CAMBRIDGE LIBRARY COLLECTION

Books of enduring scholarly value

Botany and Horticulture

Until the nineteenth century, the investigation of natural phenomena, plants and animals was considered either the preserve of elite scholars or a pastime for the leisured upper classes. As increasing academic rigour and systematisation was brought to the study of 'natural history', its subdisciplines were adopted into university curricula, and learned societies (such as the Royal Horticultural Society, founded in 1804) were established to support research in these areas. A related development was strong enthusiasm for exotic garden plants, which resulted in plant collecting expeditions to every corner of the globe, sometimes with tragic consequences. This series includes accounts of some of those expeditions, detailed reference works on the flora of different regions, and practical advice for amateur and professional gardeners.

A Botanical Arrangement of All the Vegetables Naturally Growing in Great Britain

This two-volume milestone work, published in 1776, was the first major publication of William Withering (1741–99), a physician who had also trained as an apothecary (his *Account of the Foxglove, and Some of its Medical Uses* is also reissued in this series). The first systematic botanical guide to British native plants, the present work uses and extends the Linnaean system of classification, but renders the genera and species 'familiar to those who are unacquainted with the Learned Languages'. Withering offers 'an easy introduction to the study of botany', explaining the markers by which the plants are classified in a particular genus, and giving advice on preserving specimens, but the bulk of the work consists of botanical descriptions (in English) of the appearance, qualities, varieties, common English names, and uses of hundreds of plants. The book continued to be revised and reissued for almost a century after Withering's death.

Cambridge University Press has long been a pioneer in the reissuing of out-of-print titles from its own backlist, producing digital reprints of books that are still sought after by scholars and students but could not be reprinted economically using traditional technology. The Cambridge Library Collection extends this activity to a wider range of books which are still of importance to researchers and professionals, either for the source material they contain, or as landmarks in the history of their academic discipline.

Drawing from the world-renowned collections in the Cambridge University Library and other partner libraries, and guided by the advice of experts in each subject area, Cambridge University Press is using state-of-the-art scanning machines in its own Printing House to capture the content of each book selected for inclusion. The files are processed to give a consistently clear, crisp image, and the books finished to the high quality standard for which the Press is recognised around the world. The latest print-on-demand technology ensures that the books will remain available indefinitely, and that orders for single or multiple copies can quickly be supplied.

The Cambridge Library Collection brings back to life books of enduring scholarly value (including out-of-copyright works originally issued by other publishers) across a wide range of disciplines in the humanities and social sciences and in science and technology.

A Botanical Arrangement of All the Vegetables Naturally Growing in Great Britain

With Descriptions of the Genera and Species, According to the System of the Celebrated Linnaeus

Volume 2

William Withering

CAMBRIDGE
UNIVERSITY PRESS

University Printing House, Cambridge, CB2 8BS, United Kingdom

Cambridge University Press is part of the University of Cambridge.
It furthers the University's mission by disseminating knowledge in the pursuit of education, learning and research at the highest international levels of excellence.

www.cambridge.org
Information on this title: www.cambridge.org/9781108075886

This edition first published 1776
This digitally printed version 2015

ISBN 978-1-108-07588-6 Paperback

A

Botanical Arrangement

OF ALL THE

VEGETABLES

Naturally growing in GREAT-BRITAIN.

VOL II.

A Botanical Arrangement OF ALL THE VEGETABLES Naturally growing in GREAT BRITAIN.

WITH DESCRIPTIONS OF THE

GENERA and SPECIES,

According to the Syſtem of the celebrated LINNÆUS.

Being an Attempt to render them familiar to thoſe who are unacquainted with the LEARNED LANGUAGES.

Under each SPECIES are added,

The moſt remarkable VARIETIES, the Natural PLACES of GROWTH, the DURATION, the TIME of FLOWERING, the PECULIARITIES of STRUCTURE, the common *Engliſh* NAMES; the NAMES of *Gerard*, *Parkinſon*, *Ray* and *Bauhine*.

The USES as MEDICINES, or as POISONS; as FOOD for Men, for Brutes, and for Inſects.

With their Applications in OECONOMY and in the ARTS.

WITH AN EASY

INTRODUCTION TO THE STUDY OF BOTANY.

SHEWING

The Method of inveſtigating PLANTS, and Directions how to Dry and Preſerve SPECIMENS.

The whole Illuſtrated by COPPER PLATES and a copious GLOSSARY.

By WILLIAM WITHERING, M. D.

Ornari res ipſa negat, contenta doceri.

IN TWO VOLUMES.

BIRMINGHAM: Printed by M. SWINNEY,

For T. CADEL and P. ELMSLEY in the Strand, and G. ROBINSON, in Pater-noſter-row, LONDON.

MDCCLXXVI.

CLASS XV.

FOUR CHIVES LONGER.

IN the flowers of this Claſs there are ſix Chives; *four* of them long, and *two* of them ſhort.

The ORDERS are two, and are diſtinguiſhed by the figure of the ſeed-veſſel, which in the *firſt* Order is a *Pouch*; that is, a roundiſh flat ſeed-veſſel furniſhed with a *Shaft*, which is frequently as long as the ſeed-veſſel itſelf. In the *ſecond* Order, the ſeed-veſſel is a *Pod*; that is, a very long ſeed-veſſel without any remarkable ſhaft.

The plants of this Claſs admit of the following NATURAL CHARACTER.

NATURAL CHARACTER.

EMPAL. *Cup* oblong; of four leaves. *Leaves* oblong egg-shaped; concave; blunt; approaching: standing in opposite pairs; hunched at the base.

the *Honey-cup* is formed within the empalement, and often occasions it to be hunched at the base.

BLOSS. Cross-shaped. *Petals* four; equal. *Claws* flattish awl-shaped; upright; generally longer than the *Cup*. *Border* flat. *Limbs* broadest towards the end; blunt; hardly touching one another at the edges. The petals are fixed in the same circle with the chives.

CHIVES. *Threads* six; awl shaped; upright; the two opposite ones as long as the cup: the other four somewhat longer, but shorter than the blossom. *Tips* rather oblong; taper; thickest at the base; upright, but with the top bent out-wards.

Honey cups; glands, which differ in different genera. They grow near the chives, and are mostly fixed at the base of the shorter chives, which are generally bent outwards to prevent the compression of the glands, and therefore appear shorter than the others.

POINT. *Seedbud* superior; daily growing taller. *Shaft* the length of the longest chives; but in some genera there is no shaft. *Summit* blunt.

S. VESS. *Pod* with two valves; often with two cells; opening from the base to the point. *Partition* projecting beyond the points of the valves and occupying the place of the shaft.

SEEDS. Roundish; inclining downwards; lodged in the partition length-ways and alternately. *Receptacle* narrow, surrounding the partition, and lodged in the seams of the seed-vessel.

OBS. *This Class is truly natural, and hath been considered as such by all the best Systematic Writers; nevertheless they have thrown into it one or more Genera that do not naturally belong to it; but this we have avoided. It contains the* Siliquose *plants of* Ray, *and the* Cross-shaped *flowers of* Tournefort.

The plants of this Class are universally called Antiscorbutic; *their taste is acrid and watery, they lose most of their virtues by drying. None of them are poisonous. In moist situations and wet seasons, they are most acrimonious. Thus the* Horse-radish SCURVY-GRASS, *growing near water is so very acrid that it can hardly be used; and* Turnep CABBAGE, *whose root in a dry sandy soil is so succulent and sweet, in wet stiff lands is hard and acrimonious.*

CLASS

CLASS XV.

FOUR CHIVES LONGER.

Order I. Seed-veſſel a Pouch.

* *Pouch entire, not notched at the end.*

262 WHITLOWGRASS. The *Pouch* with nearly flat valves. *Shaft* none.
263 AWLWORT. - The *Pouch* with half egg-ſhaped valves. *Shaft* ſhorter than the Pouch.
264 CAMLINE. - The *Pouch* with concave valves. *Shaft* permanent.
265 CRESSET. - The *Pouch* with valves only half the length of the partition.

** *Pouch notched at the end.*

266 ROCKWORT. - The two outermoſt petals the largeſt.
267 SCURVYGRASS. The *Pouch* heart-ſhaped. *Valves* blunt; hunched.
268 DITTANDER. - The *Pouch* heart-ſhaped; *Valves* ſharply keel-ſhaped.
269 MITHRIDATE. The *Pouch* inverſely heart-ſhaped. *Valves* in ſome ſpecies bordered; keel-ſhaped.

Order II. Seed-veſſel a Pod.

** Cup cloſed; its leaves approaching length-ways.*

270 CHARLOCK. - *Pod* jointed.
271 WORMSEED. *Pod* four-edged.
272 WALLFLOWER. *Pod* marked by a ſmall gland on each ſide the ſeed-bud.
273 DAMEWORT. - *Glands* ſituated within the ſhorter chives. *Petals* oblique.
274 TURKEYPOD. - *Glands* four, within the leaves of the cup. *Summit* ſimple.
275 CABBAGE. - *Glands* two within the ſhorter chives, and two on the outſide the longer chives.
276 TOWERER. - *Petals* upright.
277 CORALWORT. *Pod* with valves which roll back when open.

** * Cup open; its leaves ſtanding wide at the top.*

278 COLEWORT. - *Pod* deciduous; globular; like a dry berry. The four *long Chives* forked at the top.
279 WOAD. - - *Pod* deciduous; ſpear-ſhaped; containing one ſeed.
280 ROCKET. - - *Pod* deciduous; circular; covered with ſharp points.
281 LADYSMOCK. - *Pod* opening. *Valves* rolling back.
282 MUSTARD. - *Pod* opening. *Cup* expanding horizontally.
283 WATERCRESS. *Pod* opening. *Valves* nearly ſtraight. *Cup* open.

† *Navew* CABBAGE.

262 WHITLOW-

262 WHITLOWGRASS. 800 Draba.

EMPAL. *Cup* with four leaves. *Leaves* egg-ſhaped; concave; open; ſomewhat upright; deciduous.

BLOSS. Four petals, forming a croſs. *Petals* oblong; rather expanding. *Claws* very ſmall.

CHIVES. *Threads* ſix; as long as the cup. The four oppoſite chives a little longer than the other two; upright but expanding. *Tips* ſimple.

POINT. *Seedbud* egg-ſhaped. *Shaft* very ſhort. *Summit* a flat knob.

S.VESS. *Pouch* oblong oval; compreſſed; entire; without a ſhaft. Cells two; *Partition* parallel to the valves, *Valves* flat, but a little concave.

SEEDS. Many, ſmall, roundiſh.

OBS. *In the firſt ſpecies the petals are divided down to the claws. In the ſecond and third ſpecies the petals are only notched at the end.*

The eſſential character *of this genus conſiſts in the Pouch being an oblong oval; compreſſed; and without a ſhaft.*

WHITLOWGRASS. The ſtalks naked; the leaves a little ſerrated—*Bloſſoms white. At night the flowers hang down. It is difficult to find ſix chives when the flower is fully opened, for they drop off when the Pouch begins to enlarge.* Early Verna

Paronychia vulgaris. *Gerard.* 624. *Ray's Syn.* 292.

Paronychia vulgaris alſines folio. *Park.* 556.

Burſa paſtoris minor, loculo oblongo. *Bauh. pin.* 108.

Common Whitlow-graſs.

On old walls and dry banks. A. March—April.

This is one of the earlieſt flowering plants we have. It is good to eat as a ſallad.

Goats, Sheep and Horſes eat it; Cows are not fond of it; Swine refuſe it.

WHITLOWGRASS. The ſtem branched. The leaves heart ſhaped; toothed; embracing the ſtem—*Fruit-ſtalks horizontal. Bloſſoms white.* Seeds *yellow.* Speedwell leaved Muralis

Burſa paſtoris major, loculo oblongo. *Bauh pin.* 108. *Ray's Syn.* 292.

Thlaſpi veronicæ folio. *Park.* 843.

Fiſſures of rocks and high paſtures. A. May.

Linnæus makes this a variety of the *Draba nemoroſa* which differs in having yellow bloſſoms and egg-ſhaped leaves, but being ſown in a garden the bloſſoms became white and the whole plant aſſumed the form of this.

Wreathen-podded Incana WHITLOWGRASS. The ſtem leaves numerous; covered with a whitiſh down. Pouches oblong, oblique, almoſt ſitting—*Flowers in a ſmall terminating bunch*; *white*.

Lunaria contorta major. *Ray's Syn.* 291.

Creeping 1. Stem creeping.

Paronychiæ ſimilis ſed major, perennis alpina repens. *Ray's Syn.* 292.

Fiſſures of rocks and high moiſt places. B. May.

Goats eat it; Cows are not fond of it.

263 AWLWORT. 799 Subularia.

EMPAL. *Cup* four leaves. *Leaves* egg-ſhaped; concave, a little expanding, deciduous.

BLOSS. Petals four; forming a croſs. *Petals* inverſely egg-ſhaped; entire; rather larger than the cup.

CHIVES. *Threads* ſix, ſhorter than the bloſſom. Two of the threads ſtanding oppoſite ſtill ſhorter. *Tips* ſimple.

POINT. *Seedbud* egg-ſhaped. *Shaft* very ſhort. *Summit* blunt.

S.VESS. *Pouch* inverſely egg-ſhaped; ſomewhat compreſſed; entire; furniſhed with a very ſhort ſhaft. *Cells* two; *Partition* placed in a contrary direction to the valves, which are egg-ſhaped and concave.

SEEDS. Several; very minute; roundiſh.

Water Aquatica AWLWORT. As there is only one ſpecies known, Linnæus gives no deſcription of it—*Leaves ſemi-cylindrical*; *full of pith*. Bloſſoms *white*. Seeds *yellow*.

Subularia erecta, Junci foliis acutis mollibus. *Ray's Syn.* 307.

At the bottom of large lakes. A.

264 CAMLINE. 796 Myagrum.

Empal. *Cup* four leaves. *Leaves* oblong egg-ſhaped; concave; opening; coloured; deciduous.

Bloss. .Four petals forming a croſs. *Petals* flat; circular; blunt. *Claws* ſlender.

Chives. *Threads* ſix; as long as the cup. The four oppoſite threads rather longer than the other two. *Tips* ſimple.

Point. *Seedbud* egg ſhaped. *Shaft* thread-ſhaped; as long as the cup. *Summit* blunt.

S.Vess. *Pouch* inverſely egg-ſhaped, ſomewhat compreſſed; entire; rigid; terminated at the point by a rigid conical ſhaft. *Valves* two; ſome of the cells often empty.

Seeds. Roundiſh.

CAMLINE. The pouches on fruit-ſtalks; inverſely egg-ſhap- Gold
ed; containing many ſeeds—*Bloſſoms pale yellow or white.* Seeds Sativ
yellow.

Myagrum. *Gerard.* 273. *Ray's Syn.* 302. ſativum. *Baub. pin,* 109

Myagrum ſylveſtre, ſeu pſeudo-myagrum. *Park.* 868.

Gold of pleaſure.

In fields amongſt flax. A. June.

It is cultivated in Germany for the ſake of the expreſſed oil of the ſeeds, which the inhabitants uſe for Medicinal, culinary and Oeconomical purpoſes The ſeeds are a ſavourite food with Geeſe.—Horſes, Cows, Goats and Sheep eat it.

265 CRESSET. 797 Vella.

Empal. *Cup* four leaves; upright; cylindrical. *Leaves* ſtrap-ſhaped; blunt; deciduous.

Bloss. Four petals; forming a croſs. *Petals* inverſely egg-ſhaped; expanding. *Claws* as long as the cup

Chives. *Threads* ſix; as long as the cup. The four oppoſite threads a little longer than the other two. *Tips* ſimple.

Point. *Seedbud* egg-ſhaped. *Shaft* conical. *Summit* ſimple.

S.Vess. *Pouch* globular; entire; cells two. *Partition* egg-ſhaped, upright; twice as large as the pouch and extending beyond it.

Seeds. Several; roundiſh.

Annual Annua CRESSET. The leaves with winged clefts. The pouches pendant—*Bloſſoms pale yellow.*

Naſturtium ſylveſtre erucæ affine. *Bauh. pin.* 105. *Ray's Syn.* 304.

Naſturtium ſylveſtre valentinum Cluſio. *Park.* 830.

Eruca naſturtio cognata tenuifolia. *Gerard.* 247.

Creſſe Rocket.

On Saliſbury plain near Stone-henge. A. June.

266 ROCKWORT. 804 Iberis.

EMPAL. *Cup* four leaves. *Leaves* inverſely egg-ſhaped; concave; expanding; ſmall; equal; deciduous.

BLOSS. Four unequal petals; inverſely egg-ſhaped; blunt; expanding. *Claws* oblong; upright. The two *outer Petals* very large; equal. The two *inner Petals* ſmall; reflected.

CHIVES. *Threads* ſix; awl-ſhaped; upright. Two lateral threads ſhorteſt. *Tips* roundiſh.

POINT. *Seedbud*: roundiſh; compreſſed. *Shaft* ſimple; ſhort. *Summit* blunt.

S.VESS. *Pouch* upright; nearly circular; compreſſed; notched at the end; encompaſſed by a ſharp border. *Cells* two. *Partition* ſpear-ſhaped. *Valves* boat-ſhaped; keeled; compreſſed.

SEEDS. Several; ſomewhat egg-ſhaped.

Naked Nudicaulis ROCKWORT. The ſtem naked; ſimple; herbaceous. The leaves indented—*Bloſſoms white*; *terminating.*

Naſturtium petræum. *Gerard.* 251. *Ray's Syn.* 303.

Naſturtium petræum, foliis burſæ paſtoris. *Bauh. pin.* 104.

Burſa paſtoris minor foliis inciſis. *Bauh. pin.* 108.

Burſa paſtoris minor. *Park.* 866.

Rock Creſſe.

Gravelly ſoil. A. May.

267 SCURVYGRASS. 803 Cochlearia.

EMPAL. *Cup* four leaves. *Leaves* egg-ſhaped; concave; opening; deciduous.

BLOSS. Four petals, forming a croſs. *Petals* inverſely egg-ſhaped; expanding; twice as large as the cup. *Claws* narrow; ſhorter than the cup; open.

CHIVES. *Threads* ſix; awl-ſhaped; as long as the cup. The two oppoſite chives ſhorter than the others. *Tips* blunt; compreſſed.

POINT. *Seedbud* heart-ſhaped. *Shaft* ſimple; very ſhort; permanent. *Summit* blunt.

S. VESS. *Pouch* heart-ſhaped; hunched; a little compreſſed; ſlightly notched at the end; furniſhed with a ſhaft; rough; blunt at the edge. *Cells* two.

SEEDS. About four in each cell.

SCURVYGRASS. The root-leaves circular; ſtem-leaves oblong; and a little indented—*Bloſſoms white*; *terminating*. Garden Officinalis

Cochlearia. *Ray's Syn.* 302. rotundifolia. *Gerard.* 401.
Cochlearia rotundifolia, ſeu batavorum. *Park.* 285.
Cochlearia folio ſubrotundo. *Bauh. pin.* 110.

1. There is a variety with ſmaller leaves. *Park.* 286.

Common Scurvy-graſs. Scrooby Graſs. Dutch Scurvy-graſs.
Sea-coaſt, and on mountains. B. April—May.

Notwithſtanding this is a native of the ſea-coaſt, it is cultivated in gardens without any ſenſible alteration of its properties. It poſſeſſes a conſiderable degree of acrimony, and this acrimony ſeems to reſide in a very ſubtil eſſential oil. Its effects as an antiſcorbutic are univerſally known; and it is a powerful remedy in the pituitous aſthma and in what Sydenham calls the ſcorbutic rheumatiſm. A diſtilled water and a conſerve are prepared from the leaves, and its juice is preſcribed along with that of oranges by the name of antiſcorbutic juices. It may be eaten as a ſallad.—Cows eat it; Horſes, Goats and Sheep refuſe it.

SCURVYGRASS. The leaves halberd-ſhaped, and angular —*Bloſſoms white*. Daniſh Danica

Cochlearia marina folio anguloſo parvo. *Ray's Syn.* 303.
In the iſland of Walney in Lancaſhire. B. May.
The following varieties are the product of cultivation:

1. Small and upright.
2. Stems creeping.
3. Bloſſoms tinged with red.

Cows eat it; Horſes, Goats and ſheep refuſe it.

SCURVY-

English
Anglica

SCURVYGRASS. The leaves betwixt egg and ſpear-ſhaped; indented—*Bloſſoms white.*

Cochlearia folio ſinuato. *Bauh. pin.* 110. *Ray's Syn.* 303.

Cochlearia vulgaris. *Park.* 285.

Cochlearia Britannica. *Gerard.* 401.

Common Sea Scurvygraſs.

Sea coaſt. B. May.

Greenland
Groenlandica

SCURVYGRASS, with kidney-ſhaped, entire, fleſhy leaves—*Root leaves very ſmall; fleſhy; very convex on the under ſurface; without veins; very entire; on long leaf-ſtalks.* Bloſſoms *white.*

Cochlearia rotundifolia. *Ray's Syn.* 302.

Mountains in Wales. B. April—May.

All the above ſpecies partake more or leſs of the properties of the *garden* SCURVYGRASS.

Swines
Coronopus

SCURVYGRASS. The leaves with winged clefts. Stem depreſſed—*Bloſſoms white; at the baſe of the leaves.*

Coronopus ruellii. *Gerard.* 427.

Coronopus recta, vel repens ruellii. *Park.* 502.

Naſturtium ſupinum capſulis verrucoſis. *Ray's Syn.* 304.

Ambroſia campeſtris repens. *Bauh. pin.* 138.

Swines creſſes.

High roads. A. June—Auguſt.

This plant is acrid, with ſomething of the flavour of Naſturtium.

Horſe-radiſh
Armoracia

SCURVYGRASS. The root leaves ſpear-ſhaped and ſcolloped. The ſtem leaves jagged—*Bloſſoms white; terminating.*

Raphanus ruſticanus. *Gerard.* 241. *Park.* 860. *Ray's Syn.* 301. *Bauh. pin.* 98.

In ditches and amongſt rubbiſh. P. May.

The ſcraped root is in common uſe at our tables as a condiment for fiſh, roaſted beef, &c, and it is uſed for many other culinary purpoſes. An infuſion of it in cold milk, makes one of the ſafeſt and beſt coſmetics. In paralytic and dropſical caſes it is an uſeful ſtimulant and diuretic. A ſtrong infuſion of it excites vomiting. A diſtilled water is prepared from it.—Horſes, Cows, Goats, Sheep and Swine refuſe it.

The ſeveral ſpecies of SCURVYGRASS nouriſh the great white cabbage Butter-fly, *Papilio Braſſicæ.*

268 DITTANDER. 801 Lepidium.

EMPAL. *Cup* four leaves. *Leaves* egg-ſhaped; concave; deciduous.

BLOSS. Four petals, forming a croſs. *Petals* inverſely egg-ſhaped; twice as long as the cup. *Claws* narrow.

CHIVES. *Threads* ſix; awl-ſhaped; as long as the cup. The two oppoſite threads ſhorter than the others. *Tips* ſimple.

POINT. *Seedbud* egg-ſhaped. *Shaft* ſimple; as long as the chives *Summit* blunt.

S. VESS. *Pouch* ſomewhat heart-ſhaped; ſlightly notched at the end; compreſſed; ſharp at the edge. *Cells* two. *Partition* ſpear-ſhaped. *Valves* boat-ſhaped; keeled.

SEEDS. Several; egg-ſhaped; tapering; narrow at the baſe; inclining downwards.

OBS. *The third ſpecies hath only two chives, and no petals.*

DITTANDER. The leaves winged; very entire. Petals imperfect at the margin; ſmaller than the cup—*Bloſſoms white. Cup whitiſh, tipt with yellow.* Mountain Petræum

Naſturtiolum montanum annuum tenuiſſime diviſum. *Ray's Syn.* 304.

Naſturtium pumilum vernum. *Bauh. pin.* 105.

On St. Vincents rock. A. April—May.

DITTANDER. The leaves betwixt egg and ſpear-ſhaped; entire; ſerrated—*Bloſſoms white; numerous; terminating.* Pepper Latifolium

Lepidium latifolium. *Bauh. pin.* 97. *Ray's Syn.* 304.

Piperitis, ſeu lepidium vulgare. *Park.* 855.

Raphanus ſylveſtris officinarum. *Gerard.* 241.

Pepperwort. Dittander. Poor-mans-pepper.

Moiſt paſtures. P. June—July.

This is one of the acrid antiſcorbutics and was formerly uſed in the place of *Horſeradiſh* SCURVYGRASS.

DITTANDER, with only two chives in each flower, and no petals. Root leaves toothed and winged: Stem leaves ſtrap-ſhaped, very entire— Narrow-leav'd Ruderale

Naſturtium ſylveſtre, oſyridis folio. *Bauh. pin.* 105. *Park.* 829. *Ray's Syn.* 303.

Thlaſpi minus. *Gerard.* 263.

Narrow-leaved wild Creſs.

Sea-coaſt. A. June.

269 MITH-

269 MITHRIDATE. 802 Thlaſpi.

EMPAL. *Cup* four leaves. *Leaves* egg-ſhaped; concave; nearly upright; deciduous.

BLOSS. Four petals, forming a croſs. *Petals* inverſely egg-ſhaped; twice as long as the cup. *Claws* narrow.

CHIVES. *Threads* ſix; half as long as the bloſſom. The two oppoſite threads ſhorter than the others. *Tips* tapering to a point.

POINT. *Seedbud* circular; compreſſed; notched at the end. *Shaft* ſimple; as long as the chives. *Summit* blunt.

S.VESS. *Pouch* compreſſed; inverſely heart-ſhaped; notched at the end; the depth of the notch being equal to the length of the ſhaft. *Cells* two. *Partition* ſpear-ſhaped. *Valves* boat ſhaped.

SEEDS. Many. inclining; fixed to the ſeams.

OBS. *In ſome ſpecies the pouch is encompaſſed by a ſharp border; in others not.*

Muſtard
Arvenſe

MITHRIDATE. The pouches round and flat. Leaves oblong; toothed; ſmooth—*Bloſſoms white.*
Thlaſpi arvenſe ſiliquis latis. *Bauh. pin.* 105.
Thlaſpi dioſcoridis. *Gerard.* 262. *Ray's Syn.* 305.
Thlaſpi drabæ folio. *Park.* 836.
Treacle Muſtard. Penny-creſs.
Corn-fields. A. June—July.
The whole plant hath ſomething of a garlic flavour. The ſeeds have the acrimony of muſtard. When Cows eat it their milk gets a bad taſte.—Cows, Goats and Swine eat it; Sheep and Horſes refuſe it.

Perennial
Hirtum

MITHRIDATE. Pouches circular; hairy. Stem leaves arrow-ſhaped; hairy.—*Root leaves oval; a little fringed. Stem downy. Cups white. Petals white; inverſely egg-ſhaped; entire.*
Thlaſpi vaccariæ incano folio perenne. *Ray's Syn.* 305.
Thlaſpi villoſum, capſulis hirſutis. *Bauh. pin.* 106.
Perennial mithridate Muſtard.

MITH-

MITHRIDATE. Pouches circular; leaves arrow-ſhaped; Creſs
toothed; covered with a whitiſh down—*Root leaves lyre-ſhaped* Campeſtre
and indented. Bloſſoms *white.* Cups *with brown ſpots.*

Thlaſpi vulgatius. *Ray's Syn.* 305.

Thlaſpi arvenſe vaccariæ folio majus. *Bauh pin.* 106.

Thlaſpi vulgatiſſimum. *Gerard.* 262.

Thlaſpi mithridaticum ſeu vulgatiſſimum, vaccariæ folio. *Park.* 835.

Mithridate Muſtard. Baſtard Creſs.

Sandy corn-fields. A. June—July.

Goats and Swine eat it; Sheep and Horſes refuſe it; Cows are not fond of it.

MITHRIDATE. Pouches inverſely heart-ſhaped; leaves Mountain
ſmooth. Root leaves inverſely egg-ſhaped; fleſhy; very entire. Montanum
Stem leaves embracing the ſtem. Bloſſoms larger than the cups
—*nearly equal*; *white.*

Thlaſpi foliis globulariæ. *Ray's Syn.* 305.

Thlaſpi montanum, glaſti folio minus. *Park.* 842 *Bauh. pin.* 106.

Thlaſpi alpinum, bellidis cæruleæ folio. *Bauh. pin.* 106.

Thlaſpeos albi ſupini varietas. *Gerard* 268.

Mountain mithridate Muſtard.

High paſtures. P. July.

MITHRIDATE. Pouches inverſely heart-ſhaped; ſtem Perfoliate
leaves heart-ſhaped; ſmooth; a little toothed. Petals as long as Perfoliatum
the cup. Stem branched—*Bloſſoms white; but little larger than
the cups.* Chives *longer than the bloſſom; white.* Tips *yellowiſh.*

Thlaſpi perfoliatum minus. *Bauh. pin.* 106. *Park.* 837. *Ray's Syn.* 305.

Thlaſpi minus Cluſii. *Gerard.* 268.

Perfoliate Treacle-Muſtard.

On Limeſtone rocks. A. June.

MITHRIDATE. Pouches inverſely heart-ſhaped; root Purſe
leaves with winged clefts—*Bloſſoms white. In the younger plants* Burſa paſtoris
*it is not unuſual to find the leaves undivided, and the empalement
hairy.*

Burſa paſtoris. *Gerard.* 276. *Ray's Syn.* 306.

Burſa paſtoris major vulgaris. *Park.* 866.

Burſa paſtoris major, folio ſinuato. *Bauh. pin* 108.

Shepherds Purſe. Shepherds Pouch.

Among rubbiſh. Road ſides. Walls. Cornfields. A. March—June.

This

This plant is a ſtrong inſtance of the influence of ſoil and ſituation, for it grows almoſt every where, and ſometimes is not more than two inches high when it flowers and perfects its ſeeds; whilſt in other ſituations it attains the heighth of two or three feet.—Horſes, Cows, Goats, Sheep and Swine eat it.

The orange tip Butterfly, *Papilio Cardamines*; and the great yellow Underwing Moth, *Phalæna Pronuba*, feed upon the different ſpecies.

Order II. Seed-veſſel a Pod.

270 CHARLOCK. 822 Raphanus.

EMPAL. *Cup* four leaves; upright. *Leaves* oblong; parallel; approaching; deciduous; hunched at the baſe.

BLOSS. Four petals, forming a croſs. *Petals* inverſely heart-ſhaped; expanding. *Claws* a little longer than the cup.

Honey-cup Glands four; *one* betwixt each ſhorter chive and the pointal, and *one* betwixt the longer chives and the cup, on each ſide.

CHIVES *Threads* ſix; awl-ſhaped; upright. Two as long as the cup; the other four as long as the claws of the bloſſom. *Tips* oblong; upright; arrow-ſhaped.

POINT. *Seedbud* oblong; diſtended; tapering; as long as the chives. *Shaft* hardly preceptible. *Summit* a knob; entire.

S.VESS. Hunched; oblong; ending in a point; ſmooth; diſtended with ſeveral protuberances ſo as to appear almoſt jointed.

SEEDS. Roundiſh; ſmooth.

CHAR-

CHARLOCK. Pods with one cell: cylindrical; jointed; smooth—*Leaves lyre-shaped. Upper leaves oblong—Spear-shaped; scolloped and a little serrated; full of veins; thin and flat.* Stem *rough; beset with pellucid bristles.* Cups *covered with white woolly hairs, except at the base, which is smooth.* Blossoms *yellow; sometimes white and streaked with livid lines.* Corn Raphanistrum

Raphanus sylvestris. *Gerard.* 240.

Raphanistrum siliqua articulata glabra, majore et minore. *Ray's Syn.* 296.

Rapistrum album articulatum. *Park.* 863.

Rapistrum flore albo, siliqua articulata. *Bauh. pin.* 95.

Rapistrum flore luteo, siliqua glabra articulata. *Ray's Syn.* 296.

White or yellow flowered Charlock.

Corn-fields. A. June—July.

In wet seasons it grows in great quantity amongst the Barley, in Sweden, and the common people who eat barley bread, are afflicted with very violent convulsive complaints in those provinces and in those seasons wherein this plant abounds. *Rothman de Raphania. Amœn. Acad. vol.* 6. *p.* 430.

Horses eat it; Cows refuse it.

271 WORMSEED. 814 Erysimum.

EMPAL. *Cup* four leaves. *Leaves* oblong egg-shaped; parallel but approaching at the top; coloured; deciduous.

BLOSS. four petals, forming a cross. *Petals* oblong; flat; very blunt. *Claws* as long as the cup; upright.

Honey-cup Glands double; on the inner side of the shorter chives.

CHIVES. *Threads* six; as long as the cup. The two opposite threads shorter than the others. *Tips* simple.

POINT. *Seedbud* strap-shaped; four edged; as long as the chives. *Shaft* very short. *Summit* a small knob; permanent.

S. VESS. *Pod* long; narrow; stiff and straight; with four edges, two valves and two cells.

SEEDS. Many; small; roundish.

Muſtard Officinale

WORMSEED. Pods contiguous to the ſpike; leaves notched—*Bloſſoms yellow, in long ſpikes; terminating.*

Eryſimum vulgare. *Bauh. pin.* 298.

Eryſimum dioſcoridis lobelio. *Gerard.* 254.

Irio ſive eryſimum vulgare. *Park.* 833.

Eruca hirſuta, ſiliqua caule appreſſa, eryſimum dicta. *Ray's Syn.* 298.

Hedge Muſtard. Bank Creſſes.

Under walls, in roads and among rubbiſh. A. May.

It is warm and acrid to the taſte, and when cultivated is uſed as a vernal pot-herb—Birds are fond of the ſeeds.—Sheep and Goats eat it; Cows, Horſes and Swine refuſe it.

Rocket Barbarea

WORMSEED. The leaves lyre-ſhaped, with a circular ſegment at the end—*Bloſſoms yellow; terminating.*

Eruca lutea, ſive barbarea. *Bauh. pin.* 98. *Ray's Syn.* 297.

Barbarea. *Gerard.* 243. flore ſimplici. *Park.* 819.

1. There is a variety in which the leaves are ſmaller and more indented. *Ray's Hiſt.* 809. Another with double bloſſoms and a third with rough leaves.

Winter Creſſes. Winter Rockett.

Wet ditches. P. May.

The common people in Sweden uſe the leaves in ſallads, early in the ſpring and late in the autumn: they alſo boil them as Cale.—Cows eat it; Horſes and Swine refuſe it; Goats and Sheep are not fond of it.

Garlic Alliaria

WORMSEED. The leaves heart-ſhaped—*Bloſſoms white; terminating.*

Heſperis allium redolens. *Ray's Syn.* 293.

Alliaria. *Gerard.* 796. *Park.* 112. *Bauh. pin.* 110.

Jack by the hedge or ſauce alone.

Ditch-banks. P. May.

The Pruſſians eat the leaves along with ſalted meats in the Spring. They are uſeful with Lettuce and the colder ſallads—the ſeeds excite ſneezing.—Cows and Goats eat it; Horſes Sheep and Swine refuſe it.

The Garlic Weevil, *Curculio Alliaria*, feeds upon it.

Treacle Cheiranthoides

WORMSEED. The leaves ſpear-ſhaped, very entire. The pods not preſſed cloſe to the ſtem.—*Bloſſoms white; terminating.*

Myagro affinis planta ſiliquis longis. *Ray's Syn.* 298.

Myagrum ſiliqua longa. *Bauh. pin.* 109.

Camelina. *Gerard.* 273. ſeu myagrum alterum amarum. *Park.* 867.

Corn-fields. A. July.

The country people give the ſeeds to deſtroy worms, and with good effect.—Horſes, Cows, Goats, Sheep and Swine eat it.

272 WALLFLOWER. 815 Cheiranthus.

EMPAL. *Cup* four leaves ; compreſſed. *Leaves* ſpear-ſhaped ; concave ; upright ; parallel but approaching towards the top ; deciduous. The two *outer* leaves hunched at the baſe.

BLOSS. Four petals, forming a croſs. *Petals* circular ; longer than the cup, *Claws* as long as the cup.

CHIVES. *Threads* ſix ; awl-ſhaped ; parallel ; as long as the cup. Two of them ſhorter than the others and hunched at the baſe. *Tips* upright ; cloven at the baſe ; ſharp and reflected at the top.

The *Honey-cup Gland* ſurrounds the baſe of each of the ſhort chives.

POINT. *Seedbud* priſm ſhaped ; with four edges ; as long as the chives : with a ſmall tubercle on each ſide the baſe. *Shaft* very ſhort ; compreſſed. *Summit* oblong ; cloven ; reflected ; thick ; permanent.

S. VESS. *Pod* long ; compreſſed ; two oppoſite angles obliterated ; marked with a little tooth. Cells two ; valves two ; ſhaft very ſhort ; ſummit cloven ; upright.

SEEDS. Many ; pendant ; alternate ; ſomewhat egg-ſhaped ; compreſſed ; with a membranaceous border.

OBS. *A little tooth on each ſide the ſeedbud is evident enough in ſome ſpecies, but not ſo in all. In the ſecond ſpecies the pod hath three points at the end.*

Yellow Cheiri

WALLFLOWER. The leaves ſpear-ſhaped; ſharp; ſmooth; branches angular. Stem ſomewhat woody—*Bloſſoms yellow; in terminating ſpikes*.

Leucojum luteum, vulgo cheiri flore ſimplici. *Ray's Syn.* 291.

Leucojum luteum vulgare. *Bauh. pin.* 202.

Keiri, ſive leucojum vulgare luteum. *Park.* 625.

Viola lutea. *Gerard.* 456.

Bloody

1. There is a variety in which the bloſſoms are ſtained with a deep reddiſh purple.

On walls, roofs, and ruins. P. May—June.

Cultivation ſupplies us with the following varieties.

1. Double yellow.
2. Large yellow.
3. Large double yellow.
4. Double bloody Wallflower.
5. Purple and gold ditto.
6. Yellow ſtriped leaved double yellow.
7. White ſtriped leaved ditto.
8. White bloſſomed.
9. Double white ditto.
10. Straw coloured bloſſoms.
11. Double ditto.

The different varieties are very commonly found in our flower-gardens, but none have a more delightful ſmell than the wild ones.

Marine Tricuſpidatus

WALLFLOWER. The leaves lyre-ſhaped. Pods with three teeth at the end—*Bloſſoms white; terminating*.

Leucojum marinum majus. *Park.* 622. *Ray's Syn.* 291.

Leucojum maritimum ſinuato folio. *Bauh. pin.* 201.

Sea Stock Gillyflower.

On the Sea coaſt. P. June.

The angle ſhaded Moth, *Phalæna Meticuloſa*, and the great yellow underwing Moth, *Phalæna Pronuba* feed upon both ſpecies.

273 DAME-

273 DAMEWORT. 817 Hefperis.

EMPAL. *Cup* four leaves. *Leaves* betwixt ftrap and fpear-fhaped; parallel, but approaching towards the top, and open at the bafe: deciduous. The *two oppofite* leaves hunched at the bafe.

BLOSS. Four petals, forming a crofs. *Petals* oblong; the length of the cup; a little bent obliquely to the left; ending in taper *Claws* which are as long as the cup.

CHIVES. *Thread* fix; awl-fhaped; as long as the tube. Two of the threads, only half as long. *Tips* narrow; upright; reflected at the top.

Honey cup Glands tapering; placed betwixt the fhorter chives and the feedbud; furrounding the chive.

POINT. *Seedbud* as long as the cup; prifm-fhaped; with four edges. *Shaft* none. *Summit* divided; placed inwards; oblong; upright; forked at the bafe; approaching at the top; fhrivelling.

S. VESS. *Pod* long; compreffed and flat; fcored. Cells two; valves two. *Valves* as long as the partition.

SEEDS. Many; egg-fhaped; compreffed.

DAMEWORT, with a fimple upright ftem. Leaves betwixt egg and fpear-fhaped; toothed. Petals with a fharp point; imperfect at the margin.—*Bloffoms purplifh white.* Scentlefs Matronalis

Hefperis fylveftris inodora. *Ray's Syn.* 293.

Hefperis pannonica inodora. *Park.* 628.

Unfavory Dames Violet.

Banks of rivers. B. May.

The following varieties are the product of cultivation.

1. White bloffomed.
2. Double purple.
3. Double white.
4. Double ftriped.

274 TURKEYPOD. 818 Arabis.

EMPAL. *Cup* four leaves; deciduous. *Leaves* parallel but approaching at the top; *two opposite* Leaves large; oblong egg-ſhaped; ſharp; a little prominent at the baſe; hunched; concave. The *other two* ſtrap-ſhaped and upright.

BLOSS. Four petals forming a croſs. *Petals* egg-ſhaped; expanding; ending in *Claws* as long as the cup.

Honey-cups four; each compoſed of a little reflected permanent ſcale, fixed to the receptacle at the bottom, and on the inner ſide of the leaves of the cup.

CHIVES. *Threads* ſix; awl-ſhaped; upright. Two as long as the cup; four twice as long. *Tips* heart-ſhaped; upright.

POINT. *Seedbud* cylindrical; as long as the chives. *Shaft* none. *Summit* blunt; entire.

S. VESS. *Pod* compreſſed; very long; ſtrap-ſhaped; the prominencies occaſioned by the ſeeds unequal. *Valves* almoſt as long as the partition.

SEEDS. Many; roundiſh; compreſſed.

OBS. *This genus is diſtinguiſhed from the* WALLFLOWER *by the honey-cup and from the* DAMEWORT *by the ſummit.*

Mouſe-ear Thaliana — TURKEYPOD. The leaves on leaf ſtalks; ſpear-ſhaped and very entire—*Stem leaves ſitting*. Bloſſoms *white*.

Turritis vulgaris ramoſa. *Ray's Syn.* 294.

Burſæ paſtoris ſimilis ſiliquoſa major et minor. *Bauh. pin.* 108.

Paronychia major et altera minor. *Park.* 556.

Leaſy — 1. There is a variety that is ſmaller and very full of leaves.

Coded Mouſe-ear.

Walls, roofs, and dry paſtures. A. May.

Sheep are not fond of it; Swine refuſe it.

275 CAB-

275 CABBAGE. 820 Braſſica.

EMPAL. *Cup* four leaves. *Leaves* betwixt ſpear and ſtrap-ſhaped; concave; channelled; hunched at the baſe; upright; parallel; deciduous.

BLOSS. Four petals, forming a croſs. *Petals* ſomewhat egg-ſhaped; flat; expanding; entire; gradually tapering into *Claws*, which are nearly as long as the cup.

Honey-cup Glands four; egg-ſhaped. One placed betwixt each ſhort chive, and the ſeedbud; and one betwixt each pair of the longer chives, and the cup.

CHIVES. *Threads* ſix; awl-ſhaped; upright. The two oppoſite ones as long as the cup, the other four longer. *Tips* upright; tapering to a point.

POINT. *Seedbud* cylindrical; as long as the chives. *Shaft* ſhort; as thick as the ſeedbud. *Summit* a knob; entire.

S. VESS. *Pod* long; nearly cylindrical, but depreſſed on each ſide. *Partition* projecting at the end; cylindrical. *Cells* two. *Valves* two; ſhorter than the partition.

SEEDS. Many; globular.

OBS. *In the ſecond and third ſpecies the Cup is the ſame colour with the Bloſſom. In other ſpecies it is green. In the laſt ſpecies the Shaft is ſword-ſhaped.*

CABBAGE. The leaves heart-ſhaped, ſmooth, and embracing the ſtem. Root-leaves rough, very entire. Pods with four edges—*Bloſſoms white; in long ſpikes.* Perforated Orientalis

Braſſica campeſtris perfoliata, flore albo. *Bauh pin.* 112. *Ray's Syn.* 293.

Perfoliata ſiliquoſa. *Gerard.* 536.

Perfoliata ſiliquoſa vulgaris ſeu Braſſica campeſtris. *Park.* 580.

On the ſea-coaſt. A. June.

Rape
Napus

CABBAGE. The root a continuation of the ſtem; ſpindle-ſhaped—*Root leaves lyre-ſhaped; ſmooth.* Stem-leaves *oblong heart-ſhaped; embracing the ſtem; a little toothed.* Cup *expanding.* Bloſſoms *yellow.*

Napus ſylveſtris. *Bauh. pin.* 95. *Ray's Syn.* 295.

Bunias ſeu Napus ſylveſtris noſtras. *Park.* 865.

Bunias ſylveſtris lobelii. *Gerard.* 235.

Wild Navew. Rape

On Ditch Banks, and among corn. B. May.

The ſeeds furniſh a large quantity of expreſſed oil called Rape Oil. The roots may be eaten like the Turnep, but they have a ſtronger taſte.—Cows, Goats and Swine eat it.

Turnep
Rapa

CABBAGE. The root a continuation of the ſtem, round; depreſſed; fleſhy. *Root-leaves rough, deeply indented.* Bloſſoms *yellow.*

Rapa ſativa rotunda. *Bauh. pin.* 89. *Ray's Syn.* 294.

Rapum majus. *Gerard.* 232.

Knolles. Turneps.

Ditch-banks and corn-fields. B. April.

The roots are eaten either raw, boiled, or roaſted Pepper is commonly uſed with them. They relax the bowers and are ſuppoſed to ſweeten the blood. They are hurtful to pregnant or hyſterical women, and to thoſe who are ſubject to flatulencies. The rind is acrimonious.—If the roots are kept in ſand, or in a cellar during the winter, they ſend out white ſhoots and yellowiſh leaves, which being rather ſweet and not unpleaſant to the palate, are uſed as ſallad, when other eſculent plants are not to be had.—But the greateſt uſe of Turneps is in feeding Oxen and Sheep in the winter.

Cultivated
Oleracea

CABBAGE. The root a continuation of the ſtem; cylindrical; fleſhy—*Bloſſoms yellow.* Seeds *duſky purple.*

Braſſica maritima arborea, ſeu procerior ramoſa. *Ray's Syn.* 293.

On clifts on the ſea-coaſt. B. April.

OBS. *The different varieties of cultivated garden Cabbage originate from this.*

Early

Early in the ſpring the Sea Cabbage is preferred before the cultivated kinds; but when gathered on the ſea-coaſt, it muſt be boiled in two waters to take away the ſaltneſs. When old, it occaſions giddineſs. The roots may be eaten like thoſe of the preceding ſpecies, but they are not ſo tender.—The induſtry of the Gardener hath produced a great number of varieties, known by the names of *Early Cabbage*, *Winter Cabbage*, *Kale*, *Red Cabbage*, *White Cabbage*, *Borecole*, *Broccoli*, *Turnep-rooted Cabbage*, *Cauliflower*, &c. as may be ſeen in Millar's Gardener's Dictionary. They are all of them much in uſe at our tables. The Red Cabbage is chiefly uſed for pickling. In ſome countries they bury the White Cabbage when full grown in the autumn, and thus preſerve it all winter. The Germans cut them to pieces and along with ſome aromatic herbs and ſalt, preſs them cloſe down in a tub where they ſoon ferment, and are then eaten under the name of Sour Crout.—The Cabbage whilſt young, is food for the *Chryſomela Saltatoria*, and afterwards for the *Papilio Braſſica*. The former may be kept off by ſtrewing the ground with Soot; and it is ſaid the latter will not touch the plants if they are whipped with the green boughs of Elder. If Cabbages are ſowed or planted for ſeveral years together in the ſame ſoil, the heads become ſmaller and the roots knotty. This is occaſioned by the Larvæ of Flies.

CABBAGE. The leaves notched; and the ſtem covered with ſtrong hairs. Pods ſmooth. Shaft ſword-ſhaped—*Bloſſoms yellow*. Rocket Erucaſtrum

Eruca ſylveſtris. *Gerard.* 247. *Ray's Syn.* 296. vulgatior. *Park.* 818.

Eruca major lutea, caule aſpero *Bauh pin.* 98.

Wild Rocket.

On old walls and among rubbiſh. P. May.

All the parts of this plant are conſiderably acrid, and have a rank diſagreeable ſmell.

The different ſpecies of Cabbage afford nouriſhment to the following inſects.

Great White Cabbage Butterfly, *Papilio Braſſicæ.*
Small White ditto. *Papilio Rapæ.*
White Butterfly with green veins. *Papilio Napi.*
Spotted red and white Under-wing Moth. *Phalæna Fulginoſa.*
Cabbage Louſe. *Aphis Braſſicæ.*
- - - - - - *Chryſomela Hyoſcyami.*

276 TOWERER. 819 Turritis.

EMPAL. *Cup* four leaves. *Leaves* oblong egg-ſhaped; parallel, but approaching towards the top; deciduous.

BLOSS. Four petals, forming a croſs. *Petals* oblong egg ſhaped; blunt; upright; entire. *Claws* upright.

CHIVES. *Threads* ſix; awl ſhaped; upright; as long as the tube. Two of them ſhorter than the others. *Tips* ſimple.

POINT. *Seedbud* as long as the bloſſom; cylindrical; a little compreſſed. *Shaft* none. *Summit* blunt.

S. VESS. *Pod* exceedingly long; ſtiff and ſtraight; with four edges, but two of the edges which are oppoſite almoſt obliterated; ſomewhat compreſſed. *Cells* two. *Valves* two. *Valves* rather ſhorter than the partition.

SEEDS. Very numerous; roundiſh; notched.

Great Glabra

TOWERER. The root-leaves toothed and covered with ſtrong hairs. Stem leaves very entire; ſmooth; embracing the ſtem—*Bloſſoms greeniſh white; in a long terminating ſpike.*

Braſſica ſylveſtris, foliis integris et hiſpidis. *Bauh. pin.* 109.
Turritis. *Gerard.* 272. *Ray's Syn.* 293. vulgatior. *Park.* 852.
Great Tower Muſtard.
Gravelly ſoil, in pits and waſte places. A. May.
The ſeeds reduced to powder are given in Sweden for pleuritic complaints.—Cows, Goats and Sheep eat it; Horſes and Swine refuſe it.

Hairy Hirſuta

TOWERER. The leaves all covered with ſtrong hairs. Stem leaves embracing the ſtem—*Bloſſoms white; in a terminating ſpike.*

Turritis muralis minor. *Ray's Syn.* 294.
Eryſimo ſimilis hirſuta, non laciniata, alba. *Bauh. pin.* 101.
Eryſimo ſimilis hirſuta planta. *Park.* 834.
Hairy Tower Muſtard. Wall Creſs.
On rocks and old walls. B. June.
Cows refuſe it.

277 CORALWORT. 811 Dentaria.

EMPAL. *Cup* four leaves. *Leaves* oblong egg-ſhaped; parallel but approaching towards the top; blunt; deciduous.

BLOSS. Four petals, forming a croſs. *Petals* circular; blunt; ſlightly notched at the end; flat; ending in *Claws* as long as the cup.

CHIVES. *Threads* ſix; awl-ſhaped; as long as the cup. Two of them ſhorter. *Tips* oblong heart-ſhaped; upright.

POINT. *Seedbud* oblong; the length of the chives. *Shaft* very ſhort and thick. *Summit* blunt; notched at the end.

S. VESS *Pod* long; cylindrical; *Cells* two; *Valves* two; *Partition* rather longer than the valves.

SEEDS. Many; ſomewhat egg-ſhaped.

CORALWORT. The lower leaves winged: upper leaves ſimple.—*Bulbs are formed at the baſe of the leaves.* Bloſſoms *white, or reddiſh.* Bulbs *black.* Bulbous Bulbifera

Dentaria bulbifera. *Gerard.* 984.

Dentaria heptaphyllos baccifera. *Bauh. pin.* 322.

In ſhady places. P. April.

Swine refuſe it.

It nouriſheth the *Chryſomela Nemorum.*

278 COLE-

278 COLEWORT. 825 Crambe.

EMPAL. *Cup* four leaves. *Leaves* egg-ſhaped; channelled; rather expanding; deciduous.

BLOSS. Four petals; forming a croſs. *Petals* large; blunt; broad; expanding. *Claws* upright, but ſtanding open; as long as the cup.

CHIVES. *Threads* ſix; two of them as long as the cup: the other four longer than the cup, and cloven at the end. *Tips* ſimple; fixed to the outermoſt diviſion of the threads.

Honey-cup Glands placed on each ſide, betwixt the bloſſom and the longer chives.

POINT. *Seedbud* oblong. *Shaft* none. *Summit* rather thick.

S. VESS. *Berry* dry; roundiſh; of one cell; deciduous.

SEED. Single; roundiſh.

OBS. *The eſſential character of this genus conſiſts in the threads being cloven at the top.*

Sea Maritima — COLEWORT. The leaves and ſtem ſmooth—*Bloſſoms white.*
Crambe maritima Braſſicæ folio. *Ray's Syn.* 307.
Braſſica maritima monoſpermos. *Bauh. pin.* 112.
Braſſica marina monoſpermos. *Park.* 270.
Braſſica marina anglica. *Gerard.* 315.

1. There is one variety with jagged leaves and another with yellowiſh bloſſoms.

On ſandy ſea-ſhores. P. May.

The young and tender leaves are boiled as Cabbage, but when full grown they occaſion giddineſs.—Horſes, Cows, Goats, Sheep and Swine eat it.

279 WOAD. 824 Isatis.

EMPAL. *Cup* four leaves. *Leaves* egg-shaped; rather expanding; coloured; deciduous.

BLOSS. Four petals, forming a cross. *Petals* oblong; blunt; expanding; gradually tapering into *Claws*.

CHIVES. *Threads* six; upright, but expanding; as long as the blossom; but two of them not quite so long. *Tips* oblong: lateral.

POINT. *Seedbud* oblong; two-edged; compressed; aslong as the shorter chives. *Shaft* none. *Summit* a blunt knob.

S. VESS. *Pod* oblong; spear shaped; blunt; compressed; two-edged; with one cell, not opening; *Valves* two; boat-shaped; compressed; keeled; deciduous.

SEED. Single; egg-shaped; in the center of the seed-vessel.

WOAD. The root-leaves scolloped and the stem-leaves arrow-shaped. Pods oblong—*Blossoms yellow*. Dyers Tinctoria
Isatis sylvestris, seu angustifolia. *Bauh. pin.* 113.
1. Isatis sativa, seu latifolia. *Bauh. pin.* 113. *Broad-leaved*
Glastum sativum. *Gerard.* 49. *Park.* 600. *Ray's Syn.* 307.
Corn-fields and under hedges. B. July.

With the juice of this plant the Ancient Britons painted their bodies to render themselves more terrible to their enemies.—It is much used by the dyers for its blue colour, and it is the basis of many other colours.—Cows eat it; Horses, Sheep and Goats refuse it.

280 ROCKET. 823 Bunias

EMPAL. *Cup* four leaves. *Leaves* oblong egg-shaped; expanding; deciduous.

BLOSS. Four petals, forming a cross. *Petals* inversely egg-shaped; twice as long as the cup. *Claws* taper; upright.

CHIVES. *Threads* six; as long as the cup. The two opposite chives not quite so long. *Tips* upright; cloven at the base.

POINT. *Seedbud* oblong. *Shaft* none. *Summit* blunt.

S. VESS. *Pod* irregular; oblong egg-shaped; with four sides; edges with one or two sharp points; not opening; deciduous.

SEEDS. Few; roundish. One placed under each point of the pod.

Sea Cakile

ROCKET. The pods egg-shaped, smooth and two-edged —*Blossoms pale purple*.

Cakile quibusdam, aliis Eruca marina, et Raphanus marinus. *Ray's Syn.* 307.

Eruca marin . *Gerard.* 248. anglica. *Park.* 821.

Eruca maritima Italica, siliqua hastæ cuspidi simili. *Bauh. pin.* 99.

Sea-shore. A. June.

Horses eat it.

281 LADYSMOCK. 812 Cardamine.

EMPAL. *Cup* four leaves. *Leaves* oblong egg-ſhaped; blunt; rather open; bunched; ſmall; deciduous.

BLOSS. Four petals, forming a croſs. *Petals* oblong egg-ſhaped; greatly expanded; ending in *Claws*, which are upright and twice as long as the cup.

CHIVES. *Threads* ſix; awl-ſhaped. The two oppoſite threads twice as long as the cup; the other four ſtill longer than them. *Tips* ſmall; oblong heart-ſhaped; upright.

POINT. *Seedbud* ſlender; cylindrical; as long as the chives. *Shaft* none. *Summit* a blunt knob; entire.

S. VESS. *Pod* long; cylindrical but compreſſed. *Cells* two; *Valves* two. The *Valves* when open roll into a ſpiral.

SEEDS. Many; roundiſh.

OBS. *In the fifth ſpecies the two ſhorter chives are often wanting; and the third ſpecies is generally without petals. In the fixth ſpecies the claws of the petals are not longer than the cup, and the two ſhorter threads are only as long as the cup.*

* *Leaves ſimple.*

LADYSMOCK. The leaves ſimple; egg-ſhaped; very entire; on long leaf-ſtalks—*Stem leaves ſitting. This is about the ſize of the* Early Whitlow-graſs, *but after flowering the pods ſhoot out and become long.* Bloſſoms *white; ſometimes tinged with purple.* Cups *brown.* — Daiſie-leaved Bellidifolia

Cardamine pumila bellidis folio, alpina. *Gerard.* 260. *Ray's Syn.* 301.

Plantula Cardamines alterius æmula Cluſii. *Park.* 812.

Naſturtium alpinum bellidis folio minus. *Bauh. pin.* 105.

St. Vincent's Rock. P. April—May.

LADYSMOCK. The leaves ſimple; oblong and toothed *Stem ſimple.* Bloſſoms *white.* — Mountains Petræa

Naſturtium petræum. *Ray's Syn.* 300.

On the higheſt mountains. P. May.

* * *Leaves winged.*

Naked Impatiens — LADYSMOCK. The leaves winged; jagged; with props. Flowers without petals—*Linnæus ſays that in the year* 1764 *he found a ſingle flower with white petals; the petals larger than thoſe of the next ſpecies, and the plant differing therefrom.*

Cardamine impatiens, vulgo Sium minus impatiens. *Gerard.* 260. *Ray's Syn.* 299.

Sium minimum, noli me tangere dictum, ſive impatiens Naſturtii ſylveſtris folio. *Park.* 1241.

Impatient Lady-ſmock.

Moiſt hills, and near rivulets on mountains. A. April.

Swine refuſe it.

Small flowered Parviflora — LADYSMOCK. The leaves winged; without props. Little leaves ſpear-ſhaped; blunt. Flowers with petals—*Chives ſhorter than the bloſſom. Pods upright: upon horizontal fruit-ſtalks.* Bloſſoms *white.*

In meadows, and near rivulets. A. March—May.

Hairy Hirſuta — LADYSMOCK. The leaves winged; flowers with four chives—*Bloſſoms white; in naked ſpikes.*

Cardamine impatiens altera hirſutior. *Ray's Syn.* 300.

Naſturtium aquaticum minus. *Bauh. pin.* 104.

Meadows and ſhady places. A. April—Auguſt.

Common Pratenſis — LADYSMOCK. The leaves winged; little leaves of the root-leaves circular; thoſe of the ſtem-leaves ſpear-ſhaped—*Bloſſoms white, or tinged with purple.*

Cardamine. *Gerard.* 259. *Ray's Syn.* 299.

Naſturtium pratenſe magno flore. *Bauh. pin.* 104.

Naſturtium pratenſe majus, ſeu Cardamine latifolia. *Park.* 825.

b. Bloſſoms double white, or double purple.

Cuckow Flower.

Meadows and moiſt paſtures. P. April.

The virtue of the flowers in Hyſteric and Epileptic caſes, was firſt mentioned by Mr. Ray, in his letters publiſhed by himſelf; and ſince then by Dr. Baker in the *Medical Tranſ.* vol. 1. p. 442. the doſe from twenty to ninety grains twice a day. Do they not act like the *Treacle* WORMSEED in the Epilepſies of children, and cure the diſeaſe by deſtroying the worms in the ſtomach and inteſtines which were the cauſe of the fits? I have accounts of their ſucceſs in young Epileptics, from good authority; but have never been fortunate enough to ſee them of much uſe in Hyſterical Caſes.—Goats and Sheep eat it; Horſes and Swine refuſe it; Cows are not fond of it.

LADY-

LADYSMOCK. The leaves winged; flowers on ſhoots, ſpringing from the boſoms of the leaves—*Stem leaves and root leaves of the ſame ſize.* Petals *more upright than in the preceding ſpecies.* Tips *purple.* *Creeping ſhoots deſcend from the baſe of the leaves.* Bloſſoms *white.* Bitter Amara

Cardamine flore majore elatior. *Ray's Syn.* 299.
Naſturtium aquaticum majus et amarum. *Bauh. pin.* 104.
Naſturtium aquaticum amarum. *Park.* 1239.
Bitter Creſſes.
Near purls of water, and in moiſt meadows. A. April.
Sheep eat it; Cows are not fond of it.
The Orange Tip Butterfly, *Papilio Cardamines,* lives upon the different ſpecies.

282 MUSTARD. 821 Sinapis.

EMPAL. *Cup* four leaves; expanding. *Leaves* ſtrap-ſhaped; concave; channelled; ſtanding croſs-ways and expanding; deciduous.

BLOSS. Four petals, forming a croſs. *Petals* circular; flat; expanding; entire. *Claws* upright; ſtrap-ſhaped; rather ſhorter than the cup; ſitting.
Honeycup Glands four; egg-ſhaped. One placed betwixt each ſhorter chive and the pointal, and one betwixt each pair of longer chives and the cup.

CHIVES. *Threads* ſix; awl-ſhaped; upright; the two oppoſite threads as long as the cup; the other four longer. *Tips* upright; but expanding; tapering.

POINT. *Seedbud* cylindrical. *Shaft* as long as the ſeedbud, and as tall as the chives. *Summit* a knob; entire.

S. VESS. *Pod* oblong; with protuberances on the lower part; rough. *Cells* two; *Valves* two. *Partition* large; compreſſed; generally twice as long as the valves.

SEEDS. Many; Globular.

OBS. *This genus differs from the* CABBAGE *by the expanding cup, and the claws of the bloſſom being upright.*

MUSTARD

Charlock Arvensis

MUSTARD. The pods with many angles; turgid, bunched out by the feeds; fmooth; longer than the two-edged bill—*Leaves harfh; deeply indented and ferrated.* Bloffoms *yellow.* Seeds *brown.*

Rapiftrum arvorum. *Gerard.* 233. *Park.* 862. *Ray's Syn.* 295.

Rapiftrum flore luteo. *Bauh. pin.* 95.

Wild Muftard. Corn Cale.

Corn-fields, and under hedges. A. May.

The Scandinavians boil and eat it as Cabbage, and in Ireland the tender tops are collected for the fame purpofe.—Cows, Goats and Swine eat it; Sheep are very fond of it; Horfes generally refufe it.

White Alba

MUSTARD. The pods rough with ftrong hairs, furnifhed with a very long bill; oblique and fword-fhaped—*Stem branched; hairy. Leaves rough, deeply indented.* Bloffoms *yellow.* Seeds *white, or reddifh.*

Sinapi album filiqua hirfuta, femine albo vel ruffo. *Ray's Syn.* 295.

Sinapi album. *Gerard.* 244.

Sinapi apii folio. *Bauh. pin.* 99.

Road-fides and ploughed-fields. A. Auguft.

This is fown in the winter and early in the fpring to fupply ur tables with falladіng—the feeds have nearly the fame properties as thofe of the next fpecies.

Black Nigra

MUSTARD. The pods fmooth; lying contiguous to the ftem—*Lower leaves rough, deeply indented: Upper leaves fmooth; entire.* Cups *yellow.* Bloffoms *pale yellow.*

Sinapi rapi folio. *Bauh. pin.* 99.

Sinapi fativum fecundum. *Gerard.* 243. *Ray's Syn.* 295.

Common Muftard.

Corn-fields, ditch-banks and road-fides. A. June.

The feeds reduced to powder, make the common muftard, fo much in requeft at our tables—they yield a confiderable quantity of expreffed oil which partakes but little of the acrimony of the plant—the feeds when unbruifed impart but little tafte to boiling water. Taken inwardly in the quantity of a meat-fpoonful or more, they gently loofen the bowels, and are of fervice in chronic Rheumatifms and Palfies.—The powdered feeds curdle milk, and give a ftrong impregnation to boiling water. This infufion taken in confiderable quantity, vomits; in fmaller dofes it is a ufeful aperient and diuretic. Cataplafms formed with crumb of bread, vinegar and powdered muftard-feed are very commonly applied to the foles of the feet as ftimulants, in Fevers that require fuch treatment; they are ufed with advantage, topically applied,

applied, in fixed rheumatic and ſciatic pains—upon the whole, wherever we want a ſtrong ſtimulus that acts upon the nervous ſyſtem without exciting much heat, we know none preferable to muſtard ſeed. Its acrimony conſiſts in an eſſential oil.

The ſpotted red and white under wing Moth, *Phalæna fuliginoſa*, lives upon the different ſpecies.

283 WATERCRESS. 813 Siſymbrium.

EMPAL. *Cup* four leaves. *Leaves* betwixt ſpear and ſtrap-ſhaped; expanding; coloured; deciduous.

BLOSS. Four petals forming a croſs. *Petals* oblong; expanding; generally ſmaller than the cup; *Claws* very ſmall.

CHIVES. *Threads* ſix; longer than the cup. The two oppoſite threads ſomewhat ſhorter. *Tips* ſimple.

POINT. *Seedbud* oblong; thread-ſhaped. *Shaft* very ſhort. *Summit* blunt.

S. VESS. *Pod* long; crooked; hunched; cylindrical. *Cells* two; *Valves* two; ſtraight; rather ſhorter than the partition.

SEEDS. Many; ſmall.

OBS. *In the ſecond and third ſpecies the pod is hunched and very ſhort; but in the ſixth ſpecies the pod is very long and very ſlender.*

* *Pods ſhort; declining.*

WATERCRESS. The pods declining; leaves winged: little leaves nearly heart-ſhaped—*Bloſſoms white, terminating.* Common Naſturtium.

Siſymbrium cardamine, ſeu naſturtium aquaticum. *Ray's Syn.* 300.

Naſturtium aquaticum vulgare. *Park.* 1239.

Naſturtium aquaticum ſupinum. *Bauh. pin.* 104.

Naſturtium aquaticum ſeu cratevæ ſium. *Gerard.* 257.

Water-creſſes.

1. There are varieties in which the leaves are ſometimes ſmaller; and in others the winged clefts are fewer. *Ray's Syn.* 301.

Springs, brooks and rivulets. P. May.

This is very univerſally uſed as an early and wholeſome ſpring ſallad—it is an excellent Antiſcorbutic and Stomachic, with leſs acrimony than the Scurvy graſs. It is an ingredient in the Antiſcorbutic Juices.

Rocket Sylveſtre

WATERCRESS. The pods declining; oblong egg-ſhaped. Leaves ſpear-ſhaped; ſerrared—*Bloſſoms yellow.*

Eruca aquatica. *Gerard.* 248. *Park.* 1242. *Ray's Syn.* 297.

Eruca ſylveſtris minor, luteo parvoque flore. *Bauh. pin.* 98.

Water Rocket.

Marſhes and ſhallow ſtreams. P. June.

Radiſh Amphibium

WATERCRESS. The pods declining; oblong egg-ſhaped. Leaves with winged clefts; ſegments ſerrated—*In deep waters the leaves beneath the ſurface are narrow and winged; thoſe above are ſpear-ſhaped and ſerrated.* Bloſſoms *yellow.*

Raphanus aquaticus rapiſtri folio. *Bauh. pin.* 97.

Raphanus aquaticus alter. *Park.* 1229.

Rapiſtrum aquaticum. *Gerard.* 240.

1. Leaves ſerrated; with winged clefts. Stem rather ſtiff and ſtraight.

Raphanus aquaticus foliis in profundas lacinias diviſis. *Bauh. pin.* 97. *Ray's Syn.* 301.

Water Radiſh.

In ſhallow waters. P. June.

Cows refuſe it; Sheep and Goats are not fond of it.

* * *Stem with few leaves.*

Wall Murale

WATERCRESS. The ſtem with very few or no leaves. The leaves ſpear-ſhaped; indented; ſerrated; nearly ſmooth. Stalks a little rough; aſcending—*Stem a little rough, with a few ſtiff hairs. Leaves on leaf-ſtalks; with a few hairs on the under ſurface. Flowering ſtalks long; at firſt drooping; afterwards aſcending.* Cups *half expanding; vaulted at the top, ſet with a very few hairs.* Bloſſoms *yellow; very blunt,* Pods *rather thick and compreſſed.* Valves *a little keel-ſhaped.*

Eruca monenſis laciniata lutea. *Ray's Syn.* 207.

Eruca ſylveſtris minor lutea, burſæ paſtoris folio. *Bauh. pin.* 98.

Wall Rocket.

Sandy ſoil near the ſea. P. June.

Yellow Monenſe

WATERCRESS. The ſtem without leaves. Leaves winged and toothed; ſomewhat hairy—*Leaves with winged clefts; the ſegments remote; blunt; generally ſeven on each leaf. Flowering ſtalks upright; ſmooth.* Cups *cloſed.* Petals *yellow; entire.*

Yellow Rocket.

On the Iſle of Angleſea. P. June.

* * * *Leaves*

* * * *Leaves winged.*

WATERCRESS. The petals ſmaller than the cup. Leaves winged; doubly compound—*Pods long; ſtiff and crooked.* Seeds *yellow.* Bloſſoms *yellow.* Flix-weed Sophia

Eryſimum ſophia dictum *Ray's Syn.* 298.

Sophia chirurgorum. *Gerard.* 1068. *Park.* 830.

Naſturtium ſylveſtre tenuiſſime diviſum. *Bauh. pin.* 105.

Flix-weed.

On roofs and among rubbiſh. A. July.

The pods retain the ſeeds all winter, and ſmall birds feed upon them—the force of a gun is ſaid to be augmented if the gun-powder is mixed with a tenth part of theſe ſeeds—the plant is ſometimes preſcribed in Hyſteric and Dyſenteric caſes; and the ſeeds are given to deſtroy worms.—Sheep and Cows eat it; Horſes and Goats are not fond of it; Swine refuſe it.

WATERCRESS. The leaves notched; toothed and naked. Stem ſmooth. Pods upright—*Leaves very green.* Bloſſoms *yellow.* Broad-leaved Irio

Eryſimum latifolium Neapolitanum. *Park.* 834. *Ray's Syn.* 298.

Eryſimum latifolium majus glabrum. *Bauh. pin.* 101.

Broad-leaved Rocket. Hedge Muſtard.

On old walls and among rubbiſh. A. May—Auguſt.

CLASS XVI.

IN this Claſs the *Threads* are all united together at the bottom, but ſeparate at the top. The ORDERS are determined by the number of Chives. The Flowers admit of the following

NATURAL CHARACTER.

EMPAL. *Cup* always preſent; permanent; in many inſtances double.

BLOSS. *Petals* five; inverſely heart-ſhaped; the edge of one lying over the edge of the next, from the right to the left.

CHIVES. *Threads* united at the bottom; ſeparate at the top. The outer ones the ſhorteſt. *Tips* fixed ſideways to the threads.

POINT. *Receptacle of the fruit* projecting in the center of the flower. *Seedbuds* upright, ſurrounding the top of the receptacle in a jointed circle. *Shafts* united at bottom into one body with the receptacle, but ſeparated at the top into as many parts as there are ſeedbuds. *Summits* expanding, ſlender,

S. VESS. *Capſules* divided into as many cells as there are ſhafts; of various figures in different Genera; and often compoſed of the ſame number of ſeed-coats united.

SEEDS. Kidney-ſhaped.

OBS. *The plants of this claſs have generally been ſuppoſed to have only one petal; but the petals are diſtinct at the baſe, although by means of the chives they ſeem united into one body; we may therefore more properly conſider them as having* five *petals.*

The plants with MANY CHIVES *are mucilaginous and emollient; and theſe properties are common to every part of the plant. None of them are poiſonous.*

CLASS XVI.

THREADS UNITED.

Order IV. Ten Chives.

284 CRANESBILL. One pointal. *Capſule* five dry berries; with a long bill.

Order VI. Many Chives.

285 VELVETLEAF. Many pointals. *Outer Cup* with three clefts. *Seedcoats* in whorls; one ſeed in each.

286 MAULS. - - Many pointals. *Outer Cup* of three leaves. *Seedcoats* ſeveral; in whorls; one ſeed in each.

287 MARSHMALLOW. Many pointals. *Outer Cup* with nine clefts. *Seedcoats* in whorls; one ſeed in each.

284 CRANESBILL. 832 Geranium.

EMPAL. *Cup* five leaves. *Leaves* egg-ſhaped; ſharp; concave; permanent.

BLOSS. *Petals* five; inverſely heart-ſhaped; or egg-ſhaped; expanding; large.

CHIVES. *Threads* ten; awl-ſhaped; expanding towards the top; alternately longer and ſhorter; ſhorter than the bloſſom. *Tips* oblong, turning about like a vane.

POINT. *Seedbud* with five angles; furniſhed with a bill. *Shaft* awl-ſhaped, longer than the chives; permanent. *Summits* five; reflected.

S.VESS. None. *Fruit* five dry berries, furniſhed with a bill.

SEEDS. Solitary; kidney-ſhaped; often furniſhed with a dry huſky coat; and a very long awn which after ſome time rolls up in a ſpiral form.

OBS. *In ſome ſpecies only every other chive is furniſhed with a tip. The awn of the ſeed in ſome ſpecies is hairy; in others ſmooth. In the three firſt ſpecies the petals are not quite equal: there is a little gland betwixt each petal; the threads are ten, but only every other thread is furniſhed with a tip; the flowers grow in rundles; the ſeeds are naked and the awns hairy. In the fourteenth ſpecies only five of the threads have tips.*

* *Five chives furniſhed with tips.*

Hemlock-leaved Cicutarium

CRANESBILL. The fruit-ſtalks ſupporting more than two or three flowers. Leaves winged; jagged; blunt. Stem branched—*Petals entire; elevated above the cup; the two upper ones with a ſpot at the baſe: red.*

Geranium Cicutæ folio inodorum. *Gerard.* 945. *Ray's Syn.* 357.

Geranium Cicutæ folio minus et ſupinum. *Baub. pin.* 319.

Geranium Moſcharum inodorum. *Park.* 708.

1. There is a variety with a white flower. *Gerard.* 946.

On walls, road-ſides, and among rubbiſh. A. April—June. Cows and Horſes eat it; Sheep are not fond of it.

CRANESBILL.

CRANESBILL. The fruit-ſtalks ſupporting more than two flowers. Leaves winged; jagged. The lateral lobe of the ſeed with winged clefts—*Bloſſoms red, or purple.* Muſk Moſchatum

Geranium Moſchatum. *Gerard.* 941. *Park.* 706. *Ray's Syn.* 358.

Geranium Cicutæ folio, Moſchatum. *Bauh. pin.* 319.

Muſked, or Muſcovy Cranes-bill. Burnet-leaved Muſk Cranes-bill.

Dry paſtures. A. May.

The whole plant ſmells like Muſk, but it loſes this property when bruiſed.

CRANESBILL. The fruit-ſtalks ſupporting two or three flowers: leaves heart-ſhaped; ſcolloped; jagged; rough. Stems trailing—*Bloſſoms pale red.* Petals *entire. The plant lies cloſe upon the ground.* Sea Maritimum

Geranium puſillum ſupinum Maritimum altheæ aut betonicæ folio noſtras. *Ray's Syn.* 356.

On the ſea-coaſt. P. June—July.

* * *Ten chives furniſhed with Tips. Fruit-ſtalks ſupporting two flowers.*

CRANESBILL. The fruit-ſtalks ſolitary; growing oppoſite to the leaves, Cups generally furniſhed with awns; ſtem upright; petals waved at the edge—*Leaves downy; alternate; with five lobes; ſerrated.* Bloſſoms *dark purple.* Spotted Phæum

Geranium montanum fuſcum. *Bauh. pin.* 318. *Ray's Syn.* 361.

Geranium maculatum ſive fuſcum. *Gerard.* 942.

Geranium pullo flore. *Park.* 704.

Hilly places. P. May—June.

CRANESBILL. The fruit-ſtalks ſupporting two flowers; petals notched at the end. Stem leaves with three lobes; entire; ſerrated. upper leaves nearly ſitting. Stems ſpreading, compreſſed—*joints thick; generally red.* Bloſſoms *red.* Knotty Nodoſum

Geranium V nodoſum Plateau. *Gerard.* 947. *Ray's Syn.* 361.

Geranium nodoſum. *Bauh. pin.* 318. *Park.* 704.

Mountains in Cumberland. P. July—Auguſt.

Mountain Sylvaticum

CRANESBILL. The fruit-ſtalks ſupporting two flowers. Leaf-ſtalks nearly central. Leaves with five lobes; jagged; ſerrated. Stem upright; petals notched at the end—*Panicle forked; nearly level.* Bloſſom *bell-ſhaped; expanding.* Cups *with awns.* Shaft *not longer than the chives.* Cup *after flowering upright, and the awns reflected.* Bloſſoms *purpliſh red; ſometimes white; or varigated.*

Geranium batrachoides montanum noſtras. *Gerard.* 942.
Geranium batrachoides minus, *Park.* 704.
Geranium batrachoides folio aconiti. *Bauh. pin.* 317.
High paſtures in the North. P. July—Auguſt.
Cows, Goats, Sheep and Swine eat it: Horſes refuſe it.

Crowfoot Pratenſe

CRANESBILL. The fruit-ſtalks ſupporting two flowers. Leaf-ſtalks nearly central. Leaves deeply divided into many parts; wrinkled, ſharp: petals entire—*Cups with awns.* Bloſſom *flat.* Shaft *longer than the chives.* Cup *after flowering cloſed and pendant* Petals *blue.*

1. By cultivation the petals will become white or ſtriped.
Geranium batrachoides. *Gerard.* 942. *Ray's Syn.* 360.
Geranium batrachoides flore cæruleo. *Park* 704.
Geranium batrachoides, Gratia Dei Germanorum, Lobelio. *Bauh. pin.* 318.
Meadows and moiſt paſtures. P. June—July.
Horſes, Cows, Goats, Sheep and Swine eat it.

Roberts Robertianum

CRANESBILL. The fruit-ſtalks ſupporting two flowers. Cups hairy; with ten angles—*Petals entire. The whole plant ſometimes turns red. It has a rank ſmell.* Stem *branched, ſpreading, jointed.* Bloſſoms *red; by cultivation white.*

Geranium Robertianum. *Gerard.* 939. *Bauh. pin.* 319. *Ray's Syn.* 358.
Geranium Robertianum vulgare. *Park.* 710.
1. There is as a variety with ſhining ſtalks. *Ray's Syn.* 358.
Herb Robert. Stock-bill.
Walls, hedges and among rubbiſh. B. April—June.
A decoction of this plant has been known to give relief in Calculous caſes. It is conſiderably aſtingent, and is given to cattle when they make bloody water.
Horſes and Goats eat it; Sheep and Swine refuſe it.

CRANESBILL.

CRANESBILL. The fruit-ſtalks ſupporting two flowers. Cups pyramidal; the angles raiſed and wrinkled. Leaves with five lobes; roundiſh—*Stem branched, reddiſh, gloſſy.* Bloſſoms *red.* Cups *not quite equal.* Shining Lucidum

Geranium ſaxatile. *Gerard* 938. *Park.* 707. *Ray's Syn.* 361.
Geranium lucidum ſaxatile. *Bauh. pin.* 318.
Shining Doves-foot Cranes-bill.
Roofs, walls, and ſhady places. A. June—July.

CRANESBILL. The fruit-ſtalks ſupporting two flowers, and alternating with the floral leaves. Cups without awns. Stem a little upright. Petals cloven but without a ſharp point in the middle—*Cups ſhorter than the petals.* Fruit-ſtalks *declining.* Bloſſoms *purple.* Soft Molle

Geranium columbinum. *Gerard.* 938 *Ray's Syn.* 359.
Geranium columbinum vulgare. *Park.* 706.
Common Doves-foot Cranes-bill.
Sheep and Goats eat it.

CRANESBILL. The fruit-ſtalks longer than the leaves; ſupporting two flowers. Leaves deeply divided into five ſegments, and theſe again having ſeveral ſhallow clefts. Seed-coats ſmooth; cups with awns—*Petals notched at the end; with a ſharp point in the middle; reddiſh blue.* Leaves *hairy on the under ſurface.* Doves-foot Columbinum

Geranium columbinum diſſectis foliis, pediculis florum longiſſimis. *Ray's Syn.* 359.
Long-ſtalked Doves-foot Cranes-bill.
Corn-fields and high paſtures. A. July.
Sheep and Goats eat it: Swine refuſe it.

CRANESBILL. The fruit-ſtalks ſupporting two flowers. Leaves deeply divided into five parts, and theſe again into three ſhallow ſegments. Petals notched at the end; as long as the cup. Seed coats covered with ſoft hairs—*Pointal woolly.* Petals *the colour of blood; not longer than the cups.* Props *coloured.* Tips *violet colour.* Fruit-ſtalks *upright.* Cups *ſtanding open after the petals are ſhed.* Jagged Diſſectum

Geranium columbinum majus diſſectis foliis. *Gerard.* 938. *Ray's Syn.* 359.
Geranium malacoides, ſeu columbinum alterum. *Park.* 706.
Geranium columbinum maximum, foliis diſſectis. *Ray's Syn.* 360.
Jagged-leaved Doves-foot Cranes-bill.
Meadows und paſtures. A. May—July.

CRANESBILL.

Round-leaved Rotundifolium CRANESBILL. The fruit-stalks supporting two flowers; petals almost entire; as long as the cup. Stem prostrate on the ground. Leaves kidney-shaped, jagged—*Petals blunt, flesh coloured or purple; by cultivation white.* Cups *with sharp points. The whole plant rather clammy.*

Geranium folio malvæ rotundo. *Bauh. pin.* 318.

Geranium columbinum majus flore minore cæruleo. *Ray's Syn.* 358.

Walls, roofs and ditch-banks. A. July.

Horses and Sheep eat it: Cows and Swine refuse it.

Upright Perenne CRANESBILL. The fruit-stalks supporting two flowers. The lower leaves with five divisions, and these again cloven into many roundish segments. Upper leaves divided into three lobes. Stem upright—*Hudson's Flor. Anglic.* P. 265.

Perennial Doves-foot Cranes-bill.

In hilly pastures. P. May—August.

Small-flowered Pusillum CRANESBILL. Fruit-stalks supporting two flowers. Petals notched at the end. Stem depressed. Leaves kidney-shaped; divided like a hand into strap-shaped sharp segments—*Cups without sharp points.* Petals *purple.* Tips *blue. Five threads with, and five without tips.*

Geranium columbinum humile, flore cæruleo minimo. *Ray's Syn.* 359. Tab. 16. fig. 2.

Geranium malacoides, seu columbinum minimum. *Park.* 707.

Small flowered Doves-foot Cranes-bill.

Corn-fields and dry sandy places. A. June.

*** *Ten chives with tips. Fruit-stalks supporting one flower.*

Bloody Sanguineum CRANESBILL. The fruit-stalks supporting a single flower. Leaves round; deeply divided into five or seven parts; and each of these again into three segments—*Blossoms purple. After flowering the whole plant sometimes becomes purple.*

Geranium hæmatodes. *Park.* 705. *Ray's Syn.* 360.

Geranium sanguineum. *Gerard.* 945.

Geranium sanguineum maximo flore. *Bauh. pin.* 318.

1. Leaves larger, paler and more deeply divided. *Ray's Syn.* 360.
2. Blossoms elegantly striped. *Ray's Syn.* 360.
3. Stems upright.

Rocks; dry pastures and hedges. 2. In the Island of Walney in Lancashire, and on the sea-coast. P. July—August.

Horses, Cows and Goats eat it: Swine refuse it.

Order

Order VI. Many Chives.

285 VELVETLEAF. 842 Lavatera.

EMPAL. *Cup* double.
Outer Cup one leaf, with three ſhallow clefts; blunt; ſhort; permanent.
Inner Cup one leaf; with five ſhallow clefts; ſegments ſharper; upright; permanent.

BLOSS. *Petals* five; united at the baſe; inverſely heart-ſhaped; flat; expanding.

CHIVES. *Threads* numerous; united at the bottom into a cylinder; looſe at the top; fixed to the bloſſom. *Tips* kidney-ſhaped.

POINT. *Seedbud* round and flat. *Shaft* cylindrical; ſhort. *Summits* many, (from ſeven to fourteen) briſtle-ſhaped; as long as the ſhaft.

S. VESS. *Seedcoats* forming a whorl round the *Receptacle*, which ſtands in the center like a pillar. The whorl depreſſed; not jointed. Seedcoats deciduous, opening inwardly.

SEEDS. Solitary; kidney-ſhaped.

OBS. Linnæus *ſays the petals are five, united at the baſe; but in the Engliſh ſpecies the bloſſom is more properly one petal, deeply divided into five ſegments, which falls off without being ſeparated.*

VELVETLEAF. The ſtem woody. Leaves with ſeven angles; downy; plaited. Fruit-ſtalks ſupporting one flower; crowded together at the baſe of the leaves.—*Bloſſoms pale purple.* Tree Arborea

Malva arborea marina noſtras. *Ray's Syn.* 252.
Malva arborea veneta dicta, parvo flore. *Baub. pin.* 315.
Sea-tree Mallow.
On the ſea-ſhore. B. Auguſt.

286 MAULS. 841 Malva.

EMPAL. *Cup* double.

Outer Cup three leaves; narrow; heart ſhaped; ſharp; permanent.

Inner Cup one leaf, with five ſhallow clefts: large; broad; permanent.

BLOSS. *Petals* five; united at the baſe; inverſely heart-ſhaped; bitten; flat.

CHIVES. *Threads* numerous; united at the bottom into a cylinder; looſe at the top; fixed to the bloſſom. *Tips* kidney-ſhaped.

POINT. *Seedbud* round. *Shaft* cylindrical; ſhort. *Summits* many; briſtle-ſhaped; as long as the ſhaft.

S.VESS. *Seedcoats* forming a whorl round the *Receptacle*, which ſtands in the center like a pillar. The whorl depreſſed; not jointed. Seedcoats deciduous; opening inwardly.

SEEDS. Solitary; kidney-ſhaped.

OBS. *All the ſpecies are mucilaginous and emollient—The duſt is a pretty microſcopic object, being toothed like the wheel of a watch.*

Small-leaved Parviflora

MAULS. The ſtem ſpreading; leaves angular; flowers at the baſe of the leaves; ſitting; congregated. Cups ſmooth; expanding—*Purple when the fruit ripens.* Bloſſoms *purpliſh.* Seedcoats *wrinkled at the top, and toothed at the ſides.*

Malva minor, flore parvo cæruleo. *Ray's Syn.* 251.
Small Mallow.
In ſandy places. B. June.

Dwarf Rotundifolia

MAULS. The ſtem proſtrate. Leaves circular; but ſomewhat heart-ſhaped; with five ſlight lobes. Fruit-ſtalks with the fruit declining—*Bloſſoms white, or tinged with purple.*

Malva ſylveſtris pumila. *Gerard.* 930.
Malva ſylveſtris folio rotundo. *Bauh. pin.* 314.
Malva ſylveſtris minor. *Park.* 299. *Ray's Syn.* 250.
Dwarf Mallow.
Road ſides, and among rubbiſh. A. June—October.
The ancients uſed to eat the leaves as we do Cabbage.—Sheep eat it; Goats, Horſes and Swine refuſe it; Cows are not fond of it.

MAULS. The ftem upright; herbaceous; leaves with feven fharp lobes; fruit-ftalks and leaf-ftalks hairy—*Bloffoms pale purple.* Common Sylveftris

Malva fylveftris. *Gerard.* 930. folio finuato. *Bauh. pin.* 314.

Malva vulgaris. *Park.* 299. *Ray's Syn.* 251.

Common Mallow.

Hedges, foot-paths, and amongft rubbifh. B. May—Oct.

The following varieties are the effect of cultivation, foil and fituation.

1. White bloffomed.
2. Blue ditto—Found in the fields.
3. Purple ftriped.
4. White ftriped.
5. Leaves variegated—Found in the fields.

The young leaves when boiled are good to eat.—Cows eat it.

MAULS. The ftem upright. Leaves a little rough; deeply divided into many parts—*Bloffoms purple; terminating: by cultivation white.* Vervain Alcea

Alcea vulgaris. *Ray's Syn.* 252.

Alcea vulgaris major. *Bauh. pin.* 316.

Alcea vulgaris, feu Malva Verbenacea. *Park.* 301.

Malva Verbenacea. *Gerard.* 930.

Vervain Mallow.

Ditch-banks, in the counties of Warwick, Leicefter and York. P. July—September.

Horfes, Cows, Sheep and Goats eat it.

MAULS. The ftem upright. Root-leaves kidney-fhaped; jagged. Stem-leaves with five divifions, and many winged clefts—*Capfules rough with hairs.* Bloffoms *pale purple; terminating.* Mufk Mofchata

Alcea tenuifolia crifpa. *Bauh. Hift.* II. ap. 1067. *Ray's Syn.* 253.

Alcea folio rotundo laciniato. *Bauh. pin.* 316.

Jagged-leaved Vervain Mallow.

Paftures and ditch-banks. B. Auguft.

Cows and Horfes eat it; Sheep refufe it.

The Grizzle or Brown March Fritillary, *Papilio Malvæ*, and the Mallows Bug, *Cimex Apterus*, are fupported by the different fpecies.

287 MARSHMALLOW. 839 Althæa.

EMPAL. *Cup* double.

Outer Cup of one leaf, ſmall and permanent; with nine unequal clefts. *Segments* very narrow.

Inner Cup one leaf; with five ſhallow clefts. *Segments* broader; ſharper, and permanent.

BLOSS. *Petals* five; united at the baſe; inverſely heart-ſhaped; bitten; flat.

CHIVES. *Threads* numerous; united at the bottom into a cylinder; looſe at the top; fixed to the bloſſom. *Tips* nearly kidney-ſhaped.

POINT. *Seedbud* round and flat. *Shaft* cylindrical; ſhort. *Summits* numerous; (about twenty:) briſtle-ſhaped; as long as the ſhafts.

S. VESS. *Seedcoats* forming a whorl round the *Receptacle*, which ſtands in the center like a pillar: the whorls depreſſed; not jointed. *Seedcoats* deciduous, opening inwardly.

SEEDS. Solitary; kidney-ſhaped but compreſſed.

Common Officinalis MARSHMALLOW, with ſimple, downy leaves—*Bloſſoms purpliſh white*.

Althæa vulgaris. *Park.* 303. *Ray's Syn.* 252.

Althæa Ibiſcus. *Gerard.* 933.

Althæa Dioſcoridis et Plinii. *Bauh. pin.* 315.

1. Leaves deeply indented.
2. Leaves more circular.
3. Leaves ſhort, dented.

In ſalt marſhes. P. Auguſt.

The whole plant, particularly the root abounds with a mild mucilage. The root boiled is much uſed as an emollient cataplaſm, and an infuſion of it is very generally preſcribed in all caſes wherein mild mucilaginous ſubſtances are uſeful.

CLASS.

CLASS XVII.

THIS Clafs comprehends the *Butterfly-fhaped* flowers, and the *Leguminous* plants of fome authors. Linnæus takes the CLASSIC character from the *difpofition*, and the character of the ORDERS from the *Number* of the Chives. From the title of the Clafs, the young Botanift would be led to imagine, that the threads are always formed into two fets, but this is by no means the cafe; in many inftances they are *all* united into one fet. The *Butterfly-fhape* of the bloffom will therefore be a furer guide. If the ftudent will get the flower of a Garden Pea, and compare it with the following NATURAL CHARACTER, there will no longer remain any difficulty in pronouncing at firft fight, whether a plant belongs to this Clafs or not.

CLASS XVII.

NATURAL CHARACTER.

EMPAL. *Cup* one leaf, bell-ſhaped; ſhrivelling; hunched at the baſe; the lower part connected with the fruit-ſtalk; the upper part blunt; containing honey. *Rim* with five teeth; ſharp; upright; oblique; unequal. The *lower Tooth* long, the *two upper* teeth ſhorter and ſtanding further aſunder. The bottom of the cup incloſing the receptacle, is moiſtened with a liquor like honey.

BLOSS. *Butterfly-ſhaped*, unequal; each petal having a diſtinct name. Thus the *Standard* is the largeſt petal, lying upon and covering the others. It is flat and horizontal; fixed by a claw to the upper edge of the receptacle; that part of it which ſtands out of the cup is nearly circular and entire; a riſing line marking it length-ways, particularly towards the end, as if it had been preſſed down at the ſides. That part of the petal next to the baſe is ſomewhat like half a cylinder, and incloſes the parts that lie under it. The border of the petal is depreſſed on each ſide, but the ſides next to the edge are turned upwards, where the half cylinder terminates. At the unfolding of the border there are two concave impreſſions, prominent on the under ſide, and compreſſing the wings which lie beneath them.

The *wings* are two equal petals, one placed on each ſide the flower under the ſtandard. The borders incumbent; parallel; circular or oblong, broadeſt outwards; the upper edge pretty ſtraight, the lower extended and rounded. The baſe of each wing is cloven; the *lower Segment* extending into a claw which is fixed to the ſide of the receptacle and is about as long as the cup. The *upper Segment* is ſhorter and bent inwards.

The *Keel* is the lowermoſt petal, generally divided; placed under the ſtandard, and betwixt the wings. It is boat-ſhaped, concave, compreſſed at the ſides; placed in the poſition of a boat upon the water. It is diminiſhed at the baſe, the lower part extending into a claw as long as the cup, and fixed to the receptacle. The upper and lateral ſegments, which

are

are ſhorter, are interwoven with thoſe parts of the wings which reſemble them in ſhape. The ſides of the keel are ſhaped like the wings, and have a ſimilar ſituation only lower and more inwards. The line that forms the keel in this petal is ſtraight as far as the middle, and then gradually riſes in an arch; but the marginal line runs ſtraight to the extremity, until it meets with, and is loſt in that of the keel.

CHIVES. *Threads* united into two bodies, differing in ſhape. The *lower Thread* incloſing the pointal, the *upper Thread* lying upon it.

Lower Thread incloſing the ſeedbud; membranaceous below the middle, and cylindrical; opening upwards and length-ways; terminating in nine awl-ſhaped threads, bent like the keel and equal to it in length; alternately two longer and two ſhorter.

Upper Thread awl or briſtle-ſhaped. Similar in ſituation to, and lying upon the opening of the cylindrical part of the lower thread; ſimple, and a little ſhorter than that: ſeparated from the others at the baſe, ſo as to give a vent on each ſide for the honey.

Tips ten. One upon the upper thread, nine upon the lower. Small; equal in ſize; terminating.

POINT. Single; ſuperior.

Seedbud oblong; nearly cylindrical; ſlightly compreſſed; ſtraight; as long as the cylinder of the lower thread, by which it is incloſed.

Shaft awl-ſhaped or thread-ſhaped; aſcending; agreeing in length and ſituation, with the diviſions of the lower thread and placed amongſt them: ſhrivelling.

Summit downy as far as it is turned upwards: Placed directly under the tips.

S. VESS. *Shell* oblong; compreſſed; blunt; with two valves and a ſeam running length-ways both above and below; both ſeams ſtraight, but the upper ſeam falling near the baſe and the lower ſeam riſing towards the end. It opens at the upper ſeam.

SEEDS. Several; roundiſh; ſmooth: fleſhy; pendant. Marked with a prominence cauſed by the young plant near the inſertion of the eye. When the young plant is excluded, the ſide lobes retain the figure of half the ſeed.

Receptacles *proper to the Seeds* are ſmall; very ſhort; ſlender at the baſe; blunt at the part by which they are fixed. Inſerted length-ways in the upper ſeam *only* of the pod, but alternately; ſo that the valves being ſeparated, the ſeeds adhere alternately to each valve.

OBS. *This Claſs is perfectly natural and the ſtructure of the flowers extremely ſingular: their ſituation is generally obliquely pendant.*

The figure of the SHELL *is not of ſo much conſequence in aſcertaining the Genera as ſome have imagined; but the* CUP, *which hath been hitherto thought unworthy of notice, is of the greateſt uſe. The* LEAVES *never ſhould be conſidered in forming the characters of Genera.*

The SEEDS *of this Claſs furniſh food for men and other animals: they are farinaceous and flatulent. The* LEAVES *are food for cattle. None of them are poiſonous.*

CLASS XVII.

THREADS in two SETS.

Order II. Six Chives.

288 FUMITORY. *Cup* two leaves. *Bloſſom* gaping, hunched at the baſe; containing honey. Three tips on each thread.

Order III. Eight Chives.

289 MILKWORT. *Cup* with two ſegments like wings. *Bloſſ.* with a cylindrical ſtandard. *Chives* connected. *Capſule* inverſely heart-ſhaped; with two cells.

Order IV. Ten Chives.

* *Threads all united.*

290 BROOM. - - *Threads* adhering cloſe to the ſeed-bud. *Summit* woolly; growing to the upper ſide of the ſhaft.

291 GREENWOOD. *Pointal* preſſing down the keel. *Summit* rolled inwards.

292 LADIESFINGER. *Cup* ſwollen and turgid; incloſing the pod.

293 GORZE. - - *Cup* two leaves. *Shell* hardly longer than the cup.

294 RESTHARROW. *Shell* diamond-ſhaped, ſitting. Standard ſcored.

* * *Summit downy.* (*without the marks of the former division.*)

295 PEASELING. *Shaft* slender; cylindrical; woolly. on the upper, or inner side.
296 PEA. - - - *Shaft* keel-shaped above, and woolly.
297 VETCHLING. *Shaft* flat above, and woolly.
298 VETCH. - - *Shaft* bearded under the *Summit.*

*** *Shells with two cells.* (*without the marks of the former divisions*)

299 COCKLEWORT. *Shell* with two cells, rounded.

**** *Shells with one or two seeds.* (*without the marks of the former divisions.*)

300 TREFOIL. - *Shell* but little longer than the cup; with one or two seeds. Flowers in heads..

***** *Shell with the appearance of being jointed.*

301 SAINTFOIN. *Shell* with roundish compressed joints. *Keel* very blunt.
302 BIRDSFOOT. *Shell* jointed; bent like a bow.
303 HORSESHOE. *Shell* compressed; membranaceous; one of the seams hollowed out with deep notches, deeper than the middle of the shell.
304 SNAILSHELL. *Shell* spiral; membranaceous; compressed. *Pointal* pressing down the keel.

****** *Shell with one cell, and many seeds.* (*Without the marks of the former divisions.*)

305 TARE. - - *Cup* with five divisions, somewhat equal; nearly as long as the blossom.
306 CLAVER. - *Shell* cylindrical; filled with cylindrical seeds.

288 FUMI-

288 FUMITORY. 849 Fumaria.

EMPAL. *Cup* two leaves. *Leaves* oppofite; equal; lateral; upright; fharp; fmall; deciduous.

BLOSS. Oblong; tubular; gaping; palate projecting and filling up the mouth.

Upper Lip flat; blunt; notched at the end; reflected. (*The Standard.*)

The Honey-cup is the bafe of the upper lip projecting backwards; blunt.

Lower Lip altogether fimiliar to the upper lip. Towards the bafe it is keel-fhaped. (*The Keel.*)

Honey-cup at the bafe is keel-fhaped; but projecting lefs in this than in moft other genera.

Mouth with four corners; blunt; cloven perpendicularly. (*Wings.*)

CHIVES. *Threads* two; equal; broad; taper; one inclofed within each lip. *Tips.* three at the end of each thread.

POINT. *Seedbud* oblong; compreffed; tapering. *Shaft* fhort. *Summit* round; compreffed; upright.

S.VESS. Pod, with one cell.

SEEDS. Roundifh.

OBS. *The Chives are almoft the only invariable part in this genus.*

FUMITORY. The feed-veffels in bunches, each containing a fingle feed. Stem fpreading.—*Leaves doubly winged; little leaves with three lobes, and thefe again cloven into two or three parts.* Bloffoms *pink and deep purple; in long terminating fpikes.* Common Officinalis

Fumaria vulgaris. *Park.* 287. *Ray's Syn.* 204.

Fumaria purpurea. *Gerard.* 1088.

Fumaria officinarum et Diofcoridis. *Bauh. pin* 143.

Corn-fields. Ditch-banks, A. April—June.

The following varieties are the product of cultivation.

1. Pale purple.
2. White bloffomed.
3. Fine leaved.

The leaves are fucculent, faline and bitter. The expreffed juice in dofes of two or three ounces, is ufeful in hypochondriacal, fcorbutic and cachectic habits. It corrects acidity and ftrengthens the tone of the ftomach. Hoffman prefers it to all other medicines as a fweetener of the blood. There is no doubt of its utility in obftructions of the vifcera and the difeafes arifing therefrom.—Cows and Sheep eat it; Goats are not fond of it; Horfes and Swine refufe it.

Ramping Capreolata

FUMITORY The ſeed-veſſels in bunches, each containing a ſingle ſeed. Leaves climbing, furniſhed with a ſort of tendrils—*Partial fruit-ſtalks crooked.* Bloſſoms *pale red.*

By ſome this is ſuppoſed to be only a variety of the former, but it is remarkable that it is not to be met with in Sweden where the former is very common.

Fumaria major ſcandens flore pallidiore. *Ray's Syn.* 204.

Woods and hedges. A. Auguſt.

Climbing Claviculata

FUMITORY. The pods ſtrap-ſhaped; leaves furniſhed with tendrils—*The tendrils grow from the ends of the leaves.* Bloſſoms *purpliſh white, in terminating ſpikes.*

Fumaria alba latifolia. *Park.* 288. *Ray's Syn.* 335. claviculata. *Gerard.* 1088.

Fumaria claviculis donata. *Bauh. pin.* 143.

Woods and moiſt hedges. A. June.

Order III. Eight Chives.

289 MILKWORT. 850 Polygala.

EMPAL. *Cup* three leaves; ſmall. *Leaves* egg-ſhaped; ſharp: permanent; two placed beneath, and one above the bloſſom.

BLOSS. Butterfly-ſhaped. But the number of petals uncertain.

Wings permanent; ſomewhat egg-ſhaped; flat; large; placed on the outſide the other parts of the bloſſom, and formed by the teeth of the cup.

Standard generally cylindrical; tubular; ſhort. *Rim* reflected; ſmall; cloven.

Keel concave; compreſſed; diſtended towards the end.

Appendages; generally two; pencil ſhaped; with three diviſions; fixed towards the end of the keel.

CHIVES. *Threads* eight; united; incloſed in the keel. *Tips* eight; ſimple.

POINT. *Seedbud* oblong. *Shaft* ſimple; upright. *Summit* terminating; thick; cloven.

S. VESS. *Capſule* betwixt turban and heart ſhaped; compreſſed; ſharp at the edge. Cells two; valves two. *Partition* placed croſs-ways to the valves; opening at the edge on each ſide.

SEEDS. Solitary; egg-ſhaped.

OBS. *The appendix to the keel is different in different ſpecies. The wings of the Bloſſom may be conſidered as two lateral coloured leaves belonging to the empalement, and then it will be a* Cup *compoſed of five leaves.*

MILKWORT. The flowers in bunches, with pencil-ſhaped appendages. Stems herbaceous; ſimple; trailing. Leaves betwixt ſtrap and ſpear-ſhaped—*Bloſſoms blue, white, or fleſh coloured; in terminating ſpikes.* Meadow Vulgaris

Polygala. *Gerard.* 563. *Ray's Syn.* 287. minor. *Park.* 1332.

Polygala vulgaris. *Park.* 215.

On heaths and poor meadow ground. P. June—July.

Linnæus found it to poſſeſs the properties of the Senega Rattle Snake Root, (POLYGALA SENEGA,) but in an inferiour degree. Duhamel uſed it in Pleuritic Caſes with the deſired ſucceſs. *Mem. de l'Acad. des Sc. de Par.* 1740. The powdered root may be given in doſes of half a dram.—Cows, Goats and Sheep eat it: Swine refuſe it.

Order IV. Ten Chives.

290 BROOM. 858 Spartium.

EMPAL. *Cup* one leaf; heart-ſhaped. but tubular; ſmall; coloured; the upper margin very ſhort; the lower towards the end ſet with three or five little teeth.

BLOSS. Buterfly-ſhaped; petals five.

Standard inverſely heart-ſhaped; entirely reflected; large.

Wings egg-ſhaped; oblong; ſhorter than the ſtandard; connected with the threads.

Keel two petals, ſpear-ſhaped; oblong; longer than the wings: connected at the keel-ſhaped margin by ſoft hairs; fixed to the threads.

CHIVES *Threads* ten; connected; unequal; the uppermoſt the ſhorteſt and from that growing gradually longer. The lower cloven into nine parts. *Tips* rather oblong.

POINT. *Seedbud* oblong; hairy. *Shaft* awl-ſhaped; hairy; riſing upwards. *Summit* fixed to the upper ſide of the ſhaft, near the end.

S. VESS. *Shell* cylindrical; long; blunt; with one cell and two valves.

SEEDS. Many; globular, but ſomewhat kidney-ſhaped.

Common Scoparium

BROOM. The leaves growing by threes, and ſolitary. Branches without prickles; angular—*Bloſſoms yellow.*

Geniſta angulofa et ſcoparia. *Bauh. pin.* 395.

Geniſta angulofa trifolia. *Ray's Syn.* 475.

Geniſta, *Gerard.* 1311. vulgaris et ſcoparia. *Park.* 228.

Dry Paſtures. S. May—June.

The young flowers are ſometimes preſerved as pickles—the plant when burnt affords a tolerably pure Alcaline Salt—Dr. Mead relates the caſe of a dropſical patient that was cured by taking half a pint of a decoction of Green Broom tops, with a ſpoonful of whole Muſtard-ſeed, every morning and evening. The patient had been tapped three times, and tryed the uſual remedies before. (*Monita et præcept. Medica. p.* 138.) An infuſion of the ſeeds drank freely, has been known to produce ſimilar happy effects: but whoever expects theſe effects to follow in every dropſical caſe, will be greatly deceived. I knew them ſucceed in one caſe that was truly deplorable; but out of a great number of caſes in which the medicine had a fair tryal, this proved a ſingle inſtance.—Cows, Horſes and Sheep refuſe it.

The Broom Moth, *Phalæna Piſi* feeds upon it.

291 GREEN-

291 GREENWOOD. 859 Geniſta.

EMPAL. *Cup* one leaf; ſmall; tubular; moſtly with two lips. *Upper Lip* with two teeth, more deeply divided than the *Lower Lip* which hath three teeth nearly equal.

BLOSS. Butterfly-ſhaped.

Standard egg-ſhaped but ſharp; diſtant from the keel; entirely reflected.

Wings oblong; flexible; ſhorter than the other petals.

Keel ſtraight; notched at the end; longer than the ſtandard.

CHIVES. *Threads* ten; connected; riſing out of the keel. *Tips* ſimple.

POINT. *Seedbud* oblong. *Shaft* ſimple; riſing upwards. *Summit* ſharp; rolled inwards.

S. VESS. *Shell* roundiſh; turgid; with one cell, and two valves.

SEEDS. Solitary; generally kidney-ſhaped.

GREENWOOD. The leaves ſpear-ſhaped and ſmooth; the branches ſcored, cylindrical and upright.—*Bloſſoms yellow.* Dyers Tinctoria

Geniſta tinctoria germanica *Bauh. pin.* 395.
Geniſta tinctoria vulgaris. *Park.* 228.
Geniſtella tinctoria. *Gerard.* 1316. *Ray's Syn.* 474.
Dyers Weed. Wood Waxen.
Paſtures, and cornfields. S. July.

A yellow colour may be prepared from the flowers.—A dram and a half of the powered ſeeds, operates as a mild purgative. A decoction of the plant is ſometimes diuretic, and therefore has proved ſerviceable in dropſical caſes—Horſes, Cows, Goats and Sheep eat it.

BROOM. Thorns ſimple; flowering branches unarmed; leaves ſpear-ſhaped—*Bloſſoms yellow.* Prickly Anglica

Geniſta minor aſpalathoides, ſeu geniſta ſpinoſa anglica. *Bauh. pin.* 395. *Ray's Syn.* 475.
Geniſta aculeata. *Gerard.* 1320.
Geniſtella aculeata. *Park.* 1004.
Needle-furze. Petty-whin.
On moiſt heaths. S. May—June.
The Broom Moth, *Phalæna Piſi*, lives upon both ſpecies.

292 LADIESFINGER. 864 Anthyllis.

EMPAL. *Cup* one leaf; egg-shaped, but oblong; a little bladder-shaped; set with soft hairs. *Rim* with five unequal teeth, permanent.

BLOSS. Butterfly-shaped.

Standard long; with reflected sides, and a claw as long as the cup.

Wings two; oblong; shorter than the standard.

Keel compressed; as long as the wings, and like them.

CHIVES. *Threads* ten; connected; rising upwards. *Tips* simple.

POINT. *Seedbud* oblong. *Shaft* simple; ascending. *Summit* blunt.

S. VESS. *Shell* roundish; inclosed in the cup; very small; with two valves.

SEEDS. One or two.

Double-headed Vulneraria

LADIESFINGER. Herbaceous: Leaves winged, unequal. Flowers in a double head—*Blossoms yellow*; *sometimes red or scarlet*; *rarely white*.

Anthyllis leguminosa. *Gerard.* 1240. vulgaris. *Park.* 1393.
Vulneraria rustica. *Ray's Syn.* 323.
Loto affinis vulneraria pratensis. *Bauh. pin.* 332.
Kidney-Vetch.
Dry and Chalky pastures. P. July.

The country people get a yellow dye from it—it makes an excellent pasturage for Sheep.—Where the soil was a reddish clay Linnæus observed the blossoms to be red, but in white clay the blossoms are white.—Goats and Cows eat it.

293 GORZE. 881 Ulex.

EMPAL. *Cup* two leaves; permanent. *Leaves* oblong egg-ſhaped; concave; ſtraight; equal; a little ſhorter than the keel. The upper leaf with two teeth, the lower with three.

BLOSS. Butterfly-ſhaped, with five petals.

Standard inverſely heart-ſhaped; notched at the end; ſtraight; large.

Wings oblong; blunt; ſhorter than the ſtandard.

Keel two petals; ſtraight: blunt; approaching at the lower edge.

CHIVES. *Threads* ten; united. *Tips* ſimple.

POINT. *Seedbud* oblong; cylindrical; hairy. *Shaft* thread-ſhaped; riſing upwards. *Summit* blunt; ſmall.

S.VESS. *Shell* oblong; turgid; nearly incloſed by the cup; ſtraight: with one cell, and two valves.

SEEDS. Few; roundiſh; notched.

GORZE. The leaves woolly and ſharp; the thorns ſcattered —*Bloſſoms woolly; yellow; rarely white.* Common Europæus

Geniſta ſpinoſa vulgaris. *Gerard.* 1319. *Ray's Syn.* 475.

Geniſta ſpinoſa vulgaris, ſeu ſcoparius theophraſti, quem Gaza nepam tranſtulit. *Park.* 1003.

Furze. Whins.

Uncultivated ground. S. April—May.

1. It varies in the thorns being long or ſhort.

This is a very hardy plant and will make fences upon the bleakeſt mountains, and cloſe to the ſea ſide, where the ſpray of the ſea deſtroys almoſt every other ſhrub—it is cut to make faggots for heating ovens, which it does very ſoon, burning rapidly and with a great degree of heat—the aſhes are uſed to make ley.—Team Horſes may be ſupported by this plant if it is cut young and bruiſed in a mill to break the thorns.--Goats, Cows, Sheep, and Horſes feed upon the tender tops.

294 RESTHARROW. 863 Ononis.

EMPAL. *Cup* with five divisions; nearly as long as the blossom. *Segments* strap-shaped; taper; a little bowed upwards. The lowest segment placed under the keel.

BLOSS. Butterfly-shaped.

Standard heart-shaped; scored; with the sides depressed more than usual.

Wings egg-shaped; half as long as the standard.

Keel tapering; generally longer than the wings.

CHIVES. *Threads* ten, united and forming a complete undivided cylinder. *Tips* simple.

POINT. *Seedbud* oblong; woolly. *Shaft* simple; rising upwards. *Summit* blunt.

S. VESS. *Shell* diamond-shaped; turgid; a little woolly; with one cell and two valves.

SEEDS. Few; kidney-shaped.

Corn Arvensis — RESTHARROW, with flowers in bunches; one or two flowers standing together. Leaves growing by threes. Upper leaves solitary; branches without thorns, somewhat woolly—*In the beginning of summer it is smooth, but in the autumn it is said to become thorny.* Blossoms *red; purple, or white.*

Anonis non spinosa purpurea. *Gerard.* 1323. *Park.* 993. *Ray's Syn.* 332.

Anonis spinis carens purpurea. *Bauh. pin.* 389.

Thorny Spinosa — 1. Branches ending in thorns.

Anonis spinosa flore purpureo. *Bauh. pin.* 389. *Ray's Syn.* 332. *Park.* 994.

Anonis, seu resta Bovis. *Gerard.* 1323.

Ononis spinosa. *Hudson.* 273.

Cammock. Petty-whin. Ground Furze.

Barren ground. P. June—August.

Notwithstanding Linnæus makes the thorny Rest-harrow only a variety of the other, and from the observations of *Loesel, in* the *Flora Prussica*, says it becomes thorny in the autumn; yet with us they seem to be different species; they are seldom found together, and the *Corn* RESTHARROW without thorns, hath never been observed to become thorny.

The smooth sort is sometimes pickled as Samphire—A decoction of the roots has been recommended in cases of Stone and Jaundice.—Cows and Goats eat it; Sheep are very fond of it; Horses and Swine refuse it.

REST-

RESTHARROW, with ſpreading ſtems and upright branches. The upper leaves ſolitary, with egg-ſhaped props—*Flowers ſolitary; at the baſe of the leaves.* Bloſſoms *red, or almoſt white.* Creeping Repens

Anonis maritima procumbens, foliis hirſutie pubeſcentibus. *Ray's Syn.* 332.

On the ſea-coaſt. P. June—July.

295 PEASELING. 871 Orobus.

EMPAL. *Cup* one leaf; tubular; blunt at the baſe. *Rim* oblique; very ſhort; with five teeth. The three lower teeth the ſharpeſt. The two upper teeth ſhorter, deeper and more bluntly divided; ſhrivelling.

BLOSS. Buterfly-ſhaped.

Standard inverſely heart-ſhaped; long; reflected at the end, and at the ſides.

Wings two; oblong; nearly as long as the ſtandard. Riſing upwards; approaching.

Keel evidently cloven in the lower part; taper; riſing upwards. Edges approaching; parallel; compreſſed. The bottom diſtended.

CHIVES. *Threads* ten; aſcending. Nine united. *Tips* roundiſh.

POINT. *Seedbud* cylindrical; compreſſed. *Shaft* thread-ſhaped; bent upwards; upright. *Summit* narrow; downy from the middle to the end of the ſhaft, on the inner ſide.

S.VESS. *Shell* cylindrical; long; taper; with the point; aſcending. One cell; two valves.

SEEDS. Many; roundiſh.

PEASELING. The leaves winged and ſpear-ſhaped. Props half arrow-ſhaped; very entire. Stems ſimple—*drooping; but upright when in flower.* Bloſſoms *purple; three or four together.* Shells *black; pendant.* Heath Tuberoſus

Orobus ſylvaticus, foliis oblongis glabris. *Ray's Syn.* 324.

Aſtragalus ſylveſtris foliis oblongis glabris. *Bauh. pin.* 351.

Aſtragalus ſylvaticus. *Gerard.* 1237.

Lathyrus ſylveſtris lignoſior. *Park.* 1072.

Wood Peaſe. Heath Peaſe.

Paſtures, woods and hedges. P. May.

The

The roots when boiled are favory and nutritious: ground o powder they may be made into bread—They are held in high esteem by the Highlanders of Scotland, who chew them, as our people do Tobacco, and find that they prevent the uneasy sensation of hunger. They imagine that they promote expectoration, and are very efficacious in curing disorders of the Lungs. They know how to prepare an intoxicating liquor from them. *Pennant's Tour*, 1772. *p.* 310. Ray's *Histor. Plantar. p.* 916. Horses, Cows, Goats and Sheep eat it.

Wood Sylvaticus

PEASELING. The stems branched, drooping; rough with hair—*Leaves formed of about seven pair in a wing.* Shell *compressed.* Seeds *two or three.* Blossom *reddish on the outside; white within, with purple lines. Five or six flowers in a cluster.*

Orobus sylvaticus nostras. *Ray's Syn.* 324.
English Wood Vetch. Bitter Vetch.
Woods and hedges. P. July.

296 PEA. 870 Pisum.

EMPAL. *Cup* one leaf; with five clefts; sharp; permanent. The two upper segments the broadest.

BLOSS. Butterfly shaped.
Standard very broad; inversely heart-shaped; reflected; notched at the end; with a point between.
Wings two; circular; approaching; shorter than the standard.
Keel compressed; half moon-shaped; shorter than the wings.

CHIVES. *Threads* ten; *One*, simple; superiour; awl-shaped; flat. *Nine*, awl-shaped; united from the middle downwards into a cylinder, which is cloven towards the top. *Tips* roundish.

POINT. *Seedbud* oblong; compressed. *Shaft* ascending; triangular; membranaceous; keel-shaped; the sides bent outwards. *Summit* fixed to the superiour angle; oblong; woolly.

S. VESS. *Shell* large; long; somewhat cylindrical, (or else compressed below,) the point tapering upwards. One cell; two valves.

SEEDS. Many: globular.

PEA.

PEA. The leaf-ſtalks flatiſh on the upper ſide. Stems angular. Props arrow-ſhaped; fruit-ſtalks ſupporting many flowers —*Bloſſoms pale red and purple.* Sea Maritimum

Piſum marinum. *Gerard.* 1250. *Ray's Syn.* 319.

Piſum ſpontaneum maritimum anglicum. *Park.* 1059.

On the ſea.ſhore. P. July.

In the year 1655, during a time of great ſcarcity, the people about Orford in Suſſex were preſerved from periſhing by eating the ſeeds of this plant, which grew there in great abundance upon the ſea-coaſt. It has been ſuppoſed that the different ſorts of garden Peas are only varieties of this, but the ſuppoſition is not well founded.—Cows, Horſes, Sheep and Goats eat it.

It affords nouriſhment to the Broom Moth, *Phalæna Piſi.*

297 VETCHLING. 872 Lathyrus.

EMPAL. *Cup* one leaf; bell-ſhaped; with five ſhallow clefts. *Segments* ſpear-ſhaped; ſharp; the two upper ones ſhorteſt, the lower one longeſt.

BLOSS. Butterfly ſhaped.

Standard inverſely heart-ſhaped; large; reflected at the end and at the ſides.

Wings oblong; creſcent-ſhaped; ſhort; blunt.

Keel half a circle; as large as the wings, but broader; opening inwardly at the middle.

CHIVES. *Threads* ten; riſing upwards; nine united. *Tips* roundiſh.

POINT. *Seedbud* compreſſed; oblong; narrow. *Shaft* upright, flat, and broad towards the top; ſharp at the end. *Summit* woolly; extending from the middle of the ſhaft to the end, along the fore part.

S.VESS. *Shell* very long; cylindrical or compreſſed; tapering to a point. Valves two.

SEEDS. Many; either cylindrical or nearly globular.

OBS. *The chief difference betwixt this and the preceding Genus conſiſts in the ſhaft.*

* *Fruit-ſtalks bearing only one flower.*

Yellow Aphaca

VETCHLING. The fruit-ſtalks ſupporting one flower. Tendrils without leaves. Props betwixt heart and arrow-ſhaped —*The* Tendrils *have ſometimes two ſpear-ſhaped leaves joined together at the baſe, but this is very rare.* Bloſſoms *yellow.*

Aphaca. *Gerard.* 1250. *Park.* 1067. *Ray's Syn.* 320.
Vicia lutea foliis convolvuli minoris. *Bauh. pin.* 345.
Corn-fields. A. June—July.

The ſeeds both of this and of all the other ſpecies are nutritious either eaten in broth, or made into bread—

Crimſon Niſſolia

VETCHLING. The fruit-ſtalks ſupporting one flower; leaves ſimple: props awl ſhaped—*Stem upright; without tendrils.* Shells *long; pendant; hairy.* Bloſſoms *crimſon.*

Lathyrus ſylveſtris minor. *Bauh. pin.* 344.
Catanance leguminoſa quorundam. *Ray's Syn.* 322.
Ervum ſylveſtre. *Gerard.* 1249. ſeu Catanance. *Park.* 1079.
Crimſon Graſs-vetch.
Round corn-fields. A. May.

This is a very beautiful plant and merits a place in our flower-gardens.

** *Fruit-ſtalks bearing more than one flower.*

Rough Hirſutus

VETCHLING. The fruit-ſtalks generally ſupporting two or three flowers. Tendrils furniſhed with two leaves. Leaves ſpear-ſhaped: ſhells hairy: ſeeds rough—*Bloſſoms crimſon, with yellow lines within.*

Lathyrus ſiliqua hirſuta. *Ray's Syn.* 320.
Lathyrus anguſtifolius, ſiliqua hirſuta. *Bauh. pin.* 344.
Rough-codded Chickling Vetch.
Corn-fields. A. July.

Tare Pratenſis

VETCHLING. The fruit-ſtalks ſupporting many flowers. Tendrils furniſhed with two very ſimple leaves. Little leaves ſpear-ſhaped —*The* Tendrils *are ſometimes divided into three parts.* Props *very entire; halberd-ſhaped.* Fruit-ſtalks *four-cornered.* Bloſſoms *yellow.* Shells *black and ſmooth.*

Lathyrus luteus ſylveſtris dumetorum. *Ray's Syn.* 320.
Lathyrus ſylveſtris flore luteo. *Gerard.* 1231. *Park.* 1062.
Lathyrus ſylveſtris luteus, foliis Viciæ. *Bauh. pin.* 344.
Tare Everlaſting. Common yellow Vetchling.
Woods, hedges, meadows and paſtures. P. July—Auguſt.
Horſes, Cows, Sheep and Goats eat it: Swine refuſe it.

VETCHLING.

VETCHLING. The fruit-ſtalks ſupporting many flowers: Tendrils furniſhed with two leaves. Little leaves ſword-ſhaped. Stems membranaceous betwixt the joints—*Spreading wide, climbing, or drooping.* Props *awl ſhaped*; *very narrow.* Fruit-ſtalks *rough at the edges.* Little leaves *not broader than the ſtems.* Bloſſoms *red, or white*; *ſometimes by cultivation dark purple.* Narrow-leaved Sylveſtris

Lathyrus ſylveſtris major. *Bauh. pin.* 344.

Lathyrus ſylveſtris. *Park.* 1061.

Lathyri majoris ſpecies flore rubente et albido minore dumetorum ſeu Germanicus. *Ray's Syn.* 319.

Narrow-leaved Peaſe-everlaſting.

Woods and hedges. P. July—Auguſt.

Horſes, Cows, Sheep and Goats eat it.

VETCHLING. The fruit-ſtalks ſupporting many flowers: tendrils furniſhed with two leaves. Little leaves ſpear-ſhaped. Stem membranaceous betwixt the joints—*Props almoſt halberd-ſhaped*; *broader than the ſtem.* Leaves *much broader than the ſtem.* Bloſſoms *pale purple.* Broad-leaved Latifolius

Lathyrus major latifolius. *Gerard.* 1229. *Ray's Syn.* 319.

Lathyrus latifolius. *Bauh. pin.* 344.

Lathyrus major perennis. *Park.* 1061.

Broad leaved Peaſe-everlaſting.

Woods and hedges. P. July.

VETCHLING. The fruit ſtalks ſupporting many flowers. Tendrils furniſhed with many leaves. Props ſpear-ſhaped—*Bloſſoms blue and red.* Marſh Paluſtris

Lathyrus peregrinis, foliis viciæ, flore ſubcæruleo pallidive purpuraſcente. *Bauh. pin.* 344.

Lathyrus viciæ formis, ſeu vicia Lathyroides noſtras. *Ray's Syn.* 320.

Marſh chickling Vetch.

Wet paſtures. P. July—Auguſt.

Horſes, Cows, Sheep and Goats eat it.

298 VETCH. 873 Vicia.

EMPAL. *Cup* one leaf; tubular; upright; with five shallow clefts; sharp. The upper teeth shortest; approaching. All the teeth equal in breadth.

BLOSS. Butterfly-shaped.

Standard oval; with a broad oblong claw; notched at the end, with a sharp point in the middle; reflected at the sides, compressed and raised in a line running lengthways.

Wings two oblong; upright; in the shape of half a heart; with an oblong claw; shorter than the standard.

Keel with an oblong cloven claw; the hollow part compressed; in the shape of half a circle; shorter than the wings.

CHIVES. *Threads* ten; nine united. *Tips* upright; roundish; with four furrows.

Honey-cup Gland short; tapering; arising from the receptacle and situated betwixt the united threads and the seedbud.

POINT. *Seedbud* narrow; compressed; long. *Shaft* thread-shaped; rather short; bent upwards. *Summit* blunt; bearded on the under-side, at the end.

S. VESS. *Shell* long; like leather; with two valves and one cell; terminated by a point.

SEEDS. Many; roundish.

* *With long fruit-stalks.*

Wood Sylvatica — VETCH. The fruit-stalks supporting many flowers. Little leaves oval; props toothed—*Blossoms white, with blue streaks.*

Vicia sylvatica multiflora. *Ray's Syn.* 322.

Tufted Wood Vetch.

Woods and hedges. P. July—August.

Horses, Cows, Sheep and Goats eat it.

VETCH. The fruit-ſtalks ſupporting many flowers, lying one over another : little leaves ſpear-ſhaped ; downy. Props entire—*Bloſſoms purple.* Tufted Cracca

Vicia multiflora. *Bauh pin.* 345. ſeu ſpicata. *Park.* 1072.

Cracca. *Ray's Syn.* 322.

Woods and hedges. P. July—Auguſt.

Dr. Plot, in his *Nat. Hiſt. of Staffordſhire. p.* 204, ſays, that this and the preceding ſpecies advance ſtarven or weak Cattle above any thing yet known.——Horſes, Cows, Sheep and Goats eat it ; Swine are not fond of it.

The Vetch Louſe, *Aphis Craccæ* lives upon it.

* * *Flowers at the baſe of the leaves ; almoſt ſitting.*

VETCH. Shells ſitting ; generally two together ; upright. Leaves indented at the end. Props marked with a ſpot on the under-ſide—*Bloſſoms light and dark purple.* Teeth *of the cup nearly equal.* Cultivated Sativa

Vicia. *Gerard.* 1227. *Ray's Syn.* 320. vulgaris ſativa. *Park.* 1072.

Vicia ſativa vulgaris ſemine nigro. *Bauh. pin.* 344.

1. There is a variety in which the ſeeds are white. *Bauh. pin.* 344.

Common Vetch, or Tare.

Amongſt corn. A. May—June.

In Glouceſterſhire they ſow it as paſturage for Horſes, and eat it off early enough to allow of Turneps being ſown the ſame year—The ſeeds are excellent food for Pidgeons.—Horſes, Cows, Sheep and Goats eat it.

VETCH. Shells ſitting ; ſolitary ; upright ; ſmooth. Six little leaves on each leaf-ſtalk ; the lower ones inverſely heart-ſhaped—*Bloſſoms bright red.* Shells *ſmooth ; about nine ſeeds in each.* Teeth *of the cup nearly equal.* Wild Lathyroides

Vicia ſylveſtris, ſeu Cracca major. *Gerard.* 1227.

Vicia ſemine rotundo nigro. *Bauh. pin.* 345.

Aracus, ſeu Cracca major. *Park.* 1070.

1. There is a variety in which the flower is very red, and the ſeed-veſſel long and black. *Ray's Syn.* 321.

Strangle Tare.

Sandy corn-fields and dry paſtures. A. May.

Yellow
Lutea

VETCH. Shells ſitting; reflected; hairy; ſolitary; five ſeeds in each. Standard of the bloſſom ſmooth—*There are varieties in which the* leaves *are dented*; *the* ſhells *almoſt ſmooth*; *the* props *alternately entire and with three teeth*. Bloſſoms *yellow*.

Vicia ſylveſtris lutea ſiliqua hirſuta. *Bauh. pin.* 345.
Vicia luteo flore ſylveſtris. *Ray's Syn.*. 321.
Meadows and paſtures. A. June—July.

Buſh
Sepium

VETCH. Shells on little foot-ſtalks; generally four together; upright. Little leaves egg-ſhaped, very entire, growing gradually ſmaller towards the end—*Cups rough with hair*; *on very ſhort fruit-ſtalks*. Bloſſoms *purple*.

Vicia ſepium perennis. *Ray's Syn.* 320.
Vicia ſepium folio rotundiore acuto. *Bauh. pin.* 345.
Vicia maxima dumetorum. *Gerard.* 1227. *Park.* 1072.
Woods, hedges and paſtures. P. May.
Horſes, Cows, Goats, Sheep and Swine eat it.

299 COCKLEWORT. 892 Aſtragalus.

EMPAL. *Cup* one leaf; tubular; with five ſharp teeth; The lower teeth gradually growing ſmaller.

BLOSS. Butterfly-ſhaped.
Standard longer than the other petals; reflected at the ſides; notched at the end; blunt; ſtraight.
Wings oblong; ſhorter than the ſtandard.
Keel as long as the wings; notched at the end.

CHIVES. *Threads* ten; almoſt ſtraight; nine united. *Tips* roundiſh.

POINT. *Seedbud* ſomewhat cylindrical. *Shaft* awl-ſhaped; aſcending. *Summit* blunt.

S.VESS. *Shell* with two cells: the cells bending to one ſide.

SEEDS. Kidney-ſhaped.

Liquorice
Glycyphyllos

COCKLEWORT. Stems proſtrate. Shells nearly three cornered and bent like a bow. Little leaves oval; longer than the fruit-ſtalks—*Bloſſoms pale yellow*.

Glaux vulgaris, ſeu Glycyrrhiza ſylveſtris. *Park.* 1098.
Glycyrrhiza ſylveſtris floribus luteo palleſcentibus. *Bauh. pin.* 352.
Hedyſarum glycyrrhizatum. *Gerard.* 1233.
Wild Liquorice. Liquorice Vetch.
Meadows and hedges. P. June—July.
Horſes, Cows, Goats and Sheep eat it: Swine refuſe it.

COCKLEWORT. Stems trailing. Flowers in a fort of bunch; upright. Leaves downy—*Bloſſoms purple.* Purple Arenarius

Aſtragalus incanus parvus purpureus noſtras. *Ray's Syn.* 326. Tab. 12. fig. 3.

Purple mountain Milkwort.

Sandy paſtures. P. July.

300 TREFOIL. 896 Trifolium

Flowers forming a little *Rundle* or *Head*, upon a common receptacle.

EMPAL. *Cup* one leaf; tubular; with five teeth; permanent.

BLOSS. Butterfly-ſhaped; generally permanent; ſhrivelling.

Standard reflected.

Wings ſhorter than the ſtandard.

Keel ſhorter than the wings.

CHIVES. *Threads* ten; nine united. *Tips* ſimple.

POINT. *Seedbud* ſomewhat egg-ſhaped. *Shaft* awl-ſhaped, aſcending. *Summit* ſimple.

S.VESS. *Shell* ſhort; with one valve, not opening; deciduous.

SEEDS. Very few. Roundiſh.

OBS. *Perhaps nothing is more difficult than to give an abſolute and eſſential character to this Genus, notwithſtanding the general habit and the properties of the plants which compoſe it ſhew that it is a natural one; and thoſe who attempt to divide it, have not been able to fix any bounds to their labours.*

* *Shells naked, containing ſeveral ſeeds.*

TREFOIL. Shells in bunches; naked; wrinkled; ſharp; two ſeeds in each. Stem upright—*Bloſſoms yellow.* Melilot: Melilot. Officinalis.

Melilotus officinarum Germaniæ. *Baub. pin.* 331.

Melilotus vulgaris. *Park.* 719. *Ray's Syn.* 331.

Melilotus Germanica. *Gerard.* 1205.

Common Melilot. Hart-clover. King's-clover.

Corn-fields in ſtiff ſoil. B. June—July.

This is more fragrant when dry than when green. A water diſtilled from the flowers poſſeſſes but little odour in itſelf, but it improves the flavour of other ſubſtances—Horſes are extremely fond of: Cows, Goats, Sheep and Swine eat it.

Birds-foot Melilot: Ornithopodioides

TREFOIL. Shells naked; generally three together: eight seeds in each; twice as long as the cup. Stems declining—*Blossoms pale purple.*

Fænum græcum humile repens, Ornithopodii siliquis brevibus erectis. *Ray's Syn.* 331.

Sandy places. A. June.

** *Shells covered. Seeds four.*

Dutch Hybridum

TREFOIL. Flowers in heads somewhat resembling rundles. shells with four seeds. Stem ascending—*branched, furrowed; about a foot high.* Props *spear-shaped; sharp; terminating in a hair. Little leaves egg-shaped; sharp; serrated, and toothed.* Fruit-stalks *not very long, but jointed.* Receptacle *egg-shaped, concave.* Chaff *spear-shaped; narrow; very sharp;* Blossoms *gaping.*

It is very probable that this plant was at first produced by the dust of the sixth species fertilizing the seedbud of the fourth species.

Pastures. P. May—September.

White Repens

TREFOIL. Flowers in heads, somewhat resembling rundles. Shells with four seeds. Stem creeping—*Undivided; cylindrical, about a span long.* Props *round; blunt.* Little leaves *round; very blunt.* Fruit-stalks *very long.* Receptacle *narrow; solid.* Chaff *egg-shaped; oblong; blunt.* Blossoms *white. The leaves stand upright against rain.*

Trifolium pratense album. *Bauh. pin.* 327. *Park.* 1110. *Ray's Syn.* 327.

Trifolium minus pratense, flore albo, seu 2. *Gerard.* 1185.

Meadows and pastures. P. May—September.

Horses, Cows and Goats eat it: Sheep are not fond of it: Swine refuse it.

*** *Cups woolly.*

Subterranean Subterraneum

TREFOIL. Heads woolly; five flowers in each: with a bushy substance in the middle; reflected; rigid; involving the seedvessel—*Shells frequently under the surface of the earth.* Blossoms *white.*

Trifolum pumilum supinum, flosculis longis albis. *Ray's Syn.* 327.

Barren pastures. A. May.

TREFOIL.

TREFOIL. Flowers in rather woolly ſpikes, incloſed by opposite membranaceous props. Bloſſoms of one petal—*The ſpikes of flowers are ſitting betwixt two oppoſite ſitting leaves, whoſe membranaceous props expanding form a ſort of common empalement. The* Cup *of the fruit is furniſhed with five awns or briſtles, the four uppermoſt of which are expanding, but the fifth and lowermoſt ſtands upright.* Bloſſoms *red*; *purple*; *or yellowiſh white*. Honeyſuckle Pratenſe

Trifolium pratenſe. *Gerard.* 1185. purpureum. *Bauh. pin.* 327. *Ray's Syn.* 328.

Trifolium pratenſe purpureum vulgare. *Park.* 1110.

1. Small purple Trefoil, with heart-ſhaped leaves. *Ray's Syn.* 328. Tab. 13. fig. 1

2. Large cultivated purple Trefoil. *Ray's Syn.* 328. *Clover*

Purple or Honey-ſuckle Trefoil. 2 Clover.

Meadows and paſtures. P. May—Steptember.

The flowering heads are uſed in Scandinavia to dye woollen cloth green. With alum they give a light, with copperas a dark green—The variety (2) engages much of the Farmers attention in modern huſbandry: it is either grazed upon the ground, or made into hay. Swine, Goats, Sheep, Horſes and Cows are all fond of it.

TREFOIL. Flowers in ſomewhat globular ſpikes; terminating; woolly. Stem upright. Leaves ſpear-ſhaped and a little ſerrated—*Bloſſoms a deep purple. This is diſtinguiſhed from the foregoing ſpecies in having two terminating heads of flowers, whereas that hath but one. In this the* Props *are green*; *in that they are ſhining, with red veins. In this the uppermoſt props are ſpear-ſhaped on each ſide, but in that ſomewhat egg-ſhaped.* Purple Alpeſtre

Trifolium medium. *Hudſon.* 284.

Trifolium purpureum majus, foliis longioribus et anguſtioribus, floribus ſaturatioribus. *Ray's Syn.* 328.

Trifolium montanum purpureum majus. *Bauh. pin.* 328.

Long-leaved purple Trefoil.

Hilly countries and high paſtures. P. July.

Hares foot Arvenſe — TREFOIL. Flowers in woolly oval ſpikes. The teeth of the cups briſtle-ſhaped; woolly: equal—*Longer than the bloſſom; which is pale red, with a bloody ſpot on the inner ſide of each of the wings.*

Trifolium arvenſe humile ſpicatum, ſeu Lagopus. *Bauh. pin*, 328. *Ray's Syn*. 330

Lagopus vulgaris, *Park*. 1107.

Lagopodium ſive pes Leporis. *Gerard*. 1192.

In ſandy ground and on the ſea-ſhore. A. July—Auguſt.

1. A ſmall elegant variety growing on the ſea-coaſt obſerved by Dillenius who gives a fig. and a deſcription of it in *Ray's Syn*. 330. T. 14. F. 2

Lagopus perpuſillus ſupinus perelegans maritimus. Ph. Br. *Ray's Ibid.*

Teaſel-headed Maritimum — TREFOIL. Flowers in woolly globular ſpikes; ſegments of the flower-cup equal; open. Stem aſcending. Leaves wedge-ſhaped; hairy. *Hudſ. Fl. Ang*. 284—*Bloſſoms, ſmall, purple or whitiſh.*

Trifolium ſtellatum glabrum. *Gerard*. 1208. *Ray's Syn*, 329.

In ſalt marſhes, and meadows near the ſea-ſhore. P. July.

OBS. *Linnæus has a ſpecies (the* Stellatum*) which correſponds pretty nearly with Hudſon's deſcription, but they appear to be diſtinct plants.*

Yellow-flowered Ochroleucum — TREFOIL. Flowers in woolly ſpikes. Stem upright; downy. Lower leaves inverſely heart-ſhaped—*Upper leaves narrow; very entire.* Spikes *on fruit-ſtalks; oblong. The lowermoſt tooth of the cup the longeſt.* Bloſſoms *the colour of brimſtone.*

Trifolium pratenſe hirſutum majus, flore albo ſulphureo *Ray's Syn*. 328.

Dry paſtures. B. May—June.

Oval-headed Scabrum — TREFOIL. Flowers in heads; ſitting; lateral; egg-ſhaped. Segments of the cups unequal; ſtiff; reflected—*Stems but little branched.* Little leaves *oval; rather thick; ſomewhat ſcolloped at the edge.* Bloſſoms *white; ſlender; as long as the cup.*

Trifolium capitulo oblongo aſpero. *Bauh. pin*. 329.

Trifolium floſculis albis, in glomerulis oblongis aſperis, cauliculis proxime adnatis. *Ray's Syn*. 329.

On Chalk-hills. A. May—June.

Round-headed Glomeratum — TREFOIL. Flowers in hemiſpherical rigid heads. Cups ſcored; open; equal—*Bloſſoms pale red.*

Trifolium cum glomerulis ad caulium nodos rotundis. *Ray's Syn*. 329.

Sandy ground. A. June.

TREFOIL. Flowers in egg-ſhaped, almoſt lateral, ſitting heads. Cups ſcored, rounded—*Bloſſoms pale purple.* Knotted Striatum

Trifolium parvum hirſutum, floribus parvis dilute purpureis, in glomerulis mollioribus et oblongis; ſemine magno. *Ray's Syn.* 329. Tab. 12. fig. 3.

Dry paſtures. A. June.

**** *Cups bladder-ſhaped.*

TREFOIL. Flowers in roundiſh ſpikes. Cups bladder-ſhaped; with two teeth; reflected. Stems creeping—*Fruit-ſtalks longer than the leaves.* Stems *ſending out roots.* Cups *a little downy.* Bloſſoms *whitiſh; but ſometimes with a reddiſh tinge.* Strawberry Fragiferum

Trifolium Fragiferum. *Gerard.* 1208. *Ray's Syn.* 329.

Trifolium Fragiferum friſicum. *Bauh. pin.* 329. *Park.* 1109.

Meadows and moiſt paſtures. P. Auguſt.

Cows eat it.

***** *Standard of the bloſſom bent inwards.*

TREFOIL. Flowers in oval tiled ſpikes. Standards of the bloſſoms bent downwards; permanent. Cups not hairy. Stem upright—*Cups before flowering a little hairy.* Bloſſoms *yellow.* Hop Agrarium

Trifolium pratenſe luteum, capitulo lupi, vel Agrarium. *Bauh. pin.* 328. *Ray's Syn.* 330.

Trifolium luteum minimum. *Gerard.* 1186. *Park.* 111.

Gravelly ſoil. A. June.

Horſes, Cows, Sheep and Goats eat it.

TREFOIL Flowers in oval tiled ſpikes. Standards of the bloſſoms bent downwards; permanent. Stems trailing—*Flowers about ten or twelve.* Leaves *not ſcored.* Stems *long and lying entirely upon the ground.* Bloſſoms *yellow.* Trailing Procumbens

Trifolium lupulinum alterum minus. *Ray's Syn.* 330. Tab. 14. Fig. 3.

Decumbent Trefoil.

Meadows and paſtures. P. May—Auguſt.

Horſes, Cows, Sheep, and Goats eat it.

Small Filiforme

TREFOIL. Flowers in ſpikes ſomewhat tiled. Standards of the bloſſoms bent downwards; permanent. Cups ſupported on little foot-ſtalks. Stems trailing—*Little leaves notched at the end, ſcored.* Fruit-ſtalks *thread-ſhaped: longer than the leaves; not much thicker than a horſe-hair; three or five diſtinct florets in each little head.* Bloſſoms *yellow.*

Trifolium lupulinum minimum. *Ray's Syn.* 331. Tab. 14. fig. 4.

Sandy paſtures. A. May—June.

The flowers of all the ſpecies dried and powdered make bread, which in times of ſcarcity hath preſerved the inhabitants of Scotland from periſhing—the leaves of all the ſpecies fold up before rain.

The plantain fritillary Butterfly, *Papilio Cinxia,* and the Black Tuſſock Moth, *Phalæna Faſcelina,* live upon the different ſpecies.

301 SAINTFOIN. 887 Hedyſarum.

EMPAL. *Cup* one leaf; with five ſhallow clefts. *Segments* awl-ſhaped; upright; permanent.

BLOSS. Butterfly-ſhaped; ſcored.

Standard reflected and compreſſed; egg-ſhaped but oblong; notched at the end; long.

Wings oblong; narrower than the other petals; ſtraight,

Keel ſtraight; compreſſed; broadeſt at the outer part, and almoſt ſquare; cloven from the baſe to the broader part.

CHIVES. *Threads* ten; with an angular bend. *Tips* roundiſh; compreſſed.

POINT. *Seedbud* ſlender; compreſſed; ſtrap-ſhaped; *Shaft* awl-ſhaped; bent like the chives. *Summit* undivided.

S. VESS. *Shell* with roundiſh joints; compreſſed; with two valves, and one ſeed.

SEED. Kidney-ſhaped; ſolitary.

OBS *In the only Britiſh ſpecies the ſhell conſiſts of one joint, but in ſome of the foreign ſpecies it is formed of ſeveral joints connected together like the links of a chain.*

SAINTFOIN. Leaves winged; ſhells prickly, containing a ſingle ſeed. The wings of the bloſſom ſhorter than the cup. Stem long—*Bloſſoms red.* Cocks head Onobrychis

Onobrychis, ſeu caput Gallinaceum. *Gerard.* 1243. *Ray's Syn.* 327

Onobrychis vulgaris. *Park.* 1082.

Onobrychis foliis Viciæ, fructu echinato major. *Bauh. pin.* 350.

Saintfoin. Cocks-head.

Meadows and paſtures, particularly in chalky ſoils. P. July.

The following varieties ariſe in cultivation.

1. White bloſſomed.
2. Blue ditto.
3. Purple ditto.
4. Striped ditto.
5. Long-leaved hoary.

This is cultivated like Clover for feeding cattle, and is particularly advantageous in dry hilly ſituations, and chalky ſoils.

302 BIRDSFOOT. 884 Ornithopus.

Rundle ſimple.

EMPAL. *Cup* one leaf; tubular; *Rim* with five teeth; nearly equal; permanent.

BLOSS. Butterfly-ſhaped.

Standard inverſely heart-ſhaped; entire.

Wings egg-ſhaped; ſtraight; hardly ſo large as the ſtandard.

Keel compreſſed; very ſmall.

CHIVES. *Threads* ten; nine united. *Tips* ſimple.

POINT. *Seedbud* narrow. *Shaft* briſtle-ſhaped; aſcending. *Summit* a dot at the end of the ſhaſt.

S. VESS. *Shell* awl-ſhaped; cylindrical; bent like a bow; ſeparated into joints; opening joint by joint.

SEEDS. Solitary, roundiſh.

BIRDSFOOT. Leaves winged; ſhells crooked—*Bloſſoms yellow; purple, and white. The plant is ſmooth. The* Leaves *on leaf-ſtalks.* Common Perpuſillus

Ornithopodium minus. *Gerard.* 1241. *Bauh. pin.* 350.

Ornithopodium radice nodoſa. *Park.* 1093.

Ornithopodium radice tuberculis nodoſa. *Bauh. pin.* 350.

Ornithopodium majus. *Bauh. pin.* 350. *Park.* 1093.

In gravelly land. P. May—Auguſt.

303 HORSESHOE. 885 Hippocrepis.

Rundle ſimple.

EMPAL. *Cup* one leaf; with five teeth. The two upper teeth not ſo deeply divided, and nearly joined: permanent.

BLOSS. Butterfly-ſhaped.

Standard heart-ſhaped; with a claw as long as the cup.

Wings oblong egg-ſhaped; blunt.

Keel creſcent ſhaped; compreſſed.

CHIVES. *Threads* ten; nine united; aſcending. *Tips* ſimple.

POINT. *Seedbud* ſlender; oblong; ending in an awl-ſhaped *Shaft*; aſcending. *Summit* undivided.

S.VESS. *Shell* compreſſed; membranaceous; very long; crooked; deeply indented along one ſeam into roundiſh hollows, ſo that it appears as if compoſed of many three-edged joints connected together by the upper ſeam.

SEEDS. Oblong; crooked; one in each joint.

OBS. *The eſſential character of this genus conſiſts in the* Shell *being ſhaped like a horſe-ſhoe.*

Tufted Comoſa

HORSESHOE. Shells on fruit-ſtalks crowded together; bowed; ſerpentine on the outer edge—*Bloſſoms yellow.*

Ferrum Equinum Germanicum. ſiliquis in ſummitate. *Bauh. pin.* 349. *Ray's Syn.* 325.

Ferrum Equinum comoſum. *Park.* 1091.

Hedyſarum Glychirrizatum. *Gerard.* 1233.

Tufted Horſe-ſhoe Vetch.

On Chalky Hills. P. July.

304 SNAIL-SHELL. 399 Medicago.

EMPAL. *Cup* one leaf; ſtraight; cylindrical, but ſomewhat bell-ſhaped; with five ſhallow clefts; taper; equal.

BLOSS. Butterfly-ſhaped.

Standard egg-ſhaped; entire; bent inwards at the edges; the whole petal reflected.

Wings oblong egg-ſhaped: fixed to the appendage of the keel; approaching at the ſides under the keel.

Keel oblong; cloven; expanding; blunt; bent downwards by the pointal, and with the ſtandard forming a gaping mouth.

CHIVES. *Threads* ten; united almoſt the whole length. *Tips* ſmall.

POINT. *Seedbud* ſtanding on a little foot-ſtalk; oblong; bowed inwards; compreſſed; incloſed by the threads; burſting out of the keel and preſſing back the ſtandard; ending in a *ſhaft* which is ſhort; awl-ſhaped; generally ſtraight. *Summit* terminating; ſmall.

S.VESS. *Shell* compreſſed; long; bent inwards.

SEEDS. Many; kidney-ſhaped, or angular.

OBS. *The ſhell in ſome ſpecies is rolled up ſpirally like a ſnail-ſhell; in others it is bent like a bow or a ſickle.*

SNAIL-SHELL. Fruit-ſtalks in bunches; ſhells twiſted in a wreath: Stem upright; ſmooth. *Leaves numerous; ſerrated; three on each leaf-ſtalk. Bloſſoms purple.* Lucern Sativa

Meadows and paſtures. P. June—July.

The modern writers upon huſbandry ſtrongly recommend the cultivation of this plant for the purpoſe of feeding cattle, but it is not yet generally adopted.

SNAIL-SHELL. Fruit-ſtalks in bunches. Shells creſcent-ſhaped; ſtem proſtrate—*Bloſſoms yellow.* Yellow Falcata

Medica ſylveſtris. *Ray's Syn.* 333.

Medica fruteſcens flavo flore cluſii. *Park.* 1114.

Trifolium luteum ſiliqua cornuta, *Gerard.* 1191.

Trifolium luteum ſylveſtre ſiliqua cornuta, et medica fruteſcens. *Bauh. pin.* 330.

Yellow Medick. Butterjags.

Roads, and foot-ways in Norfolk. P. July.

In hot, dry, barren ſandy places it is well worth the trouble of ſowing for the purpoſe of making hay. A practice long ſince adopted in ſome parts of Sweden.—Cows, Horſes, Goats, and Sheep eat it.

SNAIL-

Melilot Lupulina SNAIL-SHELL. Flowers in oval ſpikes. Shells kidney-ſhaped; one ſeed in each. Stems trailing—*Shells ſcored; wrinkled; ſet with ſtiff hairs.* Bloſſoms *yellow.*

Trifolium luteum lupulinum. *Gerard.* 1186. *Ray's Syn.* 331.

Trifolium montanum lupulinum. *Park.* 1105.

Trifolium pratenſe luteum, capitulo breviore. *Bauh. pin.* 328.

Melilot Trefoil.

Corn-fields and ſandy paſtures. A. May—Auguſt.

Cows, Horſes, Goats and Sheep eat it; but it is leſs grateful to them than the other ſpecies.

Wreathed Polymorpha SNAIL-SHELL. With ſhells reſembling a ſnail-ſhell: props toothed: ſtem ſpreading.—

The varieties of this ſpecies are often cultivated in our flower gardens for the ſake of the curiouſly formed ſeed-veſſels, which bear ſome reſemblance to green Caterpillars, Snail-ſhells, Hedge-hogs, &c. The following are the principal varieties.

1. *Claver Arabica* With prickly ſhells; generally three together. Little leaves heart-ſhaped—*With a brown ſpot upon each.* Props *fringed.* Bloſſoms *yellow.*

Trifolium cochleatum folio cordato maculato. *Bauh pin.* 329. *Ray's Syn.* 333.

Trifolium Cordatum. *Gerard.* 1190.

Meadows, dry paſtures; and near the ſea-ſhore. A. May.

2. *Small Minima* With prickly ſhells: prickles hooked; alternate. Props entire—*Shells ſeveral together.*

Trifolium echinatum fructu minore. *Bauh. pin.* 330.

Medica echinata minima. *Ray's Syn.* 333.

Smalleſt Hedge-hog Trefoil.

In ſandy places. A. May.

3 *Soft Mollis* With ſhells rather compreſſed, and covered with ſoftiſh prickles.

Trifolium cochleatum modiol s ſpinoſis. *Ray's Syn.* 333.

Hedge-hog Trefo : with ſmall fruit like the ſegment of a cone, or nave of a cart-wheel.

Near the ſea-ſhore. A.

4. *Sea Maritima* With prickly ſhells; and the upper leaves covered with ſoft hair.

Medica marina ſupina noſtras foliis ad ſummos ramulos villoſis. *Ray's Syn.* 334.

Medicæ marinæ ſpinoſa ſpecies? *Gerard.* 1200.

Medica marina major et minor ſpinoſa? *Park. Theat.* 1115.

On the ſea-ſhore. A.

5. *Corn Arvenſis* With rough ſhells covered with ſmall tubercles; without prickles; ſlightly compreſſed: about ten growing on a fruit-ſtalk.

Medica polycarpos fructu minore compreſſo ſcabro. *Ray's Syn.* 333.

Trifolium cochleatum polycarpon ſive medica racemoſa. *Park.* 1114.

In corn-fields. A. May.

305 TARE.

305 TARE. 874 Ervum.

EMPAL. *Cup* with five divifions ; nearly as long as the bloffom. *Segments* narrow ; tapering ; nearly equal.

BLOSS. Butterfly-fhaped.

Standard flat ; a little reflected ; circular ; large.

Wings blunt ; half as long as the ftandard.

Keel fhorter than the wings ; tapering.

CHIVES. *Threads* ten ; rifing upwards ; nine united. *Tips* fimple.

POINT. *Seedbud* oblong. *Shaft* fimple ; rifing upwards ; *Summit* blunt ; without a beard.

S. VESS. *Shell* oblong ; blunt ; cylindrical ; with protuberances formed by the feeds.

SEEDS. Four ; nearly round.

OBS. *The chief difference betwixt this Genus and the* VETCH *confifts in the fummit. In the* Smooth TARE *the cup is cut into five unequal teeth ; and the fummit when viewed through a Microfcope appears bearded, fo that is ought to have been arranged with the* Vetches.

TARE. With about two flowers on a fruit-ftalk. Seeds globular ; four in each fhell—*Little leaves about ten ; generally alternate.* Fruit-ftalks *hair-like ; very flender ; fupporting one or two flowers.* Flowers *fmall ; violet or blood coloured.* Shell *oblong oval ; fmooth.* Smooth. Tetrafpermum

Vicia fegetum, fingularibus filiquis glabris. *Bauh. pin.* 345.

Cracca minor filiquis fingularibus, flofculis cærulefcentibus. *Ray's Syn.* 322.

Corn-fields. A. June.

Horfes, Cows, Goats and Sheep eat it.

TARE. With many flowers on a fruit-ftalk. Seeds globular, two in each fhell—*Bloffoms blue and white.* Hairy Hirfutum

Vicia fegetum cum filiquis plurimis hirfutis. *Bauh. pin.* 345.

Vicia fylveftris, feu cracca minima. *Gerard.* 1028.

Arachus, feu cracca minor. *Park.* 1070.

Cracca minor. *Ray's Syn.* 322.

Corn-fields. A. June.

Horfes, Cows, Goats and fheep eat it.

TARE-

Spring Soloniense

TARE. The fruit-ſtalks ſupporting about two flowers, with awns. Leaf-ſtalks tapering. Little leaves blunt—*Stem ſeven or eight inches high; a little downy.* Fruit-ſtalks *longer than the leaves; terminated by a tendril, under which are one or two flowers on little fruit-ſtalks; alternate.* Bloſſoms *purple.*

Vicia minima præcox Pariſienſium. *Ray's Syn.* 321.

In chalky ſoils. A. April—May.

306 C L A V E R. 897 Lotus.

Rundle ſimple.

EMPAL. *Cup* one leaf; cylindrical; with five ſhallow clefts. *Teeth* ſharp; equal; upright; permanent.

BLOSS. *Butterfly-ſhaped.*

Standard circular; bent downwards; with an oblong concave claw.

Wings circular; ſhorter than the ſtandard; broad; approaching upwards.

Keel hunched in the lower part; cloſed above; taper; aſcending; ſhort.

CHIVES. *Threads* ten; aſcending; nine united; rather broad at the ends. *Tips* ſmall; ſimple.

POINT. *Seedbud* cylindrical; oblong. *Shaft* ſimple; aſcending. *Summit* a dot, bending inwards.

S. VESS. *Shell* cylindrical; ſtiff and ſtraight; full; longer than the cup: with two valves, one cell, and as if tranſverſely divided into many.

SEEDS. Many; cylindrical.

Birds-foot Corniculatus

CLAVER. Flowers in depreſſed heads. Stems drooping. Shells cylindrical; expanding—*Bloſſoms yellow.* Seeds *numerous.*

Lotus corniculata glabra minor. *Ray's Syn.* 334.

Lotus, ſeu melilotus pentaphyllos minor glabra. *Bauh. pin.* 332.

Trifolium ſiliquoſum minus. *Gerard.* 1191.

Birds-foot Trefoil.

1. Leaves narrow. Stems almoſt woody. *Ray's Syn.* 334. Leſſer, buſhy, narrow-leaved Birds-foot Trefoil.
2. Flowers large; yellow; ſhining. *Bauh. pin.* 332. Greater Birds-foot Trefoil.
3. A larger variety, but little hairy. *Ray's Syn.* 334.
4. Leaves white with down on the under ſurface. *Ray's Syn.* 334.

Hedges, paſtures. 2, in woods and moiſt meadows: 4, in Chalk-pits. P. June—Auguſt.

In Hertfordſhire it is cultivated as paſturage for ſheep—The flowers become greeniſh when dried, and in this reſpect they reſemble the flowers of the plants that produce Indigo.—Cows, Goats and Horſes eat it; Sheep and Swine are not fond of it.

The Skipping Thrips, *Thrips Phyſapus* is found upon it.

CLASS

CLASS XVIII.

THREADS in many SETS.

THIS Clafs comprehends the Plants whofe Flowers have the Chives united by the Threads into three or more Sets.

Order IV. Many Chives.

307 TUTSAN. - - *Cup* with five divifions : beneath. *Blofs* : five Petals. *Shafts* one, three or five. *Capfule* with one or more Cells.

307 TUTSAN. 902 Hypericum.

EMPAL. *Cup* with five diviſions. *Segments* ſomewhat egg-ſhaped; concave; permanent.

BLOSS. *Petals* five; oblong egg-ſhaped; blunt; expanding; bending from the left to the right,

CHIVES. *Threads* numerous; hair-like; connected at the baſe into three or five ſets. *Tips* ſmall.

POINT. *Seedbud* roundiſh. *Shafts* three (ſometimes one or two or five) ſimple; diſtant; as long as the chives. *Summits* ſimple.

S. VESS. *Capſule* roundiſh; with as many cells as there are ſhafts.

SEEDS. Several; oblong.

Park-leaves Androſæmum

TUTSAN. Flowers with three ſhafts. Fruit like a berry. Stem ſomewhat woody, two edged—*Bloſſoms yellow.*

Hypericum maximum Androſæmum vulgare dictum. *Ray's Syn.* 343.

Androſæmum vulgare. *Park.* 575.

Androſæmum maximum fruteſcens. *Bauh. pin.* 280.

Clymenum Italorum. *Gerard.* 548.

Park-leaves. St. John's-wort.

Woods and moiſt hedges. P. July.

This plant is not uncommon in our flower-gardens.

St. Peter's Quadrangulum

TUTSAN. Flowers with three ſhafts. Stem four-cornered; herbaceous—*Bloſſoms yellow.*

Hypericum Aſcyron dictum, Caule quadrangulo. *Ray's Syn.* 344.

Aſcyron. *Gerard.* 542. vulgare. *Park.* 575.

St. Peter's-wort

Moiſt hedges. P. July.

Cows, Goats and Sheep eat it; Horſes and Swine refuſe it.

St. John's Perforatum

TUTSAN. Flowers with three ſhafts. Stem two edged; Leaves blunt; with ſemi-tranſparent dots—*Flowers open at night as well as in the day.* Bloſſoms *yellow.* *When cultivated, the dots on the leaves are ſometimes red.* Tips *double*; *yellow*; *furniſhed with a ſmall black gland.*

Hypericum. *Gerard.* 540. *Ray's Syn.* 342. vulgare. *Bauh. pin.* 279. *Park.* 572.

Saint John's wort.

In rough uncultivated grounds and hedges. P. July.

This

This plant has long held a place in the Medicinal catalogues, but its uſe is very much undetermined—The ſemi-tranſparent dots on the leaves are the receptacles of an eſſential oil—The leaves given in ſubſtance are ſaid to deſtroy worms—The flowers tinge ſpirits and oils of a fine purple colour, which is probably derived from the little glands upon the tips, and upon the edges of the petals.---Cows, Goats and Sheep eat it; Horſes and Swine refuſe it.

TUTSAN. Flowers with three ſhafts; at the baſe of the leaves; ſolitary. Stems two-edged; proſtrate; thread-ſhaped. Leaves ſmooth.—*Greatly reſembling the preceding ſpecies; but the leaves have no tranſparent dots.* Bloſſoms *yellow.* Trailing Humifuſum

Hypericum minus ſupinum. *Park.* 572. *Ray's Syn.* 343.

Hypericum ſupinum glabrum. *Gerard.* 541. *Bauh. pin.* 279.

1. Leaves growing by threes.

Gravelly ſoil. P. July.

TUTSAN. Flowers with three ſhafts. Cups ſerrated, and glandular. Stem cylindrical; upright. Leaves egg-ſhaped; ſmooth. *Floral leaves at the top of the Stem, ſmall, and fringed with glands.* Bloſſoms *yellow.* Mountain Montanum

Hypericum elegantiſſimum non ramoſum, folio lato. *Ray's Syn.* 343.

Aſcyron ſeu Hypericum bifolium glabrum non perforatum. *Bauh. pin.* 280.

Mountain St. John's-wort.

Woods and rough hilly ground. P. July.

TUTSAN. Flowers with three ſhafts. Cups ſerrated, and glandular. Stem cylindrical; upright. Leaves egg-ſhaped; a little downy—*with ſemi-tranſparent dots.* Flowers *cloſing at night.* Bloſſoms *yellow.* Hairy Hirſutum

Androſæmum hirſutum. *Bauh. pin.* 280.

Hairy St. John's-wort.

1. Leaves oblong egg-ſhaped; on ſhort leaf-ſtalks. Segments of the cup oblong; ſharp. *Oblong.*

Hedges and rough grounds. P. July.

Sheep eat; Horſes refuſe it.

Marſh Elodes

TUTSAN. Flowers with three ſhafts. Stem cylindrical; creeping. Leaves circular; woolly—*Bloſſoms yellow.*

Aſcyron ſupinum villoſum paluſtre. *Bauh. pin.* 280. *Park.* 574. *Ray's Syn* 344.

Aſcyron ſupinum elodes Cluſii. *Gerard.* 542.

Marſh St. John's wort.

In putrid bogs and amongſt Peat. July.

Upright Pulchrum

TUTSAN. Flowers with three ſhafts. Cups ſerrated, and glandular. Leaves heart-ſhaped; ſmooth; with ſemi tranſparent dots. Stem cylindrical—*Bloſſoms yellow, with a tinge of red.*

Hypericum minus erectum. *Bauh. pin.* 279.

Hypericum pulchrum Tragi. *Ray's Syn.* 342.

Hypericum quintum ſeu pulchrum Tragi. *Gerard.* 540.

Upright St. John's-wort.

Woods, hedges and heaths. P. July.

CLASS.

CLASS. XIX.

THIS Claſs comprehends thoſe Flowers which Mankind have very generally agreed to call *Compound.* The Eſſential character of a COMPOUND FLOWER conſiſts in the TIPS being united ſo as to form a Cylinder, and a ſingle Seed being placed upon the Receptacle under each Floret. The DANDELION and the THISTLE are compound Flowers; that is, each of theſe *Flowers* are compoſed or compounded of a number of ſmaller Flowers, called FLORETS.

Character of the FLOWER.

It is compoſed of many *Florets*, ſitting upon a COMMON RECEPTACLE, and incloſed by one COMMON EMPALEMENT. The

Surface of the RECEPTACLE is either concave; flat; convex; pyramidal; or globular. It is either

Naked, that is, marked only with little dots, as in DANDELION; or

Hairy; covered with ſoft upright hairs as in THISTLE; or

Chaffy; beſet with awl-ſhaped, narrow, compreſſed, upright chaffy ſubſtances, ſeparating the Florets, as in CHAMOMILE or YARROW.

The COMMON EMPAL. is a *Cup* which ſurrounds the Florets and the common Receptacle. (*When the Florets have bloſſomed it contracts; but when the Seeds are ripe it expands and falls back.*) It is either

Simple; when formed with only a ſingle row of ſcales or Leaves.

Tiled; when the ſcales are numerous the outer ones gradually growing ſhorter and lying upon the inner ones, like the Tiles upon a Houſe.

Leafy; when a ſingle row of equal and longer ſegments ſtands next to the florets, and another row of very ſmall ſcales ſurrounds the baſe of thoſe ſegments.

The ſtructure of the FLORETS that compoſe a compound flower, will be beſt underſtood by pulling to pieces the flower of a Thiſtle, or of Dandelion, or of the Sunflower, and comparing the Florets with the following

Natural Character of a FLORET.

EMPAL. None but the crown of the ſeed ſitting upon the top of the ſeedbud.

BLOSS. One petal. *Tube* very ſlender and long; ſitting upon the ſeedbud. (*it is either*)

1. TUBULAR. *Border* bell-ſhaped, with five clefts. *Segments* reflected and expanding.

2. NARROW. *Border* ſtrap-ſhaped; flat; turned outwards; lopped at the end, which is either entire, or marked with three or five teeth.

3. NONE. *Border* wanting; and ſometimes the petal is altogether deficient.

CHIVES. *Threads* five; hair-like; very ſhort; fixed to the neck of the bloſſom. *Tips* five; upright; ſtrap-ſhaped; connected at the ſides ſo as to form a hollow cylinder, as long as the border of the bloſſom, and marked at the top with five teeth.

POINT. *Seedbud* oblong; ſtanding under the bloſſom upon the common receptacle. *Shaft* thread-ſhaped; upright; as long as the chives; paſſing through the hollow cylinder formed by the tips. *Summit* cloven; the ſegments rolled back and expanding.

S. VESS. Properly ſpeaking, none; though in ſome foreign Genera there is a ſort of leathery cruſt over the ſeed.

SEED. Single; oblong; frequently with four edges: generally narrower towards the baſe.

Crowned with a FEATHER, which either conſiſts of many *Undivided* hair-like ſpokes, placed in a circle; or of ſpokes that are *Branched* or radiated. This *Feather* again is either ſupported upon a little *Pillar*, or elſe *Sitting* immediately upon the ſeed.

- - - - with a ſmall CUP which hath generally five teeth, and is permanent.

- - - - neither with a *Cup* nor with a feather.

OBS. *In examining the minuter Florets, the diſſecting Needles, and the Botanic Microſcope, will be found extremely uſeful.*

The

The diſpoſition of the Chives and Pointals varying, occaſions the following

Diſtinctions of Florets.

Bloss. Tubular {
1. Furniſhed with chives and a pointal.
2. Furniſhed with chives, but no pointal.
3. Furniſhed with a pointal, but no chives.
4. Without either chives, pointal or ſummit.

Bloss. Narrow {
1. Furniſhed with chives and a pointal.
2. Furniſhed with chives, but no pointal.
3. Furniſhed with a pointal, but no chives.
4. Without either chives, pointal or ſummit.

From conſidering the different ſtructure of the Florets, it is evident that the compound Flowers may be compoſed either

Of

1. Florets *Tubular* in the center, with chives and pointals. *Tubular* in the circumference, with chives and pointals.
2. Florets *Tubular* in the center, with chives and pointals. *Tubular* in the circumference, with only pointals.
3 Florets *Tubular* in the center, with chives and pointals. *Tubular* in the circumference, with neither chives nor pointals.
4. Florets *Tubular* in the center with chives and pointals. *Narrow* in the circumference, with chives and pointals.
5. Florets *Tubular* in the center, with chives and pointals. *Narrow* in the circumference, with only pointals.
6. Florets *Tubular* in the center, with chives and pointals. *Narrow* in the circumference, with neither chives nor pointals.
7. Florets *Tubular* in the center with chives and pointals. *Pointals* in the circumference without bloſſoms.
8. Florets *Tubular* in the center, with chives and imperfect pointals. *Pointals* in the circumference without bloſſoms.
9. Florets *Narrow* in the center, with chives and pointals. *Narrow* in the circumference with chives and pointals.

The

The ORDERS therefore, according to the ſyſtem we have adopted, will be as follows.

I. *Chives and Pointals equal.* That is, when all the Florets are furniſhed with Chives and Pointals. (9. 1. 4. of the preceding table.)

II. *Superfluous Pointals.* That is, when the Florets in the center have both Chives and Pointals; but the florets in the circumference have only Pointals. (2. 5. 7. of the preceding table.)

III. *Barren Florets.* That is, when the Florets in the center have both Chives and Pointals; but the Florets in the circumference neither. (3. 6. of preceding table.)

IV. *Neceſſary Pointals.* That is, when the Florets in the center have both Chives and Pointals, but from ſome defect in the Pointals produce no Seed. The Florets in the circumference have Pointals only and produce perfect Seeds. (8 of the preceding table.)

V. *Separated Florets.* That is when ſeveral Cups filled with florets, are contained in another larger cup, ſo as to form but one flower.

VI. *Flowers ſimple.* That is, when there is only one flower in a cup, and theſe not incloſed by another larger cup ſo as to form but one flower.

OBSERVATIONS.

This is a Natural Claſs except the laſt Order, which from the principles of the Syſtem is neceſſarily introduced here.

The plants of this Claſs are ſuppoſed to have various ſpecific virtues. Moſt of them are bitter. None of them are poiſonous, except perhaps the ſtrong ſcented LETTUCE *when growing in ſhady ſituations.*

The elaſticity of the Empalement in the OXTONGUE, THISTLE, *and many other Genera, is too remarkable to paſs unnoticed by the ſlighteſt obſerver. It ſeems as if the expanſion of the florets firſt burſt the Empalement open, and when theſe wither, it cloſes again. the downy hairs that crown the Seeds, before upright, now begin to expand, and by this expanſion again open the leaves of the Empalement, and bend them quite back. The Seeds now eſcape, and the Empalement becoming dry and withered, no longer retains its elaſtic power.*

The hairy appendages of the Seeds are very properly called FEATHERS; *for by means of theſe, the Seeds are wafted about in the air and diſſeminated far and wide. The ſtructure of theſe feathers deſerve our notice: there is hardly a child that is inſenſible to their beauty in the* DANDELION.

CLASS

CLASS XIX.

TIPS UNITED.

Order I. Chives and Pointals equal.

* *All the Florets* NARROW.

308 ENDIVE. - *Receptacle* a little chaffy. *Feather* with about five teeth. *Cup* double.

309 HAWKSEYE, *Receptacle* chaffy. *Feather* ſomewhat downy. *Cup* tiled.

310 GOATSBEARD. *Receptacle* naked. *Feather* downy. *Cup* ſimple.

311 OXTONGUE. *Receptacle* naked. *Feather* downy. *Cup* double.

312 DANDELION. *Receptacle* naked. *Feather* downy. *Cup* tiled with flexible ſcales.

313 SOWTHISTLE. *Receptacle* naked. *Feather* hairy. *Cup* titled, and hunched.

314 SUCCORY. *Receptacle* naked. *Feather* hairy. *Cup* double; with ſcales of different ſhapes.

315 IVYLEAF. - *Receptacle* naked. *Feather* hairy. *Cup* double; containing about five florets.

316 LETTUCE. - *Receptacle* naked. *Feather* hairy. *Cup* tiled; ſcales ſkinny at the edge.

317 HAWKWEED. *Receptacle* naked. *Feather* hairy. *Cup* tiled; egg-ſhaped.

318 NIPPLE-

318 NIPPLEWORT. *Receptacle* naked. *Feather* none. *Cup* double.

319 YELLOWEYE. *Receptacle* naked. *Feather* not hairy. *Cup* nearly equal.

* * *Flowers in globular heads.*

320 TWINGEWORT. *Cup* radiate. The *Rays* coloured.

321 BURDOCK. - *Cup* with the ſcales bent inwards at the points, and hooked.

322 THISTLE. - *Cup* with thorny ſcales; diſtended. *Receptacle* hairy.

323 ARGENTINE. *Cup* with thorny ſcales; diſtended; *Receptacle* like a honey-comb.

324 SAWWORT. *Cup* with ſcales tiled; ſharp, but without thorns; nearly cylindrical.

* * * *All the Florets* TUBULAR.

325 LIVERHEMP. *Receptacle* naked. *Feather* downy. *Cup* tiled. *Pointal* very long.

326 CUDWORT. *Receptacle* chaffy. *Feather* chaffy. *Cup* tiled.

327 DOUBLETOOTH. *Receptacle* chaffy. *Feather* with awns. *Cup* tiled,

† *Common Groundſel.* † *Commom Tanſcy.*

Order II. Superfluous Pointals.

* *All the florets* TUBULAR.

328 SOUTHERNWOOD. *Receptacle* almoſt naked. *Feather* none. *Bloſſoms* in the circumference none.

329 TANSEY. *Receptacle* naked. *Feather* only a ſort of border. Bloſſoms of the circumference with three clefts.

330 SPIKENARD. *Receptacle* naked. *Feather* hairy. Bloſſoms of the circumference with three clefts.

331 CATSFOOT. *Receptacle* naked. *Feather* downy. *Cup* with ſkinny concave ſcales.

† *Common Groundſel.*

* * *Florets*

* * *Florets of the circumference* NARROW.

332 DAISIE. - - *Receptacle* naked. *Feather* none. *Cup* ſimple, with equal ſcales.

333 FEVERFEW. *Receptacle* naked. *Feather* none. *Cup* tiled with ſharp ſcales.

334 GOLDINGS. - *Receptacle* naked. *Feather* none. *Cup* with the inner ſcales membranaceous.

335 ELECAMPANE. *Receptacle* naked. *Feather* hairy. *Tips* with two briſtles at the baſe.

336 FLEABANE. - *Receptacle* naked. *Feather* hairy. *Bloſſoms* in the circumference very ſlender.

337 GOLDENROD. *Receptacle* naked. *Feather* hairy. *Bloſſoms* of the circumference about ſix; remote.

338 FLEAWORT. - *Receptacle* naked. *Feather* hairy. *Cup* equal; ſimple.

339 GROUNDSEL. *Receptacle* naked. *Feather* hairy. *Cup* with the ſcales dead at the ends.

340 BUTTERBUR. *Receptacle* naked. *Feather* hairy. *Cup* with ſomewhat membranaceous ſcales.

341 STARWORT. - *Receptacle* naked. *Feather* hairy. *Cup* a little rough and ſcurfy.

342 CHAMOMILE. *Receptacle* chaffy. *Feather* none. *Cup* hemiſpherical.

343 YARROW. - *Receptacle* chaffy. *Feather* none. *Circumference* containing about five florets. *Cup* oblong.

Order III. Barren Florets.

344 KNAPWEED. *Receptacle* briſtly. *Feather* hairy. *Bloſſoms* of the circumference tubular.

Order IV. Neceſſary Pointals.

345 CUDWEED. *Receptacle* naked. *Feather* none. *Florets with Pointals* amongſt the ſcales of the cup.

† *Fleabane.*

Order

Order VI. Flowers Simple.

346 SCABIOUS. *Cup* common. *Blossom* five petals; regular. *Capsule* beneath; with two cells.

347 CARDINALFLOWER. *Cup* with five teeth. *Blossom* one petal; irregular. *Capsule* beneath; with two cells.

348 VIOLET. - - *Cup* five leaves. *Blossom* five petals; irregular. *Capsule* superiour; with three valves.

349 WEATHERCOCK. *Cup* two leaves. *Blossom* five petals; irregular. *Capsule* superiour; with five valves.

308 ENDIVE. 921 Cichorium.

EMPAL. *Common,* double; cylindrical. *Scales* eight; narrow; ſpear-ſhaped; equal; forming a cylinder. Five ſcales ſhorter than and lying upon the others, form the outer cup.

BLOSS. *Compound,* flat; uniform. *Florets* with chives and pointals about twenty; placed in a circle.

Individuals; one petal; narrow; lopped; deeply divided into five teeth.

CHIVES. *Threads* five; hair-like; very ſhort. *Tips* forming a hollow cylinder, with five edges.

POINT. *Seedbud* oblong. *Shaft* thread-ſhaped; as long as the chives. *Summits* two; rolled back.

S.VESS. None; but the cylindrical *Cup* cloſing at the point.

SEEDS. Solitary; compreſſed; with ſharp angles. *Crown* of the ſeed an imperfect margin, with about five teeth;

RECEPT. Chaffy.

Wild Intybus

ENDIVE. The flowers in pairs; ſitting. Leaves notched. *Scales of the outer cup fringed.* Bloſſoms *lateral; blue: by cultivation roſe-coloured. They open at eight in the morning and cloſe at four in the afternoon.*

Cichoreum ſylveſtre. *Gerard.* 284. *Park.* 776. *Ray's Syn.* 172.

Cichorium ſylveſtre, ſeu officinarum. *Bauh. pin.* 125.

Wild Succory, or Cichory.

Borders of corn-fields. B. July—Auguſt.

The leaves when blanched are eaten early in the ſpring in ſallads. They loſe their bitterneſs by cultivation—The roots gathered before the ſtem ſhoots up are eatable, and when dried will make bread.—Sheep, Goats and Swine eat it; Cows and Horſes refuſe it.

309 HAWKSEYE. 918 Hypochæris.

EMPAL. *Common*, oblong; tiled; diſtended at the baſe. *Scales* ſpear-ſhaped; ſharp.

BLOSS. *Compound* tiled; uniform. *Florets* with chives and pointals, numerous; the inner ones gradually ſhorter.

Individuals one petal; narrow; ſtrap-ſhaped; lopped; with five teeth.

CHIVES. *Threads* five; hair-like; very ſhort. *Tips* forming a hollow cylinder.

POINT. *Seedbud* egg-ſhaped. *Shaft* thread-ſhaped; as long as the chives. *Summits* two; reflected.

S. VESS. None: the *Cup* becoming globular but tapering, cloſes on the ſeeds.

SEEDS. Solitary; oblong. *Feather* ſtanding on a pillar; downy.

RECEPT. Chaffy. *Chaff* ſpear-ſhaped; narrow; as long as the ſeeds.

HAWKSEYE. The ſtem almoſt naked, generally with a single branch. Leaves oblong egg-ſhaped; entire; toothed.—*Cup hairy; compoſed of large ſcales.* Bloſſoms *yellow: they open at ſix in the morning and cloſe at four in the afternoon.* Spotted Maculata

Hieracium I latifolium. *Ray's Syn.* 167.

Hieracium alpinum latifolium hirſutie incanum, flore magno. *Bauh. pin.* 12.

Spotted Hawkweed. Broad-leaved Hungarian Hawkweed.

On high grounds. P. July.

The leaves are boiled and eaten like Cabbage—Horſes are fond of this plant when green, but they do not like it when dry.—Cows, Goats and Swine eat it; Sheep are not fond of it.

HAWKESEYE. Smooth. Cups oblong; tiled. Stem branched; naked. Leaves toothed and indented.—*Bloſſoms very ſmall, yellow. Scales of the* Cup *ſmooth.* Fruit-ſtalks *thickeſt towards the top. The flowers open at nine in the morning and cloſe again at twelve or one.* Smooth Glabra

Hieracium minus, dentis Leonis folio oblongo glabro. *Bauh. pin.* 127.

Hieracium parvum in arenoſis naſcens, ſeminum pappis denſius radiatis. *Ray's Syn.* 166.

Smooth Hawkweed

High gravelly paſtures. B. July.

Long-rooted Radicata

HAWKSEYE. The leaves notched; blunt; rough. Stem branched; naked; smooth. Fruit-stalks scaly—*Blossoms large; yellow within; reddish green on the out-side. They close at three in the afternoon. Scales of the* Cup *fringed on the back. At the base of the leaves, and in the angles formed by the branches, are certain long yellow hairs.*

Hieracium, dentis Leonis folio obtuso majus. *Bauh. pin.* 127.
Hieracium longius radicatum. *Gerard.* 295. *Park.* 790. *Ray's Syn.* 165.
Long-rooted Hawkweed.
In Pastures. P. May—September.

310 GOATSBEARD. 905 Tragopogon.

EMPAL. *Common,* simple; leaves eight; spear-shaped; equal; every other leaf standing more inwards; but all united at the base.

BLOSS. *Compound,* tiled; uniform. *Florets* with chives and pointals numerous; the outer ones rather longest.

Individuals one petal; narrow; lopped; with five teeth.

CHIVES. *Threads* five; hair-like; very short. *Tips* forming a cylinder.

POINT. *Seedbud* oblong. *Shaft* thread-shaped; as long as the chives. *Summits* two; rolled back.

S. VESS. None: the cup closing; tapering; as long as the seeds; a little distended.

SEEDS. Solitary; oblong; tapering towards each end; angular; rough; terminated by a long awl-shaped pillar supporting the *Feather,* which is downy and flat; with about thirty-two spokes.

RECEPT. Naked; flat; rough.

OBS. *In some species the seeds are straight and the cup longer than blossoms. In others the seeds are crooked and the cup shorter than the blossoms.*

Yellow Pratense

GOATSBEARD. The cup equal with the florets in the circumference. Leaves entire, stiff and straight—*The blossoms expan early in the morning but close again before noon. Yellow.*

Tragopogon luteum. *Gerard.* 735. *Park.* 412. *Ray's Syn.* 171.
Tragopogon pratense luteum majus. *Bauh. pin.* 274.
Go to Bed at Noon.
Corn-fields and pastures. B. June.
Before the stem shoots up, the roots boiled like Sparagus have the same flavour, and are nearly as nutritious--Cows, Sheep and Horses eat it; Swine devour it greedily; Goats are not fond it.

GOATS-

GOATSBEARD. The cup longer than the florets in the circumference. Leaves entire; ſtiff and ſtraight. Fruit-ſtalks thickeſt towards the top—*Bloſſoms purple.* Purple Porrifolium

Tragopogon purpuro-cæruleum porrifolio guod Artifi vulgo. *Baub. pin.* 274.

Tragopogon purpureum. *Gerard.* 735. *Park.* 412. *Ray's Syn.* 172.

Salſafy.

Meadows and marſhes. B. June.

The roots are eſculent, and when cultivated in gardens are called Salſafie—

311 OXTONGUE. 907 Picris.

EMPAL. *Common*, double. The *Outer Cup* large; with five heart-ſhaped, flat, flexible, approaching leaves. The *Inner Cup* tiled; egg-ſhaped.

BLOSS. *Compound*, tiled; uniform. *Florets* with chives and pointals; numerous.

Individuals one petal; narrow; ſtrap-ſhaped; lopped; with five teeth.

CHIVES. *Threads* five; hairlike; very ſhort. *Tips* forming a hollow cylinder.

POINT. *Seedbud* nearly egg-ſhaped. *Shaft* as long as the chives. *Summits* two; reflected.

S.VESS. None. The *Cup* unchanged contains the ſeeds.

SEEDS. Solitary; diſtended; blunt; furrowed tranſverſely. *Feather* downy.

RECEPT. Naked.

OXTONGUE. The outer cup with five leaves: larger than the inner cup, which is furniſhed with awns—*Leaves undivided; embracing the Stem.* Bloſſoms *yellow. They expand at four or five in the morning and never cloſe before noon; ſometimes they remain open till nine at night.* Common Echioides

Hieracium Echioides, capitulis Cardui benedicti. *Baub. pin.* 128. *Ray's Syn.* 166.

Bugloſſum luteum. *Gerard.* 798. ſeu Lingua bovis. *Park.* 800.

Corn-fields. A. July—Auguſt.

This is an agreeable pot-herb when young. The juice is milky, but not too acrid.

Hawkweed
Hieracium

OXTONGUE. Cups flexible. Leaves entire—*The outer cup consists of about ten leaves, only half as long as the inner ones. The* Inner Cup *hath about the same number of leaves, covered with strong hair.* Blossoms *yellow.*

Hieracium asperum, majore flore, in agrorum limitibus. *Ray's Syn.* 167.

Hieracium asperum. *Gerard.* 298.

Cichorium pratense luteum hirsutie asperum. *Bauh. pin.* 126.

Cichorium pratense luteum asperum. *Park.* 777.

Yellow Succory. Rough Hawk-weed.

Borders of corn fields. A. July

1. There is a variety with toothed and indented leaves.

Near the sides of brooks. A. September.

312 DANDELION. 912 Leontodon.

EMPAL. *Common,* tiled; oblong. The *Inner scales* strap-shaped parallel; equal. The *Outer scales* fewer, and generally reflected down to the base.

BLOSS. *Compound,* tiled; uniform. *Florets* with chives and pointals numerous; the inner ones gradually shorter.

Individuals one petal; narrow; lopped; with five teeth.

CHIVES. *Threads* five; hairlike; very short. *Tips* forming a hollow cylinder.

POINT. *Seedbud* nearly egg-shaped. *Shaft* thread-shaped; as long as the central blossoms. *Summits* two; rolled back.

S.VESS. None. *Cup* oblong; straight.

SEEDS. Solitary; oblong; rough. *Feather* supported on a little pillar; downy.

RECEPT. Naked; dotted.

OBS. *In the second species the feather is sitting.*

Common
Taraxacum

DANDELION. The scales of the cup reflected downwards. Leaves smooth, notched, and edged with little teeth—*Sometimes in moist situations the leaves are without the little teeth.* Blossoms *yellow. They generally expand at five or six in the morning and close again early in the afternoon.*

Dens Leonis. *Gerard.* 290. *Ray's Syn.* 171. vulgaris. *Park.* 780.

Dens Leonis latiore folio. *Bauh. pin.* 126.

Pissabed.

1. The leaves are sometimes narrower and a little downy. *Bauh. pin.* 129. *Ray's Syn.* 171.

Road-sides, pastures, Ditch-banks. P. April—September.

Early

Early in the ſpring whilſt the leaves are yet white, and hardly unfolded, they are an excellent ingredient in ſallads. The French eat the roots and tender leaves with bread and butter. Children that eat it in the evening experience its diuretic effects, which is the reaſon that other Europæan nations as well as ourſelves, vulgarly call it Piſſ-a-bed.—When a ſwarm of Locuſts had deſtroyed the Harveſt in the Iſland of Minorca, many of the inhabitants ſubſiſted upon this plant—The expreſſed juice has been given to the quantity of four ounces three or four times a day, and Boerhaave had a great opinion of the utility of this and other lacteſcent plants in viſceral obſtructions—Goats eat it; Swine devour it greedily; Sheep and Cows are not fond of it; Horſes refuſe it. Small Birds are fond of the ſeeds.

The Black Tuſſock Moth, *Phalæna Faſcelina*, and the *Thrips Phyſapus*, feed upon it.

DANDELION. Stem branching; fruit-ſtalks ſcaly. Leaves ſpear ſhaped; toothed; very entire; ſmooth—*The root appears as if bitten off.* Bloſſoms *yellow. They open about ſeven in the morning and cloſe about three in the afternoon.* Autumnal Autumnale

Hieracium minus præmorſa radice. *Park.* 794. *Ray's Syn.* 165.

Hieracium minus; ſeu Leporinum. *Gerard.* 296.

Hieracium Chondrillæ folio glabro, radice ſuccisa majus *Baub. pin.* 127.

Yellow Devils-bit.

1. There is a variety with jagged leaves.

Paſtures. P. Auguſt.

Horſes, Goats and Swine eat it; Cows and Sheep refuſe it.

DANDELION. Cup quite upright. Leaves toothed; hairy; very entire. Hairs-forked—*Bloſſoms yellow: greeniſh on the outſide. They open at four in the morning and cloſe at three in the afternoon.* Rough Hiſpidus

Dens Leonis hirſutus, leptocaulus, Hieracium dictus. *Ray's Syn.* 171.

Hieracium, Dentis Leonis folio hirſutum. *Gerard.* 303.

Hieracium aſperum flore magno Dentis Leonis. *Baub. pin.* 127.

Hieracium ſperum, foliis et floribus Dentis Leonis bulboſi. *Park.* 788. *Ray's Syn.* 167.

Paſtures. P. May—June.

Hairy
Hirtum

DANDELION. Cup quite upright. Leaves toothed; hairy hairs undivided—*Blossoms yellow on both sides.*

Hieracium pumilum saxatile asperum, præmorsa radice *Ray's Syn.* 167.

Hieracium Dentis Leonis folio, hirsutie asperum, minus laciniatum. *Bauh. pin.* 127.

Leontodon hispidum. *B. Hudson.* 297.

Pastures. P. June—September.

313 SOWTHISTLE. 908 Sonchus.

EMPAL. *Common*, distended and hunched; with many strap-shaped; unequal scales.

BLOSS. *Compound*, tiled; uniform. *Florets* with chives and pointals, numerous; equal.

Individuals, one petal; narrow; lopped; with five teeth.

CHIVES. *Threads* five; hair-like; very short. *Tips* forming a hollow cylinder.

POINT. *Seedbud* somewhat egg-shaped. *Shaft* thread-shaped; as long as the chives. *Summits* two; reflected.

S.VESS. None; the *Cup* closing forms a compressed globe, but tapering.

SEEDS. Solitary; rather long. *Feather* hairy.

RECEPT. Naked.

Marsh
Palustris

SOWTHISTLE. Flowers in a sort of rundle; fruit-stalks and cups rough with hair. Leaves notched; with awns at the base—*Blossoms deep yellow. They expand at six or seven in the morning and close at two in the afternoon.*

Sonchus tricubitalis, folio cuspidato. *Ray's Syn.* 163.

Sonchus arborescens alter. *Gerard.* 294.

Sonchus asper arborescens. *Bauh. pin.* 124.

Banks of rivers. P. August.

Tree
Arvensis

SOWTHISTLE. Flowers in a sort of rundle; fruit-stalks and cups rough with hair. Leaves notched; heart-shaped at the base—*Blossoms yellow. They expand betwixt six and seven and close again betwixt eleven and twelve in the morning.*

Sonchus repens; multis Hieracium majus. *Ray's Syn.* 163.

Sonchus arborescens. *Gerard.* 291.

Hieracium majus, folio Sonchi. *Bauh. pin.* 126.

Corn-fields and ditch-banks. P. August.

The flowers follow the course of the Sun, very regularly.

Cows and Goats eat it: Horses are very fond of it.

SOWTHISTLE.

SOWTHISTLE. Fruit-ſtalks downy ; cups ſmooth—*Leaves compreſſed and embracing the ſtem. The* Fruit-ſtalks *when old become ſmooth.* Bloſſoms *pale yellow. They open about five in the morning and cloſe again at eleven or twelve.* Common Oleraceus

Sonchus lævis. *Gerard.* 292. *Ray's Syn.* 162. vulgatius. *Park.* 805.

Sonchus lævis laciniatus latifolius. *Bauh. pin.* 124.

1. There are ſome varieties in the roughneſs or ſmoothneſs of the plants, and in the leaves being more or leſs jagged, which depend upon the ſoil, ſituation and time of growth.

Amongſt rubbiſh. Corn-fields. Gardens. A. June—Auguſt.

The leaves are good amongſt other pot-herbs—They are a very favorite food with Hares and Rabbits—Sheep, Goats and Swine eat it : Horſes are not fond of it.

The Sowthiſtle Louſe, *Aphis Sonchi,* lives upon it.

SOWTHISTLE. Fruit-ſtalks ſcaly ; flowers in bunches. Leaves notched—*Fruitſtalks clammy.* Cups *brown.* Bloſſoms *blue. They open at ſeven in the morning and cloſe at noon.* Blue Alpinus

Sonchus lævis laciniatus cæruleus, ſeu Sonchus Alpinus cæruleus. *Bauh. pin.* 124.

Blue Mountain Sow-thiſtle.

On the ſides of mountains. A.

The Laplanders get the ſtems before the flowers expand ; ſtrip off the bark and eat them , but they are much improved by the addition of Oil and Vinegar—It communicates a bitter taſte to the milk of Cattle that are fed with it.—Cows, Goats and Swine eat it : Horſes and Sheep are very fond of it.

Moſt of the ſpecies nouriſh the Sow-thiſtle Louſe, *Aphis Sonchi*

314 SUCCORY. 914 Crepis.

EMPAL. *Common* double.

Outer Cup very ſhort; open; deciduous.

Inner Cup egg-ſhaped; ſimple; furrowed; permanent. *Scales* ſtrap-ſhaped; approaching.

BLOSS. *Compound*, tiled; uniform. *Florets* many; equal: with chives and pointals

Individuals one petal; narrow; lopped; with five teeth.

CHIVES. *Threads* five; hair-like; very ſhort. *Tips* forming a hollow cylinder.

POINT. *Seedbud* nearly egg-ſhaped. *Shaft* thread-ſhaped; as long as the chives. *Summits* two; reflected.

S.VESS. None. *Cup* roundiſh.

SEEDS. Solitary; oblong. *Feather* ſometimes ſtanding on a pillar; hairy.

RECEPT. Naked.

OBS. *In the* ſmooth SUCCORY *the Feather of the ſeed is ſitting.*

Stinking
Fætida

SUCCORY. Leaves hairy; notched, and almoſt winged. Leaf-ſtalks toothed—*The flowers before they open hang down.* Leaves *ſmell like bitter almonds.* Bloſſoms *purpliſh on the outſide; yellow within.*

Hieracium Caſtorei odore monſpelienſium. *Ray's Syn.* 165.

Senecio hirſutus. *Bauh. pin.* 131.

Stinking Hawk-weed.

On chalky ſoil. B. June—July.

Smooth
Tectorum

SUCCORY. Leaves ſmooth; ſitting; ſpear-ſhaped; notched. The lower leaves toothed—*Stem angular and furrowed.* Cups *furrowed and ſet long-ways with clammy hairs. This plant varies greatly in its appearance according to the place of its growth.* Bloſſoms *yellow. They expand at four in the morning and cloſe about noon.*

Hieracium luteum glabrum, ſive minus hirſutum. *Ray's Syn.* 165.

Cichorium pratenſe luteum lævius. *Bauh pin.* 126. *Park.* 778.

Smooth Succory Hawk-weed.

Walls. Roofs. Paſtures. A. June—September.

The variations are chiefly in the leaves which are

1. Large and ſharp.
2. Small and ſharp.
3. Small and blunt.
4. Like Lettuce leaves.
5. It is ſometimes conſiderably branched.

Cows, Goats, Sheep and Swine eat it: Horſes are not fond of it.

SUCCORY.

SUCCORY. Leaves notched with winged clefts; rough; toothed above the base. Cups covered with sharp points—*Stem angular; rough; four feet high or more; brittle.* Blossoms *yellow.* Rough Biennis

Hieracium maximum Chondrillæ folio, asperum. *Bauh. pin.* 127. *Ray's Syn.* 166.

Rough Succory Hawk-weed.

Fields and hedges. B. July—August.

315 IVYLEAF. 911 Prenanthes.

EMPAL. *Common*, double; cylindrical; smooth. The *Scales* of the cylinder equal in number to the florets. The *Scales* of the base few; unequal; very short.

BLOSS. *Compound*, generally consisting of a single row of florets. *Florets* from five to eight; with chives and pointals; equal; standing in a circle.

Individuals one petal; narrow; lopped; with four teeth.

CHIVES. *Threads* five; hairlike; very short. *Tips* forming a hollow cylinder.

POINT. *Seedbud* nearly egg-shaped. *Shaft* thread-shaped; longer than the chives. *Summit* cloven; reflected.

S.VESS. None. *Cup* cylindrical; closing a little at the rim.

SEEDS. Solitary; heart-shaped. *Feather* hairy.

RECEPT. Naked.

IVYLEAF, with five florets in each compound flower. Leaves notched.—*Blossoms yellow. Feather of the seed supported by a little pillar.* Lettuce Muralis

Lactuca sylvestris murorum flore luteo. *Ray's Syn.* 162.

Sonchus lævis muralis. *Gerard.* 293.

Sonchus lævis alter, parvis floribus. *Park.* 805.

Sonchus lævis laciniatus muralis parvis floribus. *Bauh. pin.* 124.

Ivy-leafed wild Lettuce.

On old walls. In shady woods. P. July.

Cows, Goats and Horses eat it: Sheep are very fond of it,

316 LETTUCE. 909 Lactuca.

EMPAL. *Common*, tiled; somewhat cylindrical. *Scales* many, tapering to a point.

BLOSS. *Compound*, tiled; uniform. *Florets* with chives and pointals, many; equal.

Individuals one petal; narrow; lopped; with four or five teeth.

CHIVES *Threads* five; hairlike; very short. *Tips* forming a hollow cylinder.

POINT. *Seedbud* somewhat egg-shaped. *Shaft* thread shaped; as long as the chives. *Summits* two; reflected.

S. VESS. None. *Cup* closing; betwixt egg-shaped and cylindrical.

SEEDS. Solitary; egg-shaped; taper; compressed. *Feather* hairy; simple. *Pillar* long, tapering at the bottom.

RECEPT. Naked.

Strong-scented Virosa

LETTUCE. Leaves pointing horizontally; toothed. The middle rib set with prickles on the under-side—*Root leaves entire.* Blossoms *numerous*; *yellow*. Seeds *black*. *The blossoms open about seven and close about ten in the forenoon.*

Lactuca sylvestris major, odore opii. *Gerard.* 309. *Ray's Syn.* 161.

Lactuca sylvestris odore viroso. *Bauh. pin.* 123.

Lactuca virosa. *Park.* 813.

Strong-scented wild Lettuce.

1. There is a variety in which the leaves are not indented.

Ditch-banks. Borders of fields. B. July—August.

The Juice smells like Opium, and upon tryal has been found to possess similar properties. If it is caught in shells, and dried by a gentle heat, it may be formed into pills.

Least Saligna

LETTUCE. Leaves narrow but somewhat halberd-shaped, and sitting. The middle rib prickly on the under-side—*The lower leaves with winged clefts.* Blossoms *yellow*.

Chondrilla viscosa humilis. *Bauh. pin.* 130. *Park.* 783. *Gerard.* 287.

The Lambda Moth, *Phalæna Gamma*; the great Tyger Moth, *Phalæna Caja*; and the Lettuce Louse *Aphis Lactucæ*, live upon these species.

317 HAWK-

317 HAWKWEED. 913 Hieracium.

EMPAL. *Common*, tiled; cylindrical: *Scales* many; ſtrap-ſhaped; very unequal; lying lengthways one over another.

BLOSS. *Compound*, tiled; uniform. *Florets* with chives and pointals, numerous; equal.

Individuals one petal; narrow; lopped; with five teeth.

CHIVES. *Threads* five; hair-like; very ſhort. *Tips* forming a hollow cylinder.

POINT. *Seedbud* nearly egg-ſhaped. *Shaft* thread-ſhaped; as long as the chives. *Summits* two; bent backwards.

S. VESS. None. *Cup* cloſing; egg-ſhaped.

SEEDS. Solitary: with four blunt edges; ſhort. *Feather* hairy.

RECEPT. Naked.

* *Stalk naked; ſupporting a ſingle flower.*

HAWKWEED. Leaves oblong; entire; toothed. Stalk nearly naked, ſupporting one flower. Cup hairy—*Leaves with white upright hairs. Stalk thick, ſet with white hairs, which are brown at the baſe; and with one or two ſmall floral leaves towards the top.* Bloſſoms *yellow*. Mountain Alpinum

Hieracium alpinum pumilum, folio lanuginoſo. *Baub. pin.* 129.

Hieracium villoſum alpinum, flore magno ſingulari. *Ray's Syn.* 169. Tab. 6. fig. 2.

On Rocks. P. July.

HAWKWEED. Leaves very entire: egg-ſhaped: downy underneath. Stalk ſupporting one flower. Suckers creeping—*Bloſſom red on the outer-ſide; pale yellow within.* Cups *ſet thick with black hairs. The flowers open at eight in the morning and cloſe about two in the afternoon.* Mouſe ear Piloſella

Piloſella repens. *Gerard.* 638. *Ray's Syn.* 170.

Piloſella minor vulgaris repens. *Park.* 690.

Piloſella major repens hirſuta. *Baub. pin.* 262.

Common creeping Mouſe-ear.

Dry paſtures. A. May—July.

This

This differs from the other lactescent plants, being less bitter and more astringent—It is esteemed hurtful to Sheep—An insect of the Cochineal genus (*Coccus Polonicus*) is often found at the roots. *Act. Upsal*, 1752. Goats eat it; Sheep are not fond of it; Horses and Cows refuse it.

* * *Stalk naked, supporting several flowers.*

Green Dubium HAWKWEED. Leaves entire; oblong egg-shaped: Stalk naked; supporting several flowers. Suckers creeping—*Blossoms pale yellow.*

Pilosella major repens, minus hirsuta. *Bauh. pin.* 262.
On hills in Westmoreland. P. July—August.
Sheep eat it.

Narrowleaved Auricula HAWKWEED. Leaves very entire, spear-shaped. Stalk naked; supporting several flowers. Suckers creeping—*Stem and* Cups *beset with black hairs.* Hairs *on the leaves very small.* Blossoms *pale yellow.*

Pilosella major erecta altera. *Bauh. pin.* 262.
Hilly pastures. P. July.
Sheep eat it; Cows refuse it.

* * * *Stem leafy.*

Golden Murorum HAWKWEED. Stem branched. Root-leaves egg-shaped; toothed. Stem leaves one or two; smaller—*Toothed*; *sharp*; *reddish on the under surface.* Flowers *few*; *yellow.* *They open about six in the morning and close about two in the afternoon.*

Hieracium murorum foliis pilosissimis. *Bauh. pin.* 129. *Ray's Syn.* 168.

Hieracium murorum Bauhini, quod est Pulmonaria Gallorum Lobelii. *Park.* 801.

Pulmonaria Gallica, seu aurea latifolia. *Gerard.* 304.
French Lungwort. Golden Lungwort
Woods, old walls, and rough shady places. P. July.

1. Narrow-leaved. Pulmonarium dictum angustifolium. *Ray's Syn.* 168.
2. Round-leaved. Hieracium macrocaulon hirsutum folio rotundiore. *Ray's Syn.* 169.
3. Long-leaved. Leptocaulon hirsutum folio longiore. *Ray's Syn.* 169.

Horses eat it.

HAWK-

HAWKWEED. Stem ſupporting a panicle of flowers. Leaves embracing the ſtem; toothed; ſmooth. Cups rough with hair—*The hairs of the cups black.* Bloſſoms *yellow. They expand at ſix in the morning and cloſe at five in the afternoon.* Marſh Paludoſum

Hieracium montanum latifolium glabrum minus. *Bauh. pin.* 129.

Hieracium montanum Cichorei folio. *Ray's Syn.* 166.

Succory-leaved Hawkweed.

Moiſt meadows and banks of rivulets. B. July.

1. Cups and fruit-ſtalks ſet with yellow hairs.

HAWKWEED. Stem upright; ſupporting many flowers. Leaves betwixt egg and ſpear-ſhaped; toothed; half embracing the ſtem—*The receptacle in this and the following ſpecies is more rough than in the others.* Bloſſoms *yellow. They continue expanded from ſeven in the morning until one or two in the afternoon.* Broad leaved Sabaudum

Hieracium fruticoſum latifolium hirſutum. *Bauh. pin.* 129. *Park.* 802. *Ray's Syn.* 167.

1. There is one variety with ſmooth leaves.
2. And another with narrower leaves. *Ray's Syn.* 170.

Broad leaved buſhy Hawkweed.

Woods and hedges. P. July—Auguſt.

HAWKWEED. Leaves ſtrap-ſhaped; ſomewhat toothed; ſcattered. Flowers in a ſort of rundles—*Bloſſoms yellow.* Seeds *purple. The bloſſoms open about ſix in the morning and cloſe at five in the afternoon.* Buſhy Umbellatum

The appearance of this plant varies in different ſituations. In woods it is more branched, more leafy, the leaves broader, of a darker colour and the bloſſoms paler, than in open airy places.

Hieracium fruticoſum anguſtifolium majus. *Park.* 801. *Bauh. pin.* 129. *Ray's Syn.* 168.

Hieracium intybaceum. *Gerard.* 298.

Narrow-leaved buſhy Hawkweed.

Woods and hedges. P. July—Auguſt.

It tinges wool with a beautiful and elegant colour. *Iter. Scan.* 342.

Horſes, Cows, Goats, Sheep and Swine eat it.

318 NIPPLEWORT. 919 Lapſana.

EMPAL. *Common*, double; egg-ſhaped; angular. *Scales of the Tube* eight; equal; ſtrap-ſhaped; with a hollow channel; keeled; ſharp. *Scales of the Baſe* ſix; tiled; ſmall. Every other ſcale ſmaller.

BLOSS. *Compound*, tiled; uniform. *Florets* with chives and pointals, about ſixteen; equal.

Individuals one petal; narrow; lopped; with five teeth.

CHIVES. *Threads* five; hair-like; very ſhort. *Tips* forming a hollow cylinder.

POINT. *Seedbud* rather oblong. *Shaft* thread-ſhaped; as long as the chives. *Summit* cloven; reflected.

S. VESS. None. The *Scales* of the tube of the cup cloſe upon the ſeeds.

SEEDS. Solitary; oblong; cylindrical; but with three edges. *Feathers* various; ſometimes wanting.

RECEPT. Naked; flat.

Common
Communis

NIPPLEWORT. The cup angular when the ſeed is ripe. Fruit-ſtalks ſlender; very much branched—*Bloſſoms yellow*; *terminating*.

Lampſana. *Gerard*. 255. *Ray's Syn*. 173. vulgaris. *Park*. 810.

Soncho affinis Lampſana domeſtica. *Bauh. pin*. 124.

Dock Creſſes.

Ditch-banks and uncultivated places. A. June—July.

Before it goes into flower it is eaten raw at Conſtantinople. Boiling increaſes its bitterneſs.—Horſes, Cows, Sheep and Swine eat it; Goats refuſe it.

319 YELLOW-

319 YELLOWEYE. 916 Hyoſeris.

EMPAL. *Common*, of ten leaves. *Scales* ſpear-ſhaped; upright; equal; ſet with very ſhort ſcales at the baſe.

Proper Cup ſuperiour; with five diviſions; very ſhort; ſharp; permanent. It often ſupplies the place of the feather.

BLOSS. *Compound*, ſomewhat tiled; uniform. *Florets* with chives and pointals, diſpoſed in one or two rows.

Individuals one petal; narrow; lopped; with five teeth.

CHIVES. *Threads* five; hair-like; very ſhort. *Tips* forming a hollow cylinder.

POINT. *Seedbud* rather oblong. *Shaft* thread-ſhaped; as long as the chives. *Summits* two; reflected.

S.VESS. None. Common empalement ſtraight.

SEEDS. Solitary; nearly oblong; compreſſed; as long as the cup; crowned with a hair-like *Feather*, or with the *Proper Cup*.

RECEPT. Naked.

YELLOWEYE. Stem divided; naked. Fruit-ſtalks thickeſt towards the top—*Bloſſoms yellow*. Hawkweed Minima

Hyoſeris maſcula. *Gerard.* 288.

Hieracium minimum Cliſiu Hyoſeris Tabernæmontani et Gerardi. *Park.* 791. *Ray's Syn.* 173.

Hieracium minus folio ſubrotundo. *Bauh. pin.* 127.

Small Swine's Succory.

Corn-fields, and in gravelly ſoil. A. May—June.

320 TWINGE-

320 TWINGEWORT. 929 Carlina.

EMPAL. *Common*, diſtended; radiate; tiled. *Scales* numerous; flexible; ſharp. The inner ſcales placed in a circle, very long; expanding; ſhining; coloured; forming rays to the compound flower.

BLOSS. *Compound*, uniform; tubular. *Florets* with chives and pointals, equal.

Individuals one petal; funnel-ſhaped. *Tube* ſlender. *Border* funnel-ſhaped; with five clefts.

CHIVES. *Threads* five; hair-like; very ſhort. *Tips* forming a hollow cylinder.

POINT. *Seedbud* ſhort. *Shaft* thread-ſhaped; as long as the chives. *Summit* oblong; cloven or entire.

S. VESS. None. *Cup* unchanged.

SEEDS. Solitary; rather cylindrical. *Feather* downy; branched.

RECEPT. Flat; ſet with cloven *Chaffy* ſubſtances three of which grow together.

Carline Vulgaris

TWINGEWORT. Stem ſupporting ſeveral flowers in a ſort of broad topped ſpike. Flowers terminating. Rays of the empalement white—*Or yellowiſh.* Leaves *toothed; with two yellow thorns on each tooth.* Bloſſoms *purple.*

Carlina ſylveſtris major. *Gerard.* 1150.

Carlina ſylveſtris quibuſdam, aliis atractylis. *Ray's Syn.* 175.

Cnicus ſylveſtris ſpinoſior. *Bauh. pin.* 378.

Wild Carline Thiſtle.

Dry paſtures. B. June.

The flowers expand in dry, and cloſe in moiſt weather. They retain this property a long time and therefore are employed as Hygrometers—It is ſaid to be an excellent remedy in Hyſterical caſes. *Amœn Acad.* V. 3. p. 64.—Goats eat it; Cows refuſe it.

321 BURDOCK

321 BURDOCK. 923 Arctium.

EMPAL. *Common*, globular; tiled. *Scales* ſpear-ſhaped; ending in awl-ſhaped prickles; long and hooked at the points.

BLOSS. *Compound*, tubular; uniform. *Florets* with chives and pointals equal.

Individuals one petal; tubular. *Tube* ſlender; very long. *Border* egg-ſhaped; with five clefts. *Segments* ſtrap-ſhaped; equal.

CHIVES. *Threads* five; hair like; very ſhort. *Tips* forming a hollow cylinder, as long as the bloſſom; with five teeth.

POINT. *Seedbud* oblong; with ſoft hairs at the end. *Shaft* thread-ſhaped; longer than the chives. *Summit* cloven; reflected.

S. VESS. None. *Cup* cloſing.

SEEDS. Solitary; like an inverted pyramid: the two oppoſite angles imperfect; hunched on the outerſide. *Feather* ſimple; ſhorter than the ſeed.

RECEPT. Chaffy; flat. *Chaff* like briſtles.

BURDOCK. Leaves heart-ſhaped; without thorns; growing on leaf-ſtalks—*Bloſſoms purple*. Common Lappa

Lappa major, ſeu Arctium Dioſcoridis. *Baub. pin.* 198. *Ray's Syn.* 197.

Bardana major. *Gerard.* 809.

Bardana vulgaris major. *Park.* 1222.

1. Flowering heads large and ſmooth. *Ray's Syn.* 196.
2. Flowering heads ſmall and ſmooth. *Ray's Syn.* 197.
3. Flowering heads large and downy. *Bauh. pin.* 198.
4. Flowering heads ſmall, globular and downy.
5. The whole plant ſmall; flowering heads covered with an elegant network.

Clott-bur.

Road-ſides, and among rubbiſh. B. July—Auguſt.

Before the flowers appear, the ſtems deprived of the bark, are boiled and eat like 'Sparagus. When raw they are good with Oil and Vinegar—A decoction of the roots, is eſteemed by ſome very ſenſible Phyſicians, as equal, if not ſuperior to that of Sarſaparilla—Boys catch Bats, by throwing the prickly heads up into the air.—Cows and Goats eat it; Sheep and Horſes refuſe it; Swine are not fond of it.

The Ghoſt Moth, *Phalæna Humuli*, feeds upon the roots, and the mottled Orange Moth upon the ſtalks.

322 THISTLE. 925 Carduus.

EMPAL. *Common*, diſtended; tiled. *Scales* numerous; ſpear-ſhaped; tapering; thorny.

BLOSS. *Compound*, tubular; uniform. *Florets* with chives and pointals, nearly equal; reflected.

Individuals one petal, funnel-ſhaped. *Tube* very ſlender; *Border* upright; egg-ſhaped at the baſe; with five clefts. *Segments* ſtrap ſhaped; equal; one more deeply divided.

CHIVES. *Threads* five; hair-like; very ſhort. *Tips* forming a hollow cylinder as long as the bloſſom; with five teeth at the mouth.

POINT. *Seedbud* egg-ſhaped. *Shaft* thread-ſhaped· longer than the chives. *Summit* ſimple; awl-ſhaped; naked; notched at the end.

S. VESS. None. *Cup* cloſes a little.

SEEDS. Solitary; inverſely egg-ſhaped; with four angles, but the two oppoſite ones imperfect. *Feather* ſitting; very long.

RECEPT. Hairy; flat.

* *Leaves running along the ſtem.*

Spe r
Lanceolatus

THISTLE. Leaves running along the ſtem; with winged clefts; rough with hair; ſegments ſtraddling. Cups egg-ſhaped; thorny; woolly. Stem hairy—*The inner ſcales of the cup are briſtly; dry, and without thorns.* Bloſſoms *purple.*

Carduus lanceatus. *Gerard.* 1174. *Ray's Syn.* 195.

Carduus lanceatus latifolius. *Bauh. pin.* 385. ſeu major. *Park.* 982.

1. The ſize of the whole plant differs conſiderably according to the ſoil and ſituation. *Ray's Syn.* 195.

Road-ſides. B. July.

Few plants are more diſregarded than this, and yet its uſe is very conſiderable. If a heap of clay is thrown up, nothing would grow upon it for ſeveral years, if the ſeeds of this plant, wafted by the wind, did not fix and vegetate thereon. Under the ſhelter of this, other vegetables appear, and the whole ſoon becomes fertile—The flowers, like thoſe of the Artichoke, have the property of Rennet in curdling milk.—Sheep and Swine refuſe it; neither Horſes, Cows or Goats are fond it—The painted Lady Butterfly, *Papilio Cardui*, and the Thiſtle Ermine Moth, feed upon it.

THISTLE

THISTLE. Leaves running a little way along the ſtem; thorny. Flowers on crooked fruit-ſtalks. The ſcales of the cup expanding upwards—*Shafts bent ſide-ways.* Leaves *with a white rib.* Bloſſoms *purple.* Seeds *brown and gloſſy.* Muſk Nutans

Cirſium majus ſingulari capitulo magno. *Baub pin.* 377.

Carduus Nutans. *Ray's Syn.* 193.

1. Bloſſoms white.

Very dry ſoil. B. July.

Horſes eat it; Cows are not fond of it; Goats and Sheep refuſe it.

THISTLE. Leaves running along the ſtem; indented; thorny at the edge. Cups on fruit-ſtalks; ſolitary; upright; woolly—*The inner ſcales of the cup expanding. The* Bloſſoms *with five clefts, but the lower ſegment is almoſt divided down to the baſe, ſo that the bloſſom may be conſidered as compoſed of two lips; with the upper lip cloven into four parts and the lower lip undivided.* Shafts *twice as long as the florets.* Leaves *halberd-ſhaped; with winged clfets; the under-ſide, ſet with white ſtraight hairs.* Bloſſoms *pale red.* Welted Acanthoides

Carduus ſpinoſiſſimus, Capitulis minoribus. *Ray's Syn.* 194.

Road-ſides and ditch-banks. A. June—July.

THISTLE. Leaves running along the ſtem; indented; thorny at the edge. Flowers incorporated; terminating. Scales ſtanding open; not thorny, but ending in a ſort of awn—*Leaves with a whitiſh down and green veins on the under ſurface.* Florets *ſometimes proliferous; purple.* Criſpus

Carduus caule criſpo. *Ray's Syn.* 194.

Carduus ſpinoſiſſimus anguſtifolius vulgaris. *Bauh. pin.* 385.

Carduus Polyanthos. *Park.* 982. primus. *Gerard.* 1173.

Thiſtle upon Thiſtle.

Road-ſides and ditch-banks. A. June.

1. There is a variety with white flowers, growing in marſhy places

Horſes, Cows, Goats and Sheep eat it.

THISTLE. Leaves running along the ſtem; toothed; thorny at the edges. Flowers in bunches, upright. Fruit-ſtalks without thorns—*Cups cloſely tiled; ſmooth; the ſharp points being hardly diſcernible.* Bloſſoms *purple.* Marſh Paluſtris

Carduus paluſtris. *Bauh. pin.* 377. *Park.* 983. *Ray's Syn.* 194.

Marſhy ſhady places. P. July.

This, and almoſt all the other ſpecies of this genus, may be eaten like the Burdock, before the flowers are formed.—Swine eat it; Horſes are very fond of it; Cows refuſe it.

Gentle Diffectus THISTLE. Leaves running along the ftem: fpear-fhaped; with little teeth; without thorns. Cups thorny—*Bloffoms purple; or white.*

Cirfium Anglicum. *Gerard.* 1183. *Ray's Syn.* 193. primum *Park.* 961.

Cirfium majus fingulari capitulo magno, five incanum varie diffectum. *Bauh. pin.* 377.

Englifh foft Thiftle.

In marfhes. P. June—July.

* * *Leaves fitting.*

Milk Marianus THISTLE. Leaves embracing the ftem; halberd-fhaped; with winged clefts; thorny. Cups without any leaves near it. Thorns channelled and fet with other little thorns—*Leaves with white ftreaks, or fpots.* Bloffoms *purple.*

Carduus Mariæ. *Gerard.* 989. *Ray's Syn.* 195. vulgaris. *Park.* 979.

Carduus albis maculis notatus vulgaris. *Bauh pin.* 281.

Ladies Thiftle.

1. There is a variety that is hairy and not fpotted. *Ray's Syn.* 194.

Road-fides, ditch-banks, and the borders of corn-fields. A. July.

This is eaten when young as a fallad. The young ftalks peeled, and foaked in water to take out the bitternefs, are excellent. The fcales of the cup are as good as Artichokes. The root is good to eat early in the fpring.

Woolly headed Eriophorus THISTLE. Leaves fitting; divided into winged clefts pointing two ways. Every other fegment upright. Cups globular; woolly—*Scales of the cup ending in a yellow thorn.* Bloffoms *purple; or white.*

Carduus Eriocephalus. *Gerard.* 1152.

Carduus capite rotundo tomentofo. *Bauh. pin.* 382.

Carduus tomentofus Corona fratum dictus. *Park.* 978.

Friars Crown.

Hilly paftures. B. July.

Melancholy Helenioides THISTLE. Leaves embracing the ftem; fpear-fhaped; toothed. Little thorns unequal; fringed; ftem unarmed—*Five or fix feet high.* Leaves *all undivided; white on the under furface.* Stem *furrowed.* Floral Leaves *awl-fhaped.* Bloffoms *purple.*

Cirfium Britannicum Clufii, repens. *Ray's Syn.* 193.

Cirfium aliud Anglicum. *Park.* 961.

Cirfium fingulari capitulo fquamato, vel incanum alterum. *Bauh pin.* 377.

Hilly countries. P. June—July.

THISTLE.

THISTLE, without a ſtem. Cup ſmooth—*Bloſſoms purple.* Dwarf Carlina acaulis, minore purpureo flore. *Bauh. pin.* 380. Acaulis *Gerard.* 1158. *Ray's Syn.* 195.

Carlina acaulis Septentrionalium. *Park.* 969.

Dwarf Carline Thiſtle.

Dry hilly countries. P. July.

Cows refuſe it.

The different ſpecies of THISTLES affords nouriſhment to the following inſects.

Green Tortoiſe Beetle. *Caſſida viridis—Caſſida nebuloſa.*

Painted Lady Butterfly. *Papilio Cardui.*

Horned Froghopper. *Cicada Cornuta.*

Thiſtle Bug. *Cimex Cardui.*

Thiſtle Fly. *Muſca Solſtitialis.*

Thiſtle Louſe. *Aphis Cardui.*

323 ARGENTINE. 927 Onopordum.

EMPAL. *Common,* roundiſh; diſtended; tiled. *Scales* numerous; thorny; prominent every way.

BLOSS. *Compound,* tubular; uniform. *Florets* with chives and pointals, equal.

Individuals one petal; funnel-ſhaped. *Tube* very ſlender. *Border* upright; diſtended; with five clefts. *Segments* equal; but one more deeply divided than the reſt.

CHIVES. *Threads* five; hair-like; very ſhort. *Tips* forming a hollow cylinder as long as the bloſſom, with five teeth.

POINT. *Seedbud* egg-ſhaped. *Shaft* thread-ſhaped; longer than the chives. *Summit* crowned.

S. VESS. None. The *Cup* cloſes a little.

SEEDS. Solitary. *Feather* hair-like.

RECEPT. Divided like a honey-comb into four-cornered membranaceous cells.

Cotton Acanthium ARGENTINE. Cups scurfy; scales expanding; awl-shaped Leaves oblong egg-shaped; indented.—*Blossoms purple. The whole plant of a whitish green colour; the stem furnished with a membrane on each side, running along it up to the cup.*

Carduus tomentosus, acanthium dictus, vulgaris. *Ray's Syn.* 196.

Acanthium album. *Gerard.* 1149.

Acanthium vulgare. *Park.* 979.

Spina alba tomentosa latifolia sylvestris. *Bauh. pin.* 382.

Cotton Thistle.

Among rubbish and in uncultivated places. B. July.

The central part of the flower, (the Receptacle) and the young stems are boiled and eaten like Artichokes—The ancients thought this plant a specific in cancerous cases.—Cows, Sheep and Horses refuse it.

324 SAW-WORT. 924 Serratula.

EMPAL. *Common*, oblong; rather cylindrical; tiled. *Scales* spear-shaped; without awns; sharp.

BLOSS. *Compound*, tubular; uniform. *Florets* with chives and pointals; equal,

Individuals one petal; funnel-shaped. *Tube* bent inwards. *Border* with five clefts; distended.

CHIVES. *Threads* five; hair like; very short. *Tips* forming a hollow cylinder.

POINT. *Seedbud* egg shaped. *Shaft* thread-shaped; as long as the chives. *Summits* two; oblong; reflected.

S. VESS. None. *Cup* unchanged.

SEEDS. Solitary; inversely egg-shaped. *Feather* sitting.

RECEPT. Naked, or chaffy.

OBS. *The feather in some species is downy, but in others hairy. The* THISTLE *is distinguished from the* SAW-WORT *by the receptacle being hairy; the cup distended; its scales thorny; and the summit less cloven.*

Dyers Tinctoria SAW-WORT. Leaves lyre-shaped; with winged clefts. The terminating segment very large. Florets uniform—*Blossoms purple, sometimes white.* Feather *of the seed glossy, with a yellow tinge.*

Serratula. *Ray's Syn.* 196.

Serratula purpurea. *Gerard.* 713.

Serratula vulgaris flore purpureo. *Park.* 474.

Woods and pastures. P. July.

This is very much used by the Dyers to give a yellow colour; but it is inferior to the YELLOW-WEED, therefore its use is confined to the coarser woollen cloths.—Goats eat it; Horses are not fond of it; Sheep, Swine and Cows refuse it.

SAW-WORT.

SAW-WORT. Cups ſomewhat hairy; egg-ſhaped. Leaves undivided—*Bloſſoms purple.* Mountain Alpina

Cirſium humile montanum Cynogloſſi folio, polyanthemum. *Ray's Syn.* 193.

1. There is a variety with narrower leaves. *Ray's Syn.* 193.

On the higheſt mountains. P. July.

SAW-WORT. Leaves toothed; thorny.—*Bloſſoms pale purple.* Feather *of the ſeed very long.* Corn Arvenſis

Carduus vulgatiſſimus viarum. *Gerard.* 1173. *Ray's Syn.* 194.

Carduus ceanothos ſive viarum et vinearum repens. *Park.* 959.

Carduus vinearum repens, folio ſonchi. *Bauh. pin.* 377.

Way-thiſtle.

Corn-fields and road-ſides. P. July.

It is ſaid to yield a very pure vegetable alkaly when burnt.

Goats eat it; neither Cows, Horſes, Sheep or Swine are fond of it.

325 LIVERHEMP. 935 Eupatorium.

EMPAL. *Common*, oblong; tiled. *Scales* betwixt ſtrap and ſpear-ſhaped; upright; unequal.

BLOSS. *Compound*, uniform; tubular. *Florets* with chives and pointals, equal.

Individuals funnel ſhaped. *Border* with five clefts; open.

CHIVES. *Threads* five; hair-like; very ſhort. *Tips* forming a hollow cylinder.

POINT. *Seedbud* very ſmall. *Shaft* thread-ſhaped; very long; cloven down to the chives; ſtraight. *Summits* ſlender.

S. VESS. None. *Cup* unchanged.

SEEDS. Oblong. *Feather* downy; long.

RECEPT. Naked.

LIVERHEMP. Leaves fingered. Five florets in each cup.—*Bloſſoms pale red.* Water Cannabinum

Eupatorium Cannabinum. *Bauh. pin.* 320. *Park.* 595. *Ray's Syn.* 179.

Eupatorium Cannabinum mas. *Gerard.* 711.

Eupatorium Cannabinum folio integro, ſeu non digitato. *Ray's Syn.* 180.

Hemp Agrimony. Dutch Agrimony. Water Agrimony. Water Hemp.

Banks of rivers and brooks. P. July—Auguſt.

An infusion of a handful of it vomits and purges smartly. An ounce of the root in decoction is a full dose. In smaller doses the Dutch peasants take it as an alterative and an antiscorbutic.—Goats eat it; Cows, Horses, Sheep and Swine refuse it.

326 CUDWORT. 943 Athanasia.

EMPAL. *Common*, tiled; egg-shaped. *Scales* spear-shaped; contiguous.

BLOSS. *Compound*, uniform; longer than the cup. *Florets* with chives and pointals equal; numerous. *Individuals* funnel-shaped. *Border* with five clefts, sharp; rather upright.

CHIVES. *Threads* five; hair-like; short. *Tips* forming a hollow cylinder.

POINT. *Seedbud* rather oblong. *Shaft* thread-shaped; a little longer than the chives. *Summit* cloven; blunt.

S. VESS. None. *Cup* unchanged.

SEEDS. Solitary; oblong. *Feather* set with very short bristles.

RECEPT. Chaffy; *Chaff* spear-shaped; longer than the seeds.

OBS. *In the* British Species *the seeds have no feather.*

Cudweed Maritima

CUDWORT. One flower upon a fruit-stalk, forming a kind of broad-topped spike. Leaves spear-shaped; scolloped; blunt; downy—*Blossoms yellow and white. As the seeds have no feather it might perhaps have been arranged under the Genus* Santolina, (*Lavender Cotton.*)

Gnaphalium maritimum. *Bauh. pin.* 263. *Ray's Syn.* 180.

Gnaphalium marinum. *Gerard.* 640. sive Colonaria. *Park.* 687.

Filago maritima. *Hudson.* 328.

Sea Cudweed.

On the sea-shore. A. June—July.

327 DOUBLE-

327 DOUBLETOOTH. 932 Bidens.

Empal. *Common*, upright; ſegments generally equal; oblong; concave and channelled.

Bloss. *Compound*, uniform; tubular. *Florets* with chives and pointals, tubular.

Individuals funnel-ſhaped *Border* with five clefts; upright.

Chives, *Threads* fives; hair-like; very ſhort. *Tips* forming a hollow cylinder.

Point. *Seedbud* oblong, *Shaft* ſimple; as long as the chives. *Summits* two; oblong; reflected.

S Vess. None. *Cup* unchanged.

Seeds. Solitary; blunt; angular. *Crown of the Seed* two or more awns, oblong; ſtraight; ſharp; rough with hooks turned backwards.

Recept. Chaffy; flat. *Chaff* deciduous.

DOUBLETOOTH. Leaves cloven into three ſegments. Cups ſomewhat leafy; ſeeds upright—*Bloſſoms yellow; terminating.* Hemp-leaved Tripartita

Verbeſina, ſeu Cannabina aquatica flore minus pulchro, elatior, et magis frequens. *Ray's Syn.* 187.

Cannabina aquatica folio tripartito diviſo. *Bauh. pin.* 321.

Eupatorium Cannabinum fæmineum. *Gerard.* 711.

Eupatorium aquaticum duorum generum. *Park.* 595.

Trifid Water Hemp Agrimony.

Marſhy places. A. July—Auguſt.

It dyes a tolerably good yellow.—Cows and Sheep eat it: Horſes, Swine and Goats refuſe it.

DOUBLETOOTH. Leaves ſpear-ſhaped; embracing the ſtem. Flowers on crooked fruit-ſtalks. Seeds upright—*Bloſſom duſky yellow.* Nodding Cernua

Verbeſina pulchriore flore luteo. *Ray's Syn.* 187.

Cannabina aquatica folio non diviſo. *Bauh. pin.* 321.

Eupatorium Cannabinum fæmina, varietas altera. *Gerard.* 711.

Eupatorium aquaticum folio integro. *Park.* 596.

Whole-leaved Water hemp Agrimony.

Ditches and marſhes. A. July—Auguſt.

Goat eat it; Horſes refuſe it.

DOUBLETOOTH. Leaves ſpear-ſhaped; ſitting. Flowers and ſeeds upright— Small Minima

Verbeſina minima. *Ray's Syn.* 188. Tab. 7. fig. 2.

Leaſt Water Hemp Agrimony.

Shallow Waters. A. July—Auguſt.

Order II. Superfluous Pointals.

328 SOUTHERNWOOD. 945 Artemifia.

EMPAL. *Common*, roundifh; tiled. *Scales* rounded; approaching.

BLOSS. *Compound. Florets with Chives and Pointals*, many; tubular; placed in the center. *Florets with only Pointals* and almoft without any petal, in the circumference.

Individuals with Chives and Pointals, funnel-fhaped. *Border* with five clefts.

CHIVES. *Threads* five; hair-like; very fhort. *Tips* forming a hollow cylinder.

POINT. *Seedbud* in thofe that have chives and pointals fmall. *Shaft* thread-fhaped; as long as the chives. *Summit* cloven; rolled back.

In the florets that have only pointals the *Seedbud* is very fmall The *Shaft* thread-fhaped, and longer than in the other; the *Summit* the fame.

S. VESS. None. *Cup* hardly changed.

SEEDS. In all the florets folitary; naked

RECEPT. Flat; naked, or woolly.

* *Stems trailing before the time of flowering.*

Field
Campeftris

SOUTHERNWOOD. Leaves with many clefts, ftrap-fhaped. Stems with long flender fhoots, trailing—*Bunches fimple; alternate.* Flowers *folitary.* Receptacle *naked. Whilft in flower it ftands more upright.* Summits *glafs-fhaped. There are about nine florets with only Pointals, and eight with Chives and Pointals in each common cup.* Bloffoms *brown.*

Abrotanum campeftre. *Bauh. pin.* 136. *Gerard.* 1106. *Park.* 94. *Ray's Syn.* 191.

High dry mountainous paftures P. Auguft.

Sheep refufe it.

Sea
Maritima

SOUTHERNWOOD. Leaves deeply divided into many parts; downy. Bunches on crooked fruit-ftalks. Florets with only Pointals, three.—*Receptacle naked. The whole plant is white; except the* Bloffoms *which are brown. The leaves are more or lefs divided in different fituations.*

Abfinthium marinum album. *Gerard.* 1099. *Ray's Syn.* 188.

Abfinthium feriphicum Belgicum. *Bauh. pin.* 139.

Abfinthium marinum noftras. *Ray's Syn.* 189.

Sea-Wormwood. Roman Worm-wood.

On the Sea-coaft. P. Auguft.

This in its wild ftate fmells like Marum or Camphor, but in our gardens it is lefs grateful, though ftill much more grateful than the next fpecies. It is ufed as an ingredient in diftilled waters, and beat with thrice its weight of fine fugar it is formed into

into a conserve. Its virtues are the same with those of the next species, but in a weaker degree.—Horses eat it; Cows, Goats and Sheep refuse it.

* * *Stems upright; herbaceous.*

SOUTHERNWOOD. Leaves compound, with many clefts. Flowers somewhat globular; pendant. Receptacle woolly.— *Leaves dark green on the upper, but white on the under surface.* Blossoms *brownish white.* Wormwood Absinthium

Absinthium vulgare. *Park.* 98. *Ray's Syn.* 188.

Absinthium latifolium, seu Ponticum. *Gerard.* 1096.

Absinthium ponticum, seu Romanum officinarum, seu Dioscoridis. *Bauh. pin.* 138.

Common Worm-wood. Mug-wort.

Road-sides. Rocky places and among rubbish. P. August.

The leaves and flowers are very bitter; the roots are warm and aromatic. A considerable quantity of essential oil rises from it in distillation. This oil is used both externally and internally to destroy Worms—The leaves put into sour beer, soon destroy the acescency—They resist putrefaction, and are therefore a principal ingredient in antisceptic fomentations. An infusion of them is a good stomachic, and with the addition of fixed alkaline salt, a powerful diuretic in dropsical cases—The ashes afford a more pure alkaline salt than most other vegetables, excepting Bean-stalks, Broom, and the larger trees.—In the *Amœn. Acad. v. 2. p.* 160. Linnæus mentions two cases, wherein an essence prepared from this plant and taken for a considerable time, prevented the formation of stones in the kidneys or bladder; the patients forbearing the use of wine and acids. It may, like other bitters, weaken the action of the nervous system, but in these instances no such effect took place—An infusion of it given to a woman that suckles, makes her milk bitter—It gives a bitterness to the flesh of Sheep that eat it; Horses and Goats are not fond of it; Cows and Swine refuse it.

SOUTHERNWOOD. Leaves with winged clefts; flat; jagged; downy on the under side. Flowers in bunches; simple; much curved. Five florets in the circumference of each flower. —*Receptacle naked.* Blossoms *purplish.* Mugwort Vulgaris

Artemisia vulgaris. *Park.* 90. *Ray's Syn.* 190. *Gerard.* 1103. major. *Bauh. pin.* 137.

Mugwort. Mother-wort.

Borders of fields. Ditch-banks. P. August.

In some countries it is used as a culinary aromatic—A decoction of it is taken by the common people to cure the Ague. Sheep and Swine refuse it; neither Horses, Cows or Goats are fond of it. The Wormwood Louse, *Aphis Absinthii* and the Lambda Moth, *Phalæna Gamma*, live upon the several species.

329 TANSEY.

329 TANSEY. 944 Tanacetum.

EMPAL. *Common*, hemispherical; tiled. *Scales* sharp; compact.

BLOSS. *Compound*, tubular; convex. *Florets with Chives and Pointals*, numerous; tubular; placed in the center. *Florets with only Pointals*, a few in the circumference.

Individuals with Chives and Pointals, funnel-shaped. *Border* with five clefts; reflected.

Individuals with only Pointals with three clefts, more deeply divided on the inner side.

CHIVES. *Threads* five; hair-like; very short. *Tips* forming a hollow cylinder.

POINT. *Seedbud*, where there are chives and pointals, oblong; small. *Shaft* thread-shaped; as long as the chives. *Summit* cloven; rolled back. *Seedbud*, where there are only pointals, oblong. *Shaft* simple. *Summits* two; reflected.

S.VESS. None. *Cup* unchanged.

SEEDS. Solitary; oblong; naked.

RECEPT. Convex; naked.

Common Vulgare

TANSEY. Leaves doubly winged; jagged; serrated.—*Blossoms yellow*.

Tanacetum. *Gerard*. 650. *Ray's Syn*. 188. vulgare. *Park*. 81.

Tanacetum luteum. *Bauh. pin*. 132.

1. There is a variety with curled leaves called double Tansey. *Bauh. pin*. 132.

2. Another with variegated leaves; and a third with larger leaves and but little smell.

High pastures. P. August.

This is a warm deobstruent bitter, and its flavour not ungrateful—The tender leaves are sometimes used to give a colour and flavour to puddings.—If a dead animal substance is rubbed with this plant, the flesh fly will not attack it.—The Finlanders obtain a green dye from it.—Cows and Sheep eat it; Horses, Goats and Swine refuse it. It gives nourishment to the Tansey Loure, *Aphis Tanaceti*; and to the *Chrysomela Tanaceti*.

330 SPIKE-

330 SPIKENARD. 950 Conyza.

EMPAL. *Common*, tiled; oblong; ſcurfy. *Scales* ſharp; the outer ones a little expanded.

BLOSS. *Compound*, tubular. *Florets with Chives and Pointals* numerous; tubular; in the center. *Florets with only Pointals* and without petals, circular; in the circumference.

Individuals with Chives and Pointals funnel-ſhaped. *Border* with five clefts; open.

Individuals with only Pointals, funnel-ſhaped. *Border* cloven into three ſegments.

CHIVES. *Threads* five; hair-like; very ſhort. *Tips* forming a hollow cylinder.

POINT. *Seedbud*, where there are chives and pointals, oblong. *Shaft* as long as the chives; thread-ſhaped. *Summit* cloven.

Seedbud, where there are only pointals, oblong. *Shaft* thread-ſhaped; as long, but more ſlender than the other. *Summits* two; very ſlender.

S. VESS. None: the cup cloſing.

SEEDS. In all the florets, ſolitary; oblong *Feather* ſimple.

RECEPT. Naked; flat.

SPIKENARD. Leaves ſpear-ſhaped; ſharp. Stem herbaceous. Flowers in a broad-topped ſpike. Cups ſcurfy—*Bloſſoms duſky purple or yellowiſh.* Plowman's Squarroſa

Conyza major vulgaris. *Bauh. pin.* 265.

Baccharis monſpelienſium. *Gerard.* 792. *Park.* 114. *Ray's Syn.* 179.

Great Fleabane.

High Grounds and road-ſides. B. July—Auguſt.

331 CATSFOOT. 946 Gnaphalium.

EMPAL. *Common*, roundiſh; tiled. *Scales* egg-ſhaped; approaching; the upper ones more flexible.

BLOSS. *Compound. Florets with Chives and Pointals*, tubular; ſometimes mixed, with the other florets without either chives or petals.

Individuals, with Chives and Pointals, funnel-ſhaped. *Border* with five clefts; reflected.

Individuals with only Pointals, without petals.

CHIVES. *Threads* five; hair-like; very ſhort. *Tips* forming a hollow cylinder.

POINT. *Seedbud*, where there are chives and pointals, egg-ſhaped. *Shaft* thread-ſhaped; as long as the chives. *Summit* cloven.

Seedbud, where there are only pointals, egg-ſhaped. *Shaft* thread-ſhaped; as long as the other, *Summit* cloven; reflected.

S. VESS. None. *Cup* permanent; ſhining.

SEEDS. In all the florets ſolitary; oblong; ſmall; crowned with a *Feather*.

RECEPT. Naked.

OBS. *In ſome ſpecies the* Feather *is hair-like; in others downy, In one ſpecies the* Chives *and* Pointals *are on diſtinct plants.*

Jerſey
Luteo-album

CATSFOOT. Stem herbaceous. Leaves ſword-ſhaped; half embracing the ſtem; ſerpentine at the edge; blunt; downy on both ſurfaces. Flowers crowded cloſe together—*The whole plant is very woolly.* Cups *yellowiſh white; ſoft; with ſcales betwixt egg and ſpear-ſhaped.* Florets with only pointals, *in the circumference; numerous.*

Gnaphalium majus lato oblongo folio. *Bauh. pin.* 263.

Elichryſum ſylveſtre latifolium capitulis conglobatis. *Bauh. pin.* 264. *Ray's Syn.* 182.

Jerſey Cudweed.

In the Iſle of Jerſey, on walls and dry banks. A. July—Auguſt.

CATSFOOT. Stem herbaceous. Leaves betwixt ſtrap and ſpear-ſhaped; tapering; alternate. Stem branched towards he top. Flowers in broad-topped level ſpikes—*Bloſſoms yellowiſh green.* Pearly Margaritaceum

Gnaphalium Americanum. *Gerard.* 641. *Ray's Syn.* 182.
Gnaphalium Americanum latifolium. *Bauh. pin.* 263.
Argyrocome, ſeu Gnaphalium Americanum. *Park.* 690.
American Cudweed
Meadows, paſtures, and banks of rivers. P. Auguſt.

CATSFOOT. Runners trailing. Stem undivided. Flowers in a ſimple broad-topped ſpike. Chives and pointals on diſtinct plants—*In the barren florets the cups are almoſt globular; but in the fertile ones nearly cylindrical.* Bloſſoms *white; purple or reddiſh; terminating.* Mountain Dioicum

Gnaphalium montanum ſuave rubens. *Gerard.* 641.
Gnaphalium montanum, flore rotundiore. *Bauh. pin.* 262.
The above are the ſynonyms of the plant bearing only *barren* flowers—The following thoſe of the other bearing *fertile* flowers.
Gnaphalium montanum album. *Gerard.* 640. *Ray's Syn.* 181.
Gnaphalium montanum ſive Pes Cati. *Park.* 690.
Gnaphalium longiore folio et flore. *Bauh. pin.* 263.
It is ſomewhat remarkable that no one of our Botaniſts appear to have obſerved both theſe plants, tho' both it is evident, are abſolutely neceſſary to perpetuate the ſpecies.
Mountain Cudweed.
On heaths and dry hilly paſtures. P. May.
Horſes Sheep and Swine eat it; Cows and Goats refuſe it.

CATSFOOT. Stem herbaceous; undivided; upright. Flowers ſcattered—*Leaves narrow; ſharp; downy on the under ſurface,* Cups *white; ſhining.* Bloſſoms *yellowiſh.* Upright Sylvaticum

Gnaphalium Anglicum. *Gerard* 639. *Ray's Syn.* 180.
Gnaphalium Anglicum vulgare majus. *Park.* 685.
Gnaphalium majus, anguſto oblongo folio, alterum. *Bauh pin.* 263.
Upright Cudweed.
Sandy woods and paſtures. B. Auguſt.
Goats eat it.

CATSFOOT. Stem herbaceous; with ſpreading branches. Flowers crowded together; terminating——*Bloſſoms yellowiſh;* Cups *brown, or black.* Black Uliginoſum

Gnaphalium longifolium humile ramoſum capitulis nigris. *Ray's Syn.* 181.
Gnaphalium medium. *Bauh. pin.* 263.
Filago minor. *Park.* 686.
Black-headed Cudweed.
In watery places. A. Auguſt.
Goats and Cows refuſe it.

332 DAISIE. 962 Bellis.

EMPAL. *Common* simple; upright. *Little Leaves* from ten to twenty, placed in a double row; spear-shaped; equal.

BLOSS. *Compound*, radiate. *Florets with chives and Pointals*, tubular; numerous; placed in the centre. *Florets with only pointals*, narrow; more in number than the leaves of the cup; placed in the circumference.

Individuals with chives and pointals funnel-shaped; with five clefts.

Individuals with only pointals narrow; spear-shaped; very slightly marked with three teeth.

CHIVES. *Threads* five; hair-like; very short. *Tips* forming a hollow cylinder.

POINT. *Seedbud* where there are chives and pointals, egg-shaped. *Shaft* simple. *Summit* notched at the end.

Seedbud where there are only pointals, egg-shaped. *Shaft* thread-shaped, *Summits* two; standing wide.

S.VESS. None. *Cup* unchanged.

SEEDS. In all the florets, solitary; inversely egg-shaped; compressed. *Feather* none.

RECEPT. Naked; conical.

Common Perennis DAISIE. Stalk naked.—*Leaves oblong; blunt; notched; spread upon the ground.* Florets *in the center yellow: those in the circumference white above, pink beneath. By cultivation the blossom becomes double and either all red, or red and white.*

Bellis sylvestris minor. *Bauh. pin.* 261. *Ray's Syn.* 184.

Bellis minor sylvestris. *Gerard.* 635. simplex. *Park.* 530.

Pastures. P. March—September.

The leaves are slightly acrid. The roots have a penetrating pungency. No attention is paid to it except what it claims from the beauty of its flowers—The flowers close at night—Horses, Sheep, and Cows refuse it.

333 FEVERFEW. 967 Matricaria.

EMPAL. *Common*, hemiſpherical. *Scales* ſtrap-ſhaped; tiled; not quite equal.

BLOSS. *Compound*, radiate. *Florets with chives and pointals* tubular; numerous; placed in the center; which is hemiſpherical. *Florets with only pointals* many; placed in the circumference.

Individuals with chives and pointals, funnel-ſhaped; with five clefts; expanding.

Individuals with only pointals, oblong; with three teeth.

CHIVES *Threads* five; hair-like; very ſhort. *Tips* forming a hollow cylinder.

POINT. *Seedbud*, where there are chives and pointals, oblong; naked. *Shaft* as long as the chives; thread-ſhaped. *Summit* cloven; expanding.

Seedbud, where there are only pointals naked. *Shaft* thread-ſhaped; as long as the other. *Summits* two; rolled back.

S.VESS. None. *Cup* unchanged.

SEEDS. In all the florets, ſolitary; oblong. *Feather* none.

RECEPT. Naked; convex.

FEVERFEW. Leaves compound; flat. Little leaves egg-ſhaped jagged. Fruit-ſtalks branched.—*Naked.* Florets *yellow in the centre; white in the circumference; double by cultivation.* Common Parthenium

Matricaria. *Gerard.* 652. *Ray's Syn.* 187. vulgaris. *Park.* 83.

Matricaria vulgaris ſeu ſativa. *Bauh. pin.* 133

Ditch-banks and amongſt rubbiſh. P. or B. June.

The whole plant has a ſtrong ſmell, and a bitter taſte: it yields an eſſential Oil by diſtillation.

FEVERFEW. Receptacles hemiſpherical. Leaves doubly winged, and ſomewhat fleſhy; convex on the upper ſide, and keeled beneath.—*Florets in the center yellow; thoſe in the circumference white.* Sea Maritima

Chamæmelum maritimum perenne humilius, foliis brevibus craſſis, obſcure virentibus. *Ray's Syn.* 186. Tab . fig. 1.

On the coaſt of Suſſex. P. July.

Sweet-scented Suaveolens FEVERFEW. Receptacles conical; rays bent downwards. Scales of the cup equal at the edge—*Florets in the center yellow; those in the circumference white.*

Sandy ground. A. May.

Goats, Sheep and Horses eat it; Swine refuse it.

Chamomile Chamomilla FEVERFEW. Receptacles conical; rays expanding. Scales of the cup equal at the edge.—*Leaves more than doubly compound, with slender segments.* Florets *in the center yellow; those in the circumference white.*

Chamæmelum. *Gerard.* 754. *Ray's Syn.* 184. vulgare. *Park.* 85.

Chamæmelum vulgare, Leucanthemum Dioscoridis. *Bauh. pin.* 135.

Corn Feverfew.

Corn-fields. A. June.

Its properties resemble those of the *Sweet-scented* CHAMOMILE.

The Finlanders use an infusion of it in Consumptive cases.—Cows, Goats and Sheep eat it; Horses are not fond of it; Swine refuse it.

334 GOLDINS. 966 Chryſanthemum.

EMPAL. *Common*, hemiſpherical; tiled. *Scales* lying cloſe upon each other. The *Inner Ones* becoming gradually larger; the very innermoſt terminating in a a ſkinny ſubſtance.

BLOSS. *Compound*, radiate. *Florets with Chives and Pointals* numerous; tubular; placed in the center. *Florets with only Pointals* twelve or more; placed in the circumference.

Individuals with Chives and Pointals, funnel-ſhaped; with five clefts; open; as long as the cup.

Individuals with only Pointals narrow; oblong; with three teeth.

CHIVES. *Threads* five; hair-like: very ſhort. *Tips* forming a hollow cylinder; generally ſhorter than the bloſſom.

POINT. *Seedbud*, where there are chives and pointals, egg-ſhaped. *Shaft* thread-ſhaped; longer than the chives. *Summits* two; rolled back.

Seedbud where there are only pointals, egg-ſhaped. *Shaft* thread-ſhaped; as long as the other. *Summits* two; blunt; rolled back.

S. VESS. None. *Cup* unchanged.

SEEDS. In all the florets ſolitary; oblong. *Feather* none.

RECEPT. Naked; dotted; convex.

OBS. *In the firſt diviſion of the ſpecies, the florets that have only pointals are ſpear-ſhaped, and the membranes of the cup are narrow. But in the ſecond diviſion they are egg-ſhaped and lopped, and the membranes of the cup are egg-ſhaped.*

* *Radiate Florets white.*

GOLDINS. Leaves embracing the ſtem; oblong; upper leaves ſerrated; lower ones toothed—*Florets in the center yellow; thoſe in the circumference white.* Daiſie Leucanthemum

Leucanthemum vulgare. *Ray's Syn.* 184.
Bellis major. *Gerard.* 634.
Bellis major vulgaris, ſeu ſylveſtris. *Park.* 528.
Bellis ſylveſtris caule folioſo major. *Bauh. pin.* 261.
Greater Daiſie. Ox-Eye.
Corn-fields and dry paſtures. P. May.
The young leaves may be eaten in ſallads—Horſes, Sheep and Goats eat it; Cows and Swine refuſe it.

Scentless Inodorum

GOLDINS. Leaves winged; with many clefts. Stem branched; spreading—*Cups skinny at the edge.* Florets *in the center yellow; those in the circumference white.* Seeds *black.*

Chamæmelum inodorum annuum humilius, foliis obscure virentibus. *Ray's Syn.* 186.

Matricaria inodora. *Hudson.* 322.

Field Feverfew.

Road-sides. A. August.

* * *Florets all yellow.*

Yellow Segetum

GOLDINS. Leaves embracing the stem; upper leaves jagged; lower ones toothed and serrated.—*Blossoms yellow.*

Chrysanthemum Segetum. *Gerard.* 743. *Ray's Syn.* 182. nostras. *Park.* 1370.

Bellis lutea, foliis profunde incisis, major. *Bauh. pin.* 262.

1. It varies in the leaves being more or less jagged.

Corn Marigold. Goulans. Goldins.

Corn-fields. A. June—July.

335 ELECAMPANE. 956 Inula.

EMPAL. *Common*, tiled: *Little Leaves* flexible; open: the outer ones the largest; equal in length.

BLOSS. *Compound*, radiate; broad. *Florets with Chives and Pointals*, equal; very numerous; placed in the center. *Florets with only Pointals* narrow; numerous; crowded; placed in the circumference.

Individuals with Chives and Pointals funnel-shaped; *Border* with five clefts; somewhat upri ht.

Individuals with only Pointals, narrow; strap-shaped; very entire.

CHIVES. *Threads* five; thread-shaped; short. *Tips* five; narrow; united; forming a hollow cylinder; each tip ending at the base in two straight bristles, as long as the threads.

POINT. *Seedbud*, where there are both chives and pointals, long. *Shaft* as long as the chives; thread-shaped. *Summit* cloven: a little upright.

Seedbud, where there are only pointals, long. *Shaft* thread-shaped; a little cloven. *Summits* upright.

S. VESS. None. *Cup* unchanged.

SEEDS. In all the florets, solitary; narrow; with four angles. *Feather* simple; as long as the seed.

RECEPT. Naked; flat.

OBS. *The essential character of this genus consists in the two bristles fixed to the lower part of each tip.*

ELECAM-

ELECAMPANE. Leaves embracing the ſtem; egg-ſhaped; wrinkled; downy underneath. Scales of the cup egg-ſhaped. —*Bloſſoms yellowiſh green.* Common Helenium

Helenium. *Gerard.* 795. *Ray's Syn.* 176. vulgare. *Baub. pin.* 276.

Helenium ſive Enula campana. *Park.* 674.

Moiſt meadows. P. July—Auguſt.

The root is eſteemed a good pectoral. Dr. Hill ſays he knows from his own experience, that an infuſion of the freſh root ſweetened with Honey, is an excellent medicine in the Hooping Cough—A decoction of the root cures Sheep that have the Scab. —Horſes and Goats eat it; Cows, Sheep and Swine refuſe it.

ELECAMPANE. Leaves embracing the ſtem; oblong heart-ſhaped. Stem woolly; flowers in panicles. Scales of the cup briſtle-ſhaped—*The ſcales of the cup are ſomewhat awl-ſhaped; ſoft; bent back; coloured at the edges.* Bloſſoms *yellow.* Autumnal Dyſenterica

Conyza media. *Gerard.* 482. *Ray's Syn.* 174.

Conyza media aſteris, flore luteo, vel tertia Dioſcoridis. *Baub. pin.* 265.

Middle Fleabane.

Banks of rivers. P. Auguſt.

It hath a peculiar acrid ſmell—The Ruſſian Soldiers, in the Perſian expedition under General Keit, were much troubled with the Bloody Flux, which was cured by the uſe of this plant. —Cows are not fond of it; Goats and Sheep refuſe it.

ELECAMPANE. Leaves embracing the ſtem; waved at the edges. Stem proſtrate. Flowers nearly globular; rays very ſhort—*Bloſſoms yellow.* Small Pulicaria

Conyza minor. *Ray' Syn.* 174.

Conyza minima. *Gerard.* 482.

Conyza minor flore globoſo. *Baub. pin.* 266.

Small Fleabane.

1. There is a variety ſtill ſmaller.

Where waters ſtagnate in winter. A. Auguſt—September.

Sheep eat it; Horſes, Cows and Goats refuſe it.

ELECAMPANE. Leaves ſtrap-ſhaped, fleſhy; three-pointed —*Bloſſom yellow; terminating.* Samphire Crithmoides

Aſter maritimus flavus, Crithmum Chryſanthemum, dictus. *Ray's Syn.* 174.

Crithmum Chryſanthemum. *Gerard.* 533. *Park.* 1287.

Crithmum maritimum flore Aſteris attici. *Baub. pin.* 288.

Golden Samphire.

On the ſea-coaſt. P. Auguſt.

336 FLEABANE. 951 Erigeron.

EMPAL. *Common*, oblong; cylindrical; tiled. *Scales* awl-ſhaped; upright; gradually longer; nearly equal.

BLOSS. *Compound*, radiate. *Florets that have both Chives and Pointals*, tubular; placed in the center. *Florets that have only Pointals*, narrow; placed in the circumference.

Individuals with Chives and Pointals, funnel-ſhaped. *Border* with five clefts.

Individuals with only Pointals, narrow; betwixt ſtrap and awl-ſhaped; upright; generally very entire.

CHIVES. *Threads* five; hair-like; very ſhort. *Tips* forming a hollow cylinder.

POINT. *Seedbud*, where there are chives and pointals, very ſmall; crowned with a feather, longer than its bloſſom. *Shaft* thread ſhaped; as long as the feather. *Summits* two; oblong; rolled back.

Seedbud, where there are only pointals, very ſmall; crowned with a feather; nearly as long as its bloſſom. *Shaft* hair-like; as long as the feather. *Summits* two very ſlender.

S. VESS. None. *Cup* cloſing.

SEEDS. In all the florets, oblong; ſmall. *Feather* long.

RECEPT. Naked; flat.

OBS. *There are ſometimes florets in the central part that have only chives and no pointals.*

In one ſpecies the florets with only pointals have no bloſſoms.

Canada
Canadenſe

FLEABANE. Stem ſupporting panicles of flowers—*Florets in the center yellow; thoſe in the circumference white with a tinge of red.*

Conyza canadenſis annua, acris, alba, Linariæ foliis. *Ray's Syn.* 175.

Amongſt rubbiſh. A. Auguſt.

Blue
Acre

FLEABANE. Fruit-ſtalks alternate; each ſupporting a ſingle flower—*Florets in the center yellow; thoſe in the circumference purple.*

Conyza cærulea acris. *Gerard.* 484. *Bauh. pin.* 265.

Conyza odorata cærulea. *Park.* 126.

Aſter arvenſis, cæruleus, acris. *Ray's Syn.* 175.

Blue-flowered Fleabane.

Dry paſtures. P. July—September.

The Germans take a decoction of it to attenuate viſcid Phlegm.—Cows and Goats refuſe it.

337 GOLDENROD. 955 Solidago.

EMPAL. *Common*, oblong; tiled. *Scales* oblong; narrow; tapering; ſtraight; approaching.

BLOSS. *Compound*, radiate. *Florets with Chives and Pointals*, tubular; numerous; placed in the center. *Florets with only Pointals*, narrow; from five to ten; placed in the circumference.

Individuals with Chives and Pointals, funnel-ſhaped; *Border* with five clefts; open.

Individuals with only Pointals, narrow; ſpear-ſhaped; with three teeth.

CHIVES. *Threads* five; hair-like; very ſhort. *Tips* forming a hollow cylinder.

POINT. *Seedbud*, where there are both chives and pointals, oblong. *Shaft* as long as the chives; thread-ſhaped. *Summit* cloven; expanding.

Seedbud, where there are only pointals, oblong; *Shaft* thread-ſhaped; as long as the other. *Summits* two; rolled back.

S.VESS. None. *Cup* but little changed.

SEEDS. In all the florets; ſolitary; oblong; inverſely egg-ſhaped. *Feather* hair-like.

RECEPT. Flat; naked.

GOLDENROD. Stem ſomewhat zigzag; angular. Flowers crowded in panicled bunches; upright—*Leaves betwixt egg and ſpear-ſhaped.* Bloſſoms *yellow*. Common Virga-aurea

Virga aurea. *Gerard.* 429. *Ray's Syn.* 176. vulgaris. *Park.* 542.

Virga aurea latifolia ſerrata. *Bauh. pin.* 268.

Woundwort.

Woods, hedges and heaths. P. Auguſt.

Varieties.

1. With leaves narrower, and more ſlightly ſerrated.

Virga aurea IV, five anguſtifolia minus ſerrata. *Bauh. pin.* 268.

2. With leaves narrower; ſlightly ſerrated; growing on long leaf-ſtalks; and flowers larger, riſing from the baſe of the leaves, and collected into a kind of ſpike. *Ray's Syn.* ib.

Virga aurea vulgari humilior. *Ray's Syn.* 176.

In Ireland, and on Hampſtead-heath.

Horſes, Cows, Sheep, Goats and Swine eat it.

Welch Cambrica GOLDENROD. Leaves between ſtrap and ſpear-ſhaped; ſomewhat ſerrated; hoary: with flowers growing in a panicle like a ſort of broad-topped ſpike, and terminating the ſtem. *Hudſ. Fl. Angl.* 319. *Mart. Cat. Cant.* 158.

Virga aurea montana, folio anguſto ſub incano, floſculis conglobatis *Ray's Syn.* 177.

In mountainous paſtures in Wales. P. June.

338 FLEAWORT. 957 Cineraria.

EMPAL. *Common*, ſimple; of many leaves. *Little Leaves* equal.

BLOSS. *Compound*, radiate. *Florets with Chives and Pointals*, equal; numerous; in the center. *Florets with only Pointals* narrow; equal in number to the leaves of the empalement; in the circumference.

Individuals with Chives and Pointals, funnel-ſhaped; with five clefts; upright.

Individuals with only Pointals, narrow; ſpear-ſhaped; with teeth at the end.

CHIVES. *Threads* five; thread-ſhaped; ſhort. *Tips* forming a hollow cylinder, with five clefts at the top.

POINT. *Seedbud*, where there are both chives and pointals, oblong. *Shaft* thread-ſhaped; as long as the chives. *Summits* two; a little upright.

Seedbud, where there are only pointals, oblong. *Shaft* thread-ſhaped; ſhort. *Summits* two; oblong; rather blunt; rolled back.

S. VESS. None. *Cup* unchanged.

SEEDS. In all the florets, ſolitary; narrow; with four angles. *Feather* hairy; in large quantity.

RECEPT. Naked; rather flat.

Marſh Paluſtris FLEAWORT. Flowers in broad-topped ſpikes. Leaves broad; ſpear-ſhaped; toothed and indented. Stem woolly— *The Leaves cover the ſtem quite up to the flowers.* Bloſſoms *yellow*.

Conyza foliis laciniatis. *Gerard.* 483. *Ray's Syn.* 174.

Conyza aquatica laciniata. *Bauh. pin.* 266.

Conyza helenitis foliis laciniatis. *Park.* 126.

Othonna paluſtris. *Hudſon.* 327.

Marſh Fleabane.

1. There is a variety in which the leaves embrace the ſtem, and the teeth and indentures are not very evident.

In Marſhes and wet ditches. P. Auguſt.

FLEAWORT. Flowers in rundles. Leaves ſpear-ſhaped woolly; furniſhed with little teeth—*Fruit-ſtalks not leafy.* Bloſſoms *yellow.* Mountain Integrifolia

Jacobæa montana lanuginoſa anguſtifolia non laciniata. *Bauh. pin.* 131

Jacobæa pannonica folio non laciniata. *Ray's Syn.* 178.

Jacobæa anguſtifolia. *Gerard.* 280.

Jacobæa anguſtifolia panonica non laciniato. *Park.* 668.

Othonna integrifolia. *Hudſon.* 327.

Mountain Rag-wort.

Hilly paſtures. P. June.

339 GROUNDSEL. 953 Senecio.

EMPAL. *Common*, double; conical; lopped. *Scales* awl-ſhaped; numerous; contiguous; equal; dead at the ends; parallel; contracted above into a cylinder; the baſe tiled by a few ſcales.

BLOSS. *Compound*, taller than the cup. *Florets with Chives and Pointals* tubular; numerous; in the center. *Florets with only Pointals*, (if any ſuch there are) narrow; in the circumference.

Individuals with Chives and Pointals, funnel-ſhaped. *Border* reflected; with five clefts.

Individuals with only Pointals, (if any) oblong; with three imperfect teeth.

CHIVES. *Threads* five; hair-like; very ſmall. *Tips* forming a hollow cylinder.

POINT. *Seedbud*, in both ſorts of florets, egg ſhaped. *Shaft* thread-ſhaped; as long as the chives. *Summits* two; oblong; rolled back.

S.VESS. None. *Cup* cloſing ſo as to form a cone.

SEED. In both ſorts of florets, ſolitary; egg-ſhaped. *Feather* hairy; long.

RECEP. Naked, flat.

OBS. *In ſome ſpecies the florets are radiate in the circumference; in others they are all tubular.*

* *Flowers without Rays.*

Common Vulgaris GROUNDSEL. Flowers not radiate. Leaves indented into wings; embracing the ſtem. Flowers ſcattered.—*Bloſſoms yellow; terminating.*

Senecio vulgaris. *Park.* 671. *Ray's Syn.* 178.
Senecio minor vulgaris. *Bauh pin.* 131.
Erigeron. *Gerard.* 278.
Simpſon.
In gardens and amongſt rubbiſh. A. May.
The bruiſed leaves are a good application to Boils—The ſeeds are very acceptable to Linnets and Gold-finches when confined in cages.—Cows are not fond of it; Goats and Swine eat it; Horſes and Sheep refuſe it.

** *Flowers with Rays; the Rays rolled backwards.*

Cotton Viſcoſus GROUNDSEL. Bloſſoms in the circumference rolled backwards. Leaves with winged clefts: clammy. Scales of the empalement flexible; as long as the cup.—*The whole plant is furniſhed with hairs, which pour out a clammy liquor.* Fruit-ſtalks *lateral; ſupporting two or three flowers.* Bloſſoms *yellow.*

Senecio incanus pinguis. *Bauh. pin.* 131.
Senecio hirſutus viſcidus major odoratus. *Ray's Syn.* 178.
Senecio fætidus. *Park.* 671.
Stinking Groundſel.
Sandy places. A. June.

Wood Sylvaticus GROUNDSEL. Bloſſoms in the circumference rolled backwards. Leaves with winged clefts, and little teeth. Stem upright; ſupporting a broad-topped ſpike of flowers.—*Bloſſoms yellow.*

Senecio minor latiore folio, ſive montanus. *Bauh. pin.* 131. *Ray's Syn.* 178.
Erigerum. *Gerard.* 279.
Mountain Groundſel.
Woods, hedges and heaths. A. July.

*** *Flowers with expanding Rays. Leaves with winged Clefts.*

Hoary Erucifolius GROUNDSEL. Flowers with rays. Leaves with winged clefts and toothed; a little hairy. Stem upright—*Downy; white.* Bloſſoms *yellow.*

Jacobæa ſenecionis folio incano perennis. *Ray's Syn.* 177.
Hoary perennial Rag-wort.
Woods and hedges. P. July.

GROUNDSEL.

GROUNDSEL. Flowers with rays. Leaves lyre-shaped; almost winged. Segments jagged. Stem upright—*Blossoms yellow; on branched fruit-stalks.* Ragwort. Jacobæa

Jacobæa vulgaris. *Ray's Syn.* 177. major. *Park.* 668.

Jacobæa. *Gerard.* 280. vulgaris laciniata. *Bauh. pin.* 131.

Rag-wort. Seggrum. St. James's Wort.

1. Flowers without rays.

Jacobæa vulgaris flore nudo. *Ray's Syn.* 177.

Grows in great plenty near the sea-shore about three or four miles from Drogheda. Some few however amongst them, are found with radiated flowers. Perhaps this variety is only the effect of the sea air, as it is well known how destructive that is to the more tender parts of vegetables.

If this plant is gathered before the flowers open, and used fresh, it dyes wool of a full green, but the colour is apt to fade. If woollen cloth is boiled in Alum-water, and then in a decoction of the flowers, it takes a beautiful deep yellow.

Road-sides; pastures. P. July.

Horses and Sheep refuse it. Cows are not fond of it.

GROUNDSEL. Flowers with rays; leaves toothed: those at the root egg-shaped; those of the stem with winged clefts; the outer segments larger than the rest. *Hudf. Fl. Ang.* 317. Water Aquaticus

Jacobæa latifolia palustris sive aquatica. *Ray's Syn.* 178.

Jacobæa latifolia. *Gerard.* 280.

Jacobæa latifolia pannonica prima. *Park.* 658.

Jacobæa alpina laciniato flore Buphthalmi. *Bauh. pin.* 131.

Broad-leaved Rag-weed.

Mostly in watery places. P. July—August.

**** *Flowers with Rays. The Rays expanding. Leaves undivided.*

GROUNDSEL. Flowers with rays. Leaves sword-shaped, sharply serrated; a little woolly on the under-side. Stem stiff and straight—*Sometimes the leaves are downy on both surfaces. Blossoms yellow.* Marsh Paludosus

Virgæ aureæ sive solidagini angustifolia affinis; lingua avis Dalechampii. *Ray's Syn.* 176.

Conyza palustris. *Park.* 1231. serratifolia. *Bauh. pin.* 266.

Marsh Rag-wort. Bird's-tongue.

Ditches and Marshes. P. August.

Broad-leaved Sarracenicus GROUNDSEL. Flowers with rays; in broad-topped spikes. Leaves spear-shaped; serrated; almost smooth—*Roots creeping very much.* Blossoms *yellow.*

Virga aurea angustifolia serrata. *Bauh. pin.* 268.

Virga aurea maxima, radice repente. *Ray's Syn.* 177.

Broad-leaved Rag-wort.

Wet pastures, hedges and woods. P. July—August.

The Ruby Tyger Moth, *Phalæna Fuliginosa*; the Cinnabar Moth, *Phalæna Jacobæa*, and the great yellow underwing Moth, *Phalæna Pronuba*, live upon the several species.

340 BUTTERBUR. 952 Tussilago.

EMPAL. *Common*, cylindrical. *Scales* spear-shaped; narrow; equal; fifteen or twenty in number.

BLOSS. *Compound*, various. *Florets with Chives and Pointals*, in some species are all tubular, in others only tubular in the center.

Florets with only Pointals, in some species narrow; in others entirely wanting.

Individuals with Chives and Pointals, funnel-shaped. *Border* with four or five clefts; sharp; reflected; longer than the empalement.

Individuals with only Pointals either none at all, or very narrow; entire; longer than the empalement.

CHIVES. *Threads* five: hair-like; very short. *Tips* forming a hollow cylinder.

POINT. *Seedbud*, where there are chives and pointals, short. *Shaft* thread-shaped; longer than the chives. *Summit* thickish.

Seedbud, where there are only pointals, short. *Shaft* thread-shaped; as long as the other. *Summit* thickish; cloven.

S. VESS. None. *Cup* but little changed.

SEEDS. In all the florets, solitary; oblong; compressed. *Feather* hairy; standing on a pillar.

RECEPT. Naked.

OBS. *In the second and third species there are no narrow florets in the circumference, but there are florets with only pointals, without blossoms. The first species hath always narrow florets in the circumference, furnished only with pointals.*

BUTTERBUR.

BUTTERBUR. Stalk tiled; ſupporting a ſingle flower. Leaves ſomewhat; heart-ſhaped; angular; toothed—*Appearing after the flower.* Bloſſoms *yellow.* Coltsfoot Farfara

Tuſſilago. *Gerard.* 811. *Park.* 1220. *Ray's Syn.* 173. vulgaris. *Bauh. pin.* 197.

Common Coltsfoot.

Moiſt ſtiff marly lands, and amongſt lime-ſtone rubbiſh. P. March.

The downy ſubſtance on the under ſurface of the leaves, makes good Tinder—The leaves are the baſis of the Britiſh Herb Tobacco—They are ſomewhat auſtere, bitteriſh, and mucilaginous to the taſte—They were formerly much uſed in Coughs and Conſumptive complaints; and perhaps not without reaſon, for Dr. CULLEN has found them to do conſiderable ſervice in Scrophulous caſes. he gives a decoction of the dried leaves. and finds it ſucceed where ſea-water has failed—*Cullens Mat. Med.* p. 458.—Goat and Sheep eat it: Cows are fond of it: Horſes and Swine refuſe it.

BUTTERBUR. Flowers in an oblong cluſter. Florets with only pointals numerous; without bloſſoms—*Florets tubular.* Tall Hybrida

Petaſites major, floribus pediculis longis inſidentibus. *Ray's Syn.* 179.

Long ſtalked Butter-bur.

Moiſt paſtures. P. March.

BUTTERBUR. Flowers in an egg-ſhaped cluſter. Florets with only pointals few, and without bloſſoms—*Florets all tubular.* *The flowers appear before the leaves, which is likewiſe the caſe with the firſt ſpecies.* Bloſſoms *pale red.* Common Petaſites

Petaſites. *Gerard.* 814. *Ray's Syn.* 179. vulgaris. *Park.* 419.

Petaſites major et vulgaris. *Bauh. pin.* 197.

Moiſt ſandy places. P. March—April.

The roots abound with a reſinous matter. They have a ſtrong ſmell and a bitteriſh acrid taſte.—Horſes, Cows, Goats and Sheep eat it.

341 STARWORT 954 Aster.

EMPAL. *Common*, tiled with ſcales. The inner ſcales ſtanding out at the points.

BLOSS. *Compound*, radiate. *Florets with Chives and Pointals*, numerous; placed in the center. *Florets with only Pointals*, ten or more; narrow; placed in the circumference.

Individuals with Chives and Pointals, funnel ſhaped. *Border* with five clefts; open.

Individuals with only Pointals, narrow; ſpear-ſhaped; with three teeth.

CHIVES. *Threads* five; hair-like; very ſhort. *Tips* forming a hollow cylinder.

POINT. *Seedbud*, where there are both chives and pointals, oblong. *Shaft* thread-ſhaped; as long as the chives. *Summit* cloven expanding.

Seedbud, where there are only pointals, oblong. *Shafts* the ſame as the other. *Summits* two: oblong; rolled back.

S.VESS. None. *Cup* but little changed.

SEEDS. In all the florets, ſolitary; oblong; egg-ſhaped. *Feather* hairy.

RECEPT. Naked; almoſt flat.

Sea Tripolium — STARWORT. Leaves ſpear-ſhaped; very entire; fleſhy; ſmooth. Branches flatted. Flowers in broad topped ſpikes.— *Florets in the center yellow: thoſe in the circumference blue.*

Aſter maritimus cæruleus Tripolium dictus. *Ray's Syn.* 175.

Tripolium vulgaris, majus et minus. *Gerard.* 413.

Tripolium majus cæruleum et minus. *Bauh. pin.* 267.

Tripolium majus, ſeu vulgare et minus. *Park.* 673.

1. It varies in being ſmaller, and in the florets being ſometimes white.

Sea Starwort.

On the ſea-ſhore, and in ſalt marſhes in the interiour parts of the Kingdom, (viz) near Ingeſtree in Staffordſhire. P. Auguſt.

Goats and Horſes eat it: Cows and Swine refuſe it: Sheep are not fond of it.

342 CHAMOMILE.

342 CHAMOMILE. 970 Anthemis.

EMPAL. *Common*, hemiſpherical. *Scales* ſtrap-ſhaped nearly equal.

BLOSS. *Compound*, radiate. *Florets with Chives and Pointal* tubular; numerous; placed in the center, which i convex. *Florets with only Pointals*, many; placed in the circumference.

Individuals with Chives and Pointals, funnel-ſhaped with five teeth upright.

Individuals with only Pointals, narrow; ſpear-ſhaped ſometimes with three teeth,

CHIVES. *Threads* five; hair-like; very ſhort. *Tips* forming a hollow cylinder.

POINT. *Seedbud*, where there are both chives and pointals oblong. *Shaft* as long the chives; thread-ſhaped *Summits* two; reflected.

Seedbud where there ar only pointals, oblong. *Shaf* the ſame as the other. *Summits* two; rolled back.

S.VESS. None. *Cup* unchanged.

SEEDS. In all the florets, ſolitary; oblong. *Feather* none

RECEPT. Chaffy; conical.

* *Rays white.*

CHAMOMILE. Leaves winged, edged with little teeth; fleſhy: naked; dotted. Stem proſtrate; cups a little downy.— *Florets in the center yellow; thoſe in the circumference white.* Marine Maritima

Chamæmelum maritimum latifolium ramoſiſſimum, flore albo. *Ray's Syn.* 186.

Matricaria maritima. *Bauh. pin.* 134.

Sea Chamomile.

Paſtures near the ſea. P. Auguſt.

CHAMOMILE. Leaves winged and compound; ſtrap-ſhaped; ſharp; ſomewhat hairy—*Florets in the center yellow; thoſe in the circumference white; bent outwards and downwards.* Sweet-ſcented Nobilis

Chamæmelum nobile ſeu Leucanthemum odoratius. *Bauh pin.* 135.

Chamæmelum odoratiſſimum repens, flore ſimplici. *Ray Syn.* 185.

Chamæmelum Romanum. *Gerard.* 755.

Trailing perennial Chamomile. Roman Chamomile.

1. By cultivation the flowers become double: that is the flore of the circumference become more numerous, and thoſe in th center diſappear in proportion.

Warm paſtures. P. July Auguſt.

The leaves and flowers have a ſtrong, not ungrateful ſmell, and a bitter nauſeous taſte. They afford an eſſential oil.—An infuſion of the flowers is often uſed as a ſtomachic and as an antiſpaſmodic. In large quantities it excites vomiting—The powdered flowers, in large doſes, have cured Agues, even when the Bark had failed—Both the leaves and flowers poſſeſs very conſiderable antiſceptic properties, and are therefore uſed in antiſcepticaomentations, and poultices. From their antiſpaſmodic powers, they are frequently found to relieve pain, eirther applied externally, or taken internally.

Corn Arvenſis

CHAMOMILE. Receptacles conical. Chaff briſtly. Seeds crowned with a border—*The whole plant is hoary, and ſpreads wide. The inner ſcales of the cup broad at the end, and membranaceous.* Florets *in the center yellow; thoſe in the circumference white.*

Chamæmelum inodonem. *Bauh. pin.* 135. *Ray's Syn.* 184.

Corn-field and road-ſides. B. July.

Cows and Sheep eat it: Horſes are not fond of it.

May-weed Cotula

CHAMOMILE. Receptacles conical. Chaff briſtly. Seeds naked—*The chaff is very ſmall. The florets are ſometimes proliferous: thoſe in the center yellow; thoſe in the circumference white; with three teeth.*

Chamæmelum fœtidum. *Bauh. pin.* 135. *Ray's Syn.* 185.

1. The variety with double bloſſoms. *Ray's Syn.* 185.

Stinking May-weed.

Corn-fields and road-ſides. A. June—July.

Toads are ſaid to be fond of this plant—It is very ungrateful and diſpleaſing to Bees—Goats and Sheep are not fond of it: Horſes, Cows and Swine refuſe it.

* * *Rays yellow*

Ox-eye. Tinctoria

CHAMOMILE. Leaves doubly winged; ſerrated, downy underneath. Stem ſupporting a broad-topped ſpike of flowers.—*Bloſſoms yellow.*

Bupthalmum Tanaceti minoris folio. *Bauh. pin.* 134.

Bupthalmum vulgare. *Gerard.* 747. *Ray's Syn.* 182.

Bupthalmum Matthioli, ſeu vulgare Millefolii foliis. *Park.* 1370.

Common Ox-eye.

Sunny paſtures. P. July.

The flowers afford a remarkably clear and good yellow dye. The flowers of the *yellow* GOLDINS reſemble them much in appearance, but experience proves they cannot be ſubſtituted in their place.

Horſes and Goats eat it: Sheep are not fond of it: Cows and Swine refuſe it.

343 YARROW.

343 YARROW. 971 Achillea.

EMPAL. *Common*, oblong; egg-ſhaped; tiled. *Scales* egg-ſhaped; ſharp; approaching.

BLOSS. *Compound*, radiate. *Florets with Chives and Pointals*, from five to ſixteen: tubular; placed in the center. *Florets with only Pointals*, from five to ten; narrow; placed in the circumference.

Individuals with Chives and Pointals, funnel-ſhaped; with five clefts; open.

Individuals with only Pointals narrow; inverſely heart-ſhaped; expanding. cloven into three ſegments, the middlemoſt the ſmalleſt.

CHIVES. *Threads* five; hair-like; very ſhort. *Tips* forming a hollow cylinder.

POINT. *Seedbud*, where there are both chives and pointals, ſmall. *Shaft* thread-ſhaped; as long as the chives. *Summit* blunt notched at the end.

Seedbud where there are only pointals, ſmall. *Shaft* thread-ſhaped; as long as the other. *Summits* two; blunt: reflected.

S.VESS. None. *Cup* but little changed. The *Receptacle* thread-ſhaped; lengthens out into the diſc of the ſeeds; egg-ſhaped, and twice as long as the empalement.

SEEDS. In all the florets, ſolitary; egg-ſhaped; woolly. *Feather* none.

RECEPT. Chaffy; elevated. *Chaff* ſpear-ſhaped; as long as the florets.

YARROW. Leaves ſpear-ſhaped; tapering; ſharply ſerrated. *Bloſſoms white: ſometimes double.* Gooſe-tongue Ptarmica
Ptarmica. *Gerard.* 606. *Ray's Syn.* 183. vulgaris. *Park.* 859.
Dracunculus pratenſis, ſerrato folio. *Bauh. pin.* 98.
Sneeze-wort. Gooſe-tongue. Baſtard Pellitory.
Moiſt woods. P. Auguſt.
The roots have a hot biting taſte—The young tops are ſharp and pleaſant in ſpring ſallads—The powdered leaves excite ſneezing—Horſes, Cows, Sheep, Goats and Swine eat it.

Milfoil Millefolium

YARROW. Leaves doubly winged; naked. Segments strap-shaped; toothed. Stem furrowed towards the top—*Blossoms white; sometimes tinged with red or purple.*

Millefolium vulgare. *Park.* 693 *Ray's Syn.* 183. album. *Bauh. pin.* 140.

Millefolium terrestre vulgare. *Gerard.* 1072.

Common Yarrow or Millfoil.

1. There is a variety with purple blossoms.

Dry pastures. P. May—July.

The flowers yield an essential oil. The leaves are celebrated by the Materia Medica writers for a variety of purposes, but they are little attended to at present.—Sheep and Swine eat it. Horses, Cows and Goats are not fond of it.

Order III. Barren Florets.

344 KNAPWEED. 984 Centaurea.

EMPAL. *Common*, tiled; roundish. *Scales* often terminating variously.

BLOSS. *Compound*, florets all tubular; but of different shapes. *Florets with Chives and Pointals*, many: placed in the center. *Florets with only Pointals*, not so many; larger; more flexible; placed in the circumference.

Individuals with Chives and Pointals, one petal. *Tube* thread-shaped *Border* distended; oblong; upright; terminating in five strap-shaped, upright segments.

Individuals with only Pointals, one petal. *Tube* slender; gradually becoming wider; bent backwards. *Border* oblong; oblique; unequally divided.

CHIVES. *Threads* five; hair-like; very short. *Tips* forming a hollow cylinder as long as the blossom.

POINT. *Seedbud*, where there are chives and pointals, small. *Shaft* thread-shaped; as long as the chives. *Summit* very blunt; (in many cloven;) with a projecting point.

Seedbud, where there are only pointals, very small. *Shaft*, next to none. *Summit* none.

S. VESS. None. *Cup* unchanged, closing.

SEEDS. In the fertile florets of the center, solitary. *Feather* generally downy, or hairy.

RECEPT. Bristly.

OBS. *The scales of the cup, and the feathers of the seeds are different in different species.*

* Scales

Scales of the cup fringed.

KNAPWEED. Cups ferrated. Leaves ftrap-fhaped; very entire. The lower leaves toothed—*Bloffoms generally blue, fometimes red, or white, or purple, or flefh-colour.* Blue-bottle Cyanus

Cyanus. *Ray's Syn.* 198. vulgaris. *Gerard.* 732. *Park.* 482.

Cyanus fegetum. *Baub. pin.* 273.

Blue-bottles. Corn-flower. Hurt-fickle.

Corn-fields. A. July.

The varieties are very numerous if they are eftimated by the colour of the bloffom, *viz.*

1. White.
2. Flefh coloured.
3. Purple.
4. Red.
5. Pale violet.
6. Dark purple.
7. Blue and white.
8. Flefh colour and white.
9. Violet colour and white.
10. Purple and white.
11. Purplifh and blue.
12. Double blue.
13. Double purple.
14. Double purple and white.

The expreffed juice of the petals is a good blue ink, it ftains linen of a beautiful blue, but the colour is not permanent in the mode it has hitherto been applied. Mr. Boyle fays, the juice of the Central Florets with the addition of a very fmall quantity of alum, makes a lafting tranfparent blue not inferior to ultramarine. *Gent. Mag.* 1748.

Cows, Goats and Sheep eat it; Horfes and Swine refufe it.

KNAPWEED. Cups fringed. Leaves with winged clefts. Segments fpear-fhaped.—*Bloffoms purple, fometimes white; on long naked fruit ftalks.* Great Scabiofa

Jacea major. *Gerard.* 728. *Ray's Syn.* 198.

Jacea nigra vulgaris laciniata. *Park.* 470.

Scabiofa major, fquamatis capitulis. *Baub. pin.* 269.

Marfellon.

Borders of corn fields. P. June. July.

Horfes, Sheep, Goats and Swine eat it; Cows refufe it.

Common Jacea

KNAPWEED. Cups ſkinny; ragged. Leaves ſpear-ſhaped. Root-leaves indented and toothed. Branches angular. —*The leaves ſometimes are white with down.* Bloſſoms *purple.*

Jacea nigra. *Gerard.* 726. *Ray's Syn.* 198. vulgaris. *Park.* 468.

Jacea nigra pratenſis latifolia. *Bauh. pin.* 271.

1. There is a variety with downy leaves. *Ray's Syn.* 199.

Matfellon. Knopweed. Horſeknops.

Meadows and paſtures. P. July—Auguſt.

The following varieties are produced by cultivation;

1. White bloſſomed.
2. Small, with jagged leaves.
3. Stems rough; leaves narrow.
4. Stems ſmooth; leaves narrow.

This plant affords a yellow dye, but inferior to that from the *Dyers* SAWWORT.—Cows, Goats and Sheep eat it; Horſes and Swine refuſe it.

The Knapweed Louſe, *Aphis Jaceæ* is found upon it.

* * *Scales of the Cups thorny.*

Starry Calcitrapa

KNAPWEED. Cups with thorns, ſomewhat double; ſitting. Leaves with winged clefts; ſtrap-ſhaped; toothed. Stem hairy—*Bloſſoms purple; ſeveral upon the ſame ſtem; ſometimes deep purple; rarely white.*

Carduus ſtellatus. *Gerard.* 1166. *Ray's Syn.* 196.

Carduus ſtellaris, ſeu Calcitrapa vulgaris. *Park.* 989.

Carduus ſtellatus, foliis Papaveris erratici. *Bauh. pin.* 387.

Star Thiſtle.

Road-ſides and barren ground. A. July.

Thiſtle Solſtitialis

KNAPWEED. Cups with double thorns; ſolitary. Leaves ſpear-ſhaped, without thorns; running along the branches. Root-leaves lyre-ſhaped, with winged clefts—*Fruit-ſtalks very long; leafy.* Thorns *of the cup only toothed at the baſe.* Bloſſoms *yellow.*

Carduus ſtellatus luteus, foliis Cyani. *Bauh. pin.* 387. *Ray's Syn.* 196.

Carduus ſolſtitialis. *Gerard.* 1166. *Park.* 989.

St. Barnaby's Thiſtle.

Hedges. A. July.

Order IV. Neceſſary Pointals.

345 CUDWEED. 995 Filago.

EMPAL. *Common*, compoſed of chaffy ſcales; tiled; con taining in the center many florets with chives and pointals, and amongſt the lower ſcales *ſolitary Florets* with only pointals.

BLOSS. *Individuals with Chives and Pointals*, funnel-ſhaped. *Border* with four clefts, upright.

Individuals with only Pointals, not very conſpicuous; thread-ſhaped; very ſlender; mouth cloven.

CHIVES. *Threads* four; hair-like; ſmall. *Tips* forming a hollow cylinder, with four teeth at the top.

POINT. *Seedbud*, where there are chives and pointals, hardly perceptible. *Shaft* ſimple. *Summit* ſharp; cloven.

Seedbud, where there are only pointals, inverſely egg-ſhaped; rather large; depreſſed. *Shaft* thread-ſhaped. *Summit* cloven; ſharp.

S. VESS. None.

SEEDS. In thoſe florets that have both chives and pointals, none. In the florets with only pointals, inverſely egg-ſhaped; compreſſed; ſmooth; ſmall. *Feather* none.

RECEPT. The center naked; but at the ſides the chafy ſcales of the empalement ſeparate the florets.

CUDWEED. Flowers in a forked panicle; round; in the diviſions of the branches; hairy. Leaves ſharp—*Cups with five corners; the ſcales gloſſy and ſharp.* Bloſſoms *brown.* Common Germanica

Gnaphalium vulgare majus. *Bauh. pin.* 263.

Gnaphalium minus, ſeu Herba impia. *Park.* 685. *Ray's Syn.* 180.

Filago, ſeu Herba impia. *Gerard.* 641.

Chafeweed.

Barren paſtures and road-ſides. A. June—July.

It is given to cattle that have the bloody flux; and hath been tried with ſucceſs in ſimilar complaints of the human body.

Least Montana

CUDWEED. Stem upright; a little forked. Flowers conical; some terminating: others in the divisions of the branches —*Leaves numerous; downy.* Blossoms *yellowish brown.*

Filago minor. *Gerard.* 641.

Gnaphalium minimum. *Ray's Syn.* 181.

Gnaphalium minus repens. *Bauh. pin.* 363.

Sandy pastures. A. June—July.

Sheep eat it; Cows and Goats refuse it.

Corn Gallica

CUDWEED. Stem upright; forked. Flowers awl-shaped; in the divisions of the branches. Leaves thread-shaped—*Leaves not hairy, but smooth and a little downy.*

Gnaphalium parvum ramosissimum foliis angustissimis polyspermon. *Ray's Syn.* 181.

Sandy corn-fields. A. June—July.

Order VI. Flowers simple.

346 SCABIOUS. 1005 Jasione

EMPAL. *Common Cup* of many leaves; permanent; alternate; the inner ones narrower; inclosing several flowers upon very short fruit-stalks.

Proper Cup with five clefts; superior; permanent.

BLOSS. *Individuals* with five petals. *Petals* spear-shaped; upright; connected at the base.

CHIVES. *Threads* five; awl-shaped; short. *Tips* five; oblong; connected at the base.

POINT. *Seedbud* beneath; roundish. *Shaft* thread-shaped; as long as the blossom. *Summit* cloven.

S.VESS. *Capsule* roundish; with five angles and two cells: crowned by the *proper Cup.*

SEEDS. Many; somewhat egg-shaped,

OBS. *The central florets are frequently barren, and the summit club-shaped and undivided.*

Sheeps Montana

SCABIOUS. As there is only one species known, *Linnæus* gives no description of it—Blossoms *blue.* Fruit-stalks *naked.*

Rapunculus scabiosæ capitulo cæruleo. *Bauh. pin.* 92. *Park.* 646. *Ray's Syn.* 278.

Scabiosa minima hirsuta. *Gerard.* 723.

Hairy sheeps Scabious.

Heaths and hilly pastures. A. June—July.

The flowers are very grateful to bees.

347 CARDINALFLOWER. 1006 Lobelia.

EMPAL. *Cup* one leaf; with five teeth; very ſmall; embracing the ſeedbud; ſhrivelling. *Teeth* nearly equal; the two upper ones a little raiſed.

BLOSS. One petal; rather gaping. *Tube* cylindrical; longer than the cup; divided on the upper part lengthways. *Border* with five diviſions. *Segments* ſpear-ſhaped; the *two upper ones* ſmaller; more reflected; more deeply divided; forming the upper lip. The *three lower ones* generally larger, and more expanding.

CHIVES. *Threads* five; awl-ſhaped; as long as the tube of the bloſſom. *Tips* connected ſo as to form an oblong cylinder, opening at the baſe in five different directions.

POINT. *Seedbud* beneath: tapering to a point. *Shaft* cylindrical; as long as the chives. *Summit* blunt; rough with hair.

S. VESS. *Capſule* egg-ſhaped; with two cells; opening at the top; encompaſſed by the cup.

SEEDS. Many; very ſmall.

CARDINALFLOWER. Leaves ſtrap-ſhaped; very entire; with a double hollow running lengthways. Stem almoſt naked — *The whole plant, and even the leaves that lye beneath the ſurface of the water are milky.* Bloſſoms *pale purple, or bluiſh white.* Water Dortmanna

Gladeolus lacuſtris Dortmanni. *Ray's Syn.* 287.

Gladeolus lacuſtris Cluſii, ſeu Leucojum paluſtre, flore ſubcæruleo. *Park.* 1250.

Leucojum paluſtre flore ſubcæruleo. *Bauh. pin.* 41.

In lakes in hilly countries. P. July—Auguſt.

348 VIOLET. 1007 Viola.

EMPAL. *Cup* five leaves; ſhort; permanent. *Little Leaves* oblong egg-ſhaped; rather ſharp at the end; blunt below; fixed above the baſe; equal; but variouſly diſpoſed. Two ſupport the upper petal; two ſupport the two lateral petals; and one ſupports the two lower petals.

BLOSS. Petals five; irregular unequal; the *Upper Petal* ſtraight; facing downwards; broader and more blunt than the reſt; notched at the end: Terminating at the baſe in a blunt *Honey-cup* reſembling a horn, and projecting betwixt the leaves of the cup.

Lateral Petals two; oppoſite; blunt; ſtraight.

Lower Petals two; larger; reflected upwards.

CHIVES. *Threads* five; very ſmall. The two near the uppermoſt petal, are furniſhed with little appendages which enter the honey-cup. *Tips* generally united; blunt; with membranes at the end.

POINT. *Seedbud* roundiſh. *Shaft* thread-ſhaped; extending beyond the tips. *Summit* oblique.

S. VESS. *Capſule* egg-ſhaped; three-edged; blunt; with one cell and three valves.

SEEDS. Many; egg-ſhaped; furniſhed with appendages fixed to the valves. *Receptacle* narrow; running like a line along each valve.

OBS. *In ſome ſpecies the* SUMMIT *is a ſimple reflected hook; in others it is a little concave knob, perforated at the end.*

* *Without Stems.*

Hairy Hirta — VIOLET, without a ſtem. Leaves heart-ſhaped; rough with hair—*Bloſſoms pale blue.*
Viola martia major hirſuta inodora. *Ray's Syn.* 365.
Woods and hedges. P. March.
Horſes, Cows, Goats and Sheep eat it.

Marſh Paluſtris — VIOLET, without a ſtem. Leaves kidney-ſhaped—*Fruitſtalks cylindrical, or only a little flatted upon one ſide. The upper petal is marked with black lines which extend quite to the top of it.* Bloſſoms *pale blue.*
Viola paluſtris rotundifolia glabra. *Ray's Syn.* 364.
1. There is a variety in which the bloſſom is ſtriped with purple. *Park.* 755.
Cows eat it; Sheep are not fond of it.

VIOLET

VIOLET, without a ſtem. Leaves heart-ſhaped. Suckers creeping along—*Bloſſoms deepiſh purple.* Sweet Odorata

Viola Martia purpurea. *Ray's Syn.* 364. flore ſimplici. *Bauh. pin.* 199.

Viola nigra ſive purpurea. *Gerard.* 850.

Viola ſimplex Martia. *Park.* 755.

1. There is a variety with white bloſſoms. *Bauh. pin.* 199.

2. Cultivation produces double flowers, blue and white, or pale purple.

Ditch-banks, in moiſt warm lanes. P. March.

The flowers and the ſeeds are ſaid to be mild laxatives. The petals give the colour to the ſyrup of Violets. This ſyrup is very uſeful in many Chemical enquiries, to detect an acid or an alkaly: the former changing the blue colour to a red, and the latter turning it green—Slips of white paper ſtained with the juice of the petals, and kept from the air and the light, anſwer the ſame purpoſe.

* * *With Stems.*

VIOLET. Stem aſcending as it attains its full growth. Leaves oblong heart-ſhaped—*At the firſt opening of the flower it hath no ſtalk, but afterwards the ſtalk grows up and fruit-ſtalks proceed from it.* Bloſſoms *pale blue.* Dogs Canina

Viola Martia inodora ſylveſtris. *Bauh. pin.* 199. *Ray's Syn.* 364.

Viola canina ſylveſtris. *Gerard.* 851.

Viola ſylveſtris. *Park.* 755.

1. Bloſſom white. *Ray's Syn.* 364.

2. Very ſmall Violet. *Ray's Syn.* 364. Tab. 24. fig. 1. ſmaller in every reſpect, and the ſpur of the bloſſom of a pale yellow.

On heaths and ditch-banks. P. April.

Cows, Sheep, Swine and Goats eat it; Horſes refuſe it.

VIOLET. Stem ſupporting two or three flowers. Leaves kidney-ſhaped, ſerrated—*Bloſſoms yellow.* Welch Biflora

Viola alpina rotundifolia lutea. *Bauh. pin.* 199.

Viola alpina. *Hudſon.* 331.

Viola Martia alpina folio tenello circinnato. *Ray's Syn.* 366.

On mountains in Wales. P.

Goats eat it.

*** *Props with winged Clefts.*

Panfie Tricolor

VIOLET. Stem three cornered; fpreading. Leaves oblong; jagged. Props with winged clefts—*Fruit-ftalks compreffed.* Floral leaves *two*; *halberd-fhaped*; *with two teeth on each fide.* Summit *globular, open and hollow*; *fringed on the lower part, fo that the little fibres of the fringe are nearly in contact with the tips.* Bloffoms *purple, yellow, and light blue.*

Viola tricolor. *Gerard.* 854. *Ray's Syn.* 365. major et vulgaris. *Park.* 756.

Viola tricolor hortenfis repens. *Bauh. pin.* 199.

1. With only two colours. *Bauh. pin.* 200.

Hearts-eafe. Panfies. Three Faces under a Hood.

Ditch-banks and corn-fields. A. May—September.

This is very frequently cultivated in flower gardens. The duft appears angular in the Microfcope, but when wet with water it becomes globular. This alteration in figure is not peculiar to the duft of this plant only—Cows and Goats eat it; Swine are not fond of it; Horfes and Sheep refufe it.

Yellow Grandiflora

VIOLET. Stem upright; three cornered. Leaves fomewhat oblong; props with winged clefts—*Bloffoms very large; yellow.* Petals *egg-fhaped*; *as long as the leaves.* Spur *twice as long as the cup, but fhorter than the petals.* Summit *globular, open and hollow.*

Viola montana lutea grandiflora noftras. *Ray's Syn.* 365.

Viola flammea lutea. *Gerard.* 851.

Viola montana lutea grandiflora. *Bauh. pin.* 200.

Viola lutea. *Hudfon.* 331.

Yellow Panfies.

In bogs upon mountains, and hilly paftures. P. May—Auguft.

The different fpecies of Violets afford nourifhment to the great Fritillary Butterfly with filver fpots; *Papilio Aglaja*; and to the high brown Fritillary Butterfly, *Papilio Adippe.*

349 WEATHERCOCK. 1008 Impatiens.

EMPAL. *Cup* two leaves; very ſmall. *Little Leaves* circular, but tapering towards a point; equal; placed at the ſides of the bloſſom; coloured; deciduous.

BLOSS. Petals five; gaping; unequal; the

Upper Petal circular; flat; upright; with three ſhallow ſegments; tapering to a point; forming the *Upper Lip*.

Lower Petals two; reflected; large; broadeſt on the outer part; blunt; irregular; forming the *Lower Lip*.

Intermediate Petals two; oppoſite; ariſing from the baſe of the upper petal.

Honey cup one leaf like a hood, receiving the bottom of the flower. Mouth oblique; riſing outwards. Baſe ending in a horn.

CHIVES, *Threads* five; very ſhort; narrower towards the baſe; bent inwards. *Tips* five; united; but ſeparate at the baſe.

POINT. *Seedbud* egg-ſhaped; but tapering. *Shaft* none. *Summit* ſimple; ſhorter than the tips.

S. VESS. *Capſule* one cell, with five valves, which opening with a ſpring roll up into a ſpiral.

SEEDS. Many; roundiſh; fixed to a pillar-like receptacle.

OBS. *In ſome ſpecies the intermediate petals are wanting; in others the honey-cup hath no horn. The figure of the capſule is various.*

WEATHERCOCK. Fruit-ſtalks ſupporting many flowers; ſolitary. Leaves egg-ſhaped. Stem ſwelling at the joints—*When the ſeeds are ripe, upon touching the capſule, they are thrown out with conſiderable force. In the day time the leaves are expanded; but at night they hang pendant.* Bloſſoms *yellow; the lateral petals ſpoted with red: by cultivation they are changed to pale yellow or purpliſh.* Impatient Noli tangere

Balſamine lutea, ſeu Noli me tangere. *Bauh. pin.* 306. *Ray's Syn.* 316.

Perſicaria ſiliquoſa. *Gerard.* 440.

Mercurialis ſylveſtris, Noli me tangere dicta, ſive perſicaria ſiliquoſa. *Park.* 296.

Quick in the Hand. Touch me not. Balſamine.

Moiſt ſhady places. Banks of rivulets. A. Auguſt.

The whole plant is conſiderably acrid—Goats eat it; Horſes, Cows and Sheep refuſe it.

The Elephant Moth, *Sphinx Elpenor*, lives upon it.

CLASS.

CLASS XX.

THIS Clafs is diftinguifhed by the CHIVES being placed upon the SHAFT, or upon the *Receptacle* lengthened out into the form of a SHAFT, fupporting both the *Pointal* and the *Chives*, and entering into part of the pointal.

The firft Order of this Clafs is natural, and its Genera are diftinguifhed by the Honey-cup alone. The ftructure of the flowers is very fingular.

SEEDBUD always below the bloffom, twifted like a fcrew.

PETALS five; the two inner ones generally approaching fo as to refemble a helmet. A *Honey-cup* forms the lower lip and occupies the place of the pointal, and of a fixth petal.

SHAFT fixed to the inner edge of the honey-cup, fo as to be fcarce diftinguifhable; as is likewife the *Summit*.

THREADS always two; very fhort; fupporting two TIPS; more flender below; without any coat; divifible like the pulp of Citron. They are inclofed in little cells opening downwards, and fixed to the inner edge of the honey-cup.

CAPSULE one cell, with three valves; opening at the angles under the keel-fhaped ribs.

SEEDS like faw-duft; very numerous; fixed to a narrow receptacle upon each valve.

The

The reader in examining and comparing the characters of this clafs, fhould always attend to the Pointal before the Chives, in order to attain a diftinct idea of the fituation of the latter.

The flowers of this Clafs have a very fingular appearance, owing to the unfual difpofition of their component parts.

A very ingenious gentleman at Worcefter, to whofe judgment and accuracy I have been much indebted in the courfe of this work, has favoured me with the following remarks upon the ftructure of the *Chives* in the firft Order of this Clafs.—The CHIVES are evidently two: but each of them appears to be compofed of a number of elaftic *Fibres* united together; each *Fibre* fupporting its own proper *Tip*. Thefe *Fibres* I call elaftic; for if you prefs down the bottom of the Chive with a needle or any other inftrument, and draw another needle along the Chive, preffing it with fome force, you may extend the Chive to twice its former length: but no fooner do you remove the force applied, than the fibres contract with an elaftic force, and the Chive returns to its former fhape. This you may repeat, and the effect will be conftantly the fame.

Thefe fibres branch out into leffer ones, each fupporting at its point an extremely minute *Tip*. Thefe tips I have examined in the microfcope, and found fome of them roundifh and others nearly triangular. The *Duft* which thefe probably contained I was not able to difcover.

Order I. Two Chives.

350 ORCHIS. - - *Honey-cup* like a little horn.
351 SATYRION. - *Honey-cup* ſhaped like a double purſe.
352 TWAYBLADE. *Honey cup* ſomewhat keel-ſhaped.
353 HELLEBORINE. *Honey-cup* egg-ſhaped; hunched on the under-ſide.
354 SLIPPER. - - *Honey-cup* diſtended and bladder-ſhaped.

Order VIII. Many Chives.

355 CUCKOWPINT. In a ſheath. *Cup* none. *Bloſſ.* none. *Chives* above the pointals.
356 GRASSWRACK. In a leaf. *Cup* none. *Bloſſ.* none. *Seeds* alternate, naked.

† *Gatteridge Spindle.*

350 ORCHIS.

350 ORCHIS. 1009 Orchis.

EMPAL. *Sheaths* scattered. *Fruit-stalk* simple.
Cup none.

BLOSS. Petals five; three *outer* ones; two *inner* ones, approaching upwards so as to form a helmet.
Honey-cup one leaf, fixed by the lower side to the receptacle betwixt the division of the petals. *Upper Lip* upright; very short. *Lower Lip* large; expanding; broad. *Tube* standing behind, shaped like a horn; hanging a little down.

CHIVES. *Threads* two; very slender and very short; fixed on the pointal. *Tips* inversely egg-shaped; upright; covered by a fold of the upper lip of the honey-cup, forming two cells.

POINT. *Seedbud* beneath; oblong; twisted. *Shaft* fixed to the upper lip of the honey-cup; very short. *Summit* compressed; blunt.

S. VESS. *Capsule* oblong; with one cell; three keels; three valves; opening in three places under the keels; connected at the base and at the end.

SEEDS. Numerous; very small; like saw-dust.

* *Bulbs of the root undivided.*

ORCHIS. Bulbs undivided. Lip of the honey-cup spear-shaped; very entire. Horn very long. Petals expanding—*Blossoms yellowish white, smelling very sweet in the night.* Leaves *generally two.* Butterfly Bifolia

Orchis alba bifolia minor, calcari oblongo. *Bauh. pin.* 83. *Ray's Syn.* 380.

Orchis psychodes. *Gerard.* 211.

Orchis sphegodes, seu Testiculis vulpinus primus. *Park.* 1351.

Pastures and uncultivated rough grounds. P. June.

ORCHIS. Bulbs undivided. Lip of the honey-cup with two horns; cloven into three segments; equal; very entire. Horn long. Petals somewhat spear-shaped—*Bulbs joined together.* Stem *about a foot high; jointed; with five or six sharp, spear-shaped, membranaceous leaves.* Floral Leaves *spear-shaped; coloured; as long as the seedbud. Two petals expanding, three approaching.* Blossoms *pale purple.* Pyramid Pyramidalis

Orchis purpurea, spica congesta pyramidali. *Ray's Syn.* 377. tab. 18.

Purple late-flowering Orchis.

Dry pastures. P. June—July.

Lizzard Coriophora

ORCHIS. Bulbs undivided. Lip of the honey-cup cloven into three ſegments; reflected; ſcolloped. Horn ſhort. Petals approaching.—*Bloſſoms greeniſh white or tinged with purple.*

Orchis odore hirci minor. *Bauh. pin.* 82.

Orchis barbata fætida minor flore albo. *Ray's Syn.* 376.

Leſſer Lizard Flower.

Fools Morio

ORCHIS. Bulbs undivided. Lip of the honey-cup cloven into four ſegments; ſcolloped. Horn blunt; aſcending. Petals blunt; approaching.—*Bloſſoms white, or red mottled with white, or violet coloured.*

Orchis morio fæmina. *Bauh. pin.* 82. *Park.* 1347. *Ray's Syn.* 377.

Cynoſorchis morio fæmina. *Gerard.* 208.

Female Fool-ſtones.

Moiſt paſtures. P. May—June.

Goats eat it; Horſes refuſe it.

Bulbous Maſcula

ORCHIS. Bulbs undivided. Lip of the honey-cup divided into four lobes; ſcolloped. Horn blunt. Petals on the back reflected—*Bloſſoms purple. This differs from the preceding ſpecies by the outer petals being longer and more ſharp, and the middle lobe of the lip cloven and longer than the lateral lobes. The flowers are likewiſe more numerous, and the ſtem twice as tall. Floral leaf coloured.*

Orchis morio mas, foliis maculatis. *Bauh. pin.* 81. *Park.* 1346.

Cynoſorchis morio mas. *Gerard.* 208.

Male Fool-ſtones.

Meadows and paſtures. P. April—May.

Mr. Moult, in a letter to Dr. Percival inſerted in the *Philoſ. Tranſ.* vol. 59. p. 1. deſcribes his method of making Salep. The beſt time to gather the roots, he obſerves, is when the ſeed is formed, and the ſtalk is going to fall; for then the new bulb, of which Salep is made, is arrived at its full ſize. This new root being ſeparated from the ſtalk, is to be waſhed in water, and the outer thin ſkin taken off. They are then to be ſet on a tin plate in an oven heated to the degree of a bread-oven. In ſix, eight or ten minutes they will have acquired a tranſparency like horn, but without being diminiſhed in ſize. They are then to be removed into another room to dry and harden, which will be done in a few days: or they may be finiſhed in a very ſlow heat in a few hours.

Salep thus prepared, may be ſold for leſs than a ſhilling a pound, and affords a mild nutriment which in times of ſcarcity, in caſes of Dyſentery and Strangury, and on ſhip-board may be extremely uſeful. *See Percival's Eſſays. part.* 2. *p.* 37.

Mr. Moult made his experiments upon the roots of this ſpecies only. The preceding ſpecies is undoubtedly equally proper for

for the purpose, and it is highly probable that every species of Orchis may be used indiscriminately. Salep has been hitherto imported from Turkey at a considerable price, but it is to be hoped we shall no longer be supplied from foreign markets, with an article that our own country can supply us with in almost any quantity. If ever plantations of it are made, the plants must be propagated by roots, for the seeds seldom come to perfection. Dr. Percival observes that he got some seeds to all appearance perfect, but yet they would not vegetate. Is it that the warmth of our climate being insufficient fully to expand the blossom, the tips are never released from their confinement in the cells formed by the upper lip of the honey-cup, and therefore the dust can never escape to fertilize the seedbud?

ORCHIS. Bulbs undivided. Lip of the honey-cup cloven into four segments, with rough dots. Horn blunt. Petals distinct.—*Blossoms purple; mottled with white.* Purple Ustulata

Orchis pannonica. *Ray's Syn.* 377.
Cynosorchis militaris pratensis humilior. *Bauh. pin.* 81.
Cynosorchis militaris panonnica. *Park.* 1345.
Cynosorchis minor pannonica. *Gerard.* 207.
Little purple-flowered Orchis.
Dry barren pastures. P. May—June.

ORCHIS. Bulbs undivided. Lip of the honey-cup cloven into five segments, with rough dots. Horn blunt. Petals thronging together.—*Blossom dark purple, or greyish white.* Soldier Militaris

Orchis galea et alis fere cinereis. *Ray's Syn.* 378.
Cynosorchis latifolia hiante cucullo major. *Bauh. pin.* 80.
Cynosorchis latifolia minor. *Park.* 1 44.
Cynosorchis major altera. *Gerard.* 205.
Man Orchis.

1. Bulbs undivided. Lip of the honey-cup bearded; divided into five lobes; the lateral ones blunt, and a little scolloped; the middlemost very short, and sharp. Petals thronging together;—*purple.* *Purple*

Orchis magna, latis foliis, Galea fusca vel nigricante. *Ray's Syn.* 373. tab. 19. fig. 2.
Orchis purpurea. *Hudson.* 334.
Chalk hills. P. June.

** *Bulbs*

* * *Bulbs of the root hand-shaped.*

Broad-leaved Latifolia ORCHIS. Bulbs somewhat hand-shaped; straight. Horn of the honey-cup conical. Lip divided into three lobes; the lateral ones reflected. Floral leaves longer than the blossom—*Stem hollow.* Root *not much branched, being divided only into two or three fingers.* Leaves, *especially the lower ones, a little spotted. The two lateral petals bent upwards.* Blossoms *purple, or flesh-coloured and sometimes white.*

Orchis palmata pratensis latifolia, longis calcaribus. *Bauh. pin.* 85. *Ray's Syn.* 380.

Orchis palmata mas, seu Palma Christi mas. *Park.* 1356.

Palma Christi mas. *Gerard.* 220.

Male-handed Orchis.

Moist meadows and marshy ground. P. May—June.

Cows eat it; Horses refuse it.

Spotted Maculata ORCHIS. Bulbs hand-shaped; expanding. Horn of the honey-cup shorter than the seedbud; lip flat. Petals on the back upright.—*Stem not hollow. Lateral lobes of the honey-cup large, and scolloped; the middle lobe very narrow and entire.* Leaves *spotted with black.* Blossoms *purplish red; sometimes flesh-coloured, or white.*

Orchis palmata pratensis maculata. *Bauh. pin.* 85. *Ray's Syn.* 381.

Orchis palmata fæmina, seu Palma Christi fæmina, foliis maculatis. *Park.* 1357.

Palma Christi fæmina. *Gerard.* 220.

Female-handed Orchis.

Woods, rich meadows, and sometimes in barren pastures. P. June.

Sheep eat it; Goats and Horses refuse it; Cows are not fond of it.

Red Conopsea ORCHIS. Bulbs hand-shaped. Horn of the honey-cup thread-shaped; as long as the seedbud. Lip cloven into three segments; very entire. The two outer petals greatly expanded. — *The others approaching.* Spike *of flowers very long.* Blossoms *red; sometimes whitish.*

Orchis palmata rubella cum longis calcaribus rubellis. *Ray's Syn.* 381.

Orchis palmata minor, calcaribus oblongus. *Bauh. pin.* 85.

Orchis palmata minor, flore rubro. *Park.* 1358.

Serapias minor, nitente flore. *Gerard.* 222.

Red-handed Orchis.

Meadows and pastures. B. June.

Goats and Cows eat it; Horses refuse it.

* * * *Roots bundled.*

ORCHIS. Roots fibrous; bundled; thread-ſhaped. Lip of the honey-cup egg-ſhaped; very entire. Stem deſtitute of leaves.—*Purple.* Bloſſoms *violet.* Bird's-Neſt Abortiva

Orchis abortiva violacea. *Baub. pin.* 86.
Limodorum Auſtriacum. *Ray's Syn.* 383.
Nidus avis purpureus. *Park.* 1362.
Nidus avis flore et caule purpuro violaceo. *Gerard.* 228.
Purple Birds Neſt.
Dry paſtures. P. May.

351 SATYRION. 1010 Satyrium.

Empal. *Sheaths* ſcattered. *Fruit-ſtalk* ſimple. *Cup* none.

Bloss. *Petals* five; oblong egg-ſhaped. *Three outer* ones; and *two inner* ones, which approach ſo as to form a helmet.

Honey-cup one leaf; connected by the lower ſide to the receptacle betwixt the diviſion of the petals. *Upper Lip* very ſhort; upright. *Lower Lip* flat; pendant; with a bag like a double purſe projecting behind.

Chives. *Threads* two; very ſlender; very ſhort; fixed on the pointal. *Tips* inverſely egg-ſhaped; covered by a fold of the upper lip of the honey-cup forming two cells.

Point. *Seedbud* beneath: oblong; twiſted. *Shaft* very ſhort; fixed to the upper lip of the honey-cup. *Summit* compreſſed; blunt.

S.Vess. *Capſule* oblong; with one cell; three keels; three valves; opening in three places under the keels; connected at the baſe and at the end.

Seeds. Numerous; very ſmall; like ſaw-duſt.

SATYRION. Bulbs undivided. Leaves ſpear-ſhaped. Lip of the honey-cup cloven into three ſegments: the middle ſegment ſtrap-ſhaped; oblique; bitten—*Bloſſoms greeniſh white; purple within.* Goat Hircinum

Orchis barbata odore hirci, breviore latioreque ſolio. *Baub. pin.* 82.
Orchis barbata fætida. *Ray's Syn.* 376.
Tragorchis maxima et T vulgaris. *Park.* 1348.
Tragorchis maximus et T. mas. *Gerard.* 225.
Lizzard flower or Goatſtones.
In chalky ſoil. P. June—July.

Green Viride SATYRION. Bulbs hand-ſhaped. Leaves oblong; blunt. Lip of the honey-cup ſtrap-ſhaped; cloven into three ſegments; the middle ſegment imperfect—*Stem angular; leafy.* Leaves *ſpear-ſhaped; alternate; downy; embracing the ſtem.* Floral leaves *ſpear-ſhaped; longer than the bloſſom.* Bloſſom *greeniſh yellow.*

Orchis palmata flore viridi. *Bauh. pin.* 86.
Orchis palmata minor, flore luteo-viridi. *Ray's Syn.* 381.
Serapias batrachites altera. *Gerard.* 224.
Frog Satyrion. Frog Orchis.
Dry paſtures. P. May—June.
Goats eat it.

Brown Fuſcum SATYRION. Bulbs hand-ſhaped. Leaves oblong. Lip of the honey-cup cloven into three equal ſegments.—
Near Kendal in Weſtmoreland. P. Auguſt.

White Albidum SATYRION Bulbs bundled. Leaves ſpear-ſhaped. Lip of the honey-cup cloven into three ſegments; ſharp; the middle ſegment blunt—*Spur very ſhort.* Petals, *three white; two greeniſh yellow.*

Orchis palmata thyrſo ſpecioſo, longo, denſe ſtipato, ex viridi albente. *Ray's Syn.* 382.
Moiſt paſtures. P. June.

Orchis Repens SATYRION. Bulbs fibrous. Root-leaves egg-ſhaped. Flowers all pointing one way—*Leaves on leave-ſtalks. Flowers with four hairy petals. Three petals downy; glued together.* Bloſſoms *greeniſh yellow.*

Pſeudo-Orchis. *Bauh. pin.* 84.

352 TWAYBLADE. 1011 Ophrys.

EMPAL. Sheaths ſcattered. *Fruit-ſtalk* ſimple. *Cup* none.
BLOSS. Petals five; oblong; approaching upwards; equal. two of them placed outwards.

Honey-cup longer than the petals; hanging down; keeled on the back part only.

CHIVES. *Threads* two; very ſhort; fixed to the pointal. *Tips* upright; covered by the inner edge of the honey-cup.

POINT. *Seedbud* beneath: oblong; twiſted. *Shaft* fixed to the inner edge of the honey-cup. *Summit* imperfect.

S. VESSS. *Capſule* ſomewhat egg-ſhaped; three edged; blunt; ſcored; with three valves; and one cell: opening at the keel-ſhaped angles,

SEEDS. Numerous; like ſaw-duſt. *Receptacle* ſtrap-ſhaped; growing to each valve of the ſeed-veſſel.

* *Bulbs branched.*

TWAYBLADE. Bulbs fibrous and bundled. Stem ſheathed; but without leaves. Lip of the honey-cup cloven—*Bloſſoms brown.* Birds-Neſt Nidus avis

Nidus avis. *Ray's Syn.* 382.
Orchis abortiva fuſca. *Bauh. pin.* 86.
Orchis abortiva rufa, ſeu Nidus avis. *Park.* 1362.
Satyrion abortivum, ſeu Nidus avis. *Gerard.* 228.
Woods and ſhady places. P. May.

TWAYBLADE. Bulbs incorporated; oblong. Stem ſomewhat leafy. Flowers placed ſpirally, but pointing one way. Lip of the honey-cup undivided, but a little ſcolloped.—*Leaves in general betwixt ſtrap and ſpear-ſhaped. The three outer petals glued together.* Bloſſoms *yellowiſh white.* Triple Spiralis

Orchis ſpiralis alba odorata. *Ray's Syn.* 378.
Triorchis. *Gerard.* 218. alba odorata minor. *Bauh. pin.* 87. *Park.* 1354.
Triple Ladies Traces.
Barren paſtures. P. Auguſt.

TWAYBLADE. Bulb fibrous. Stem with two leaves. Leaves egg-ſhaped. Lip of the honey-cup cloven.—*Bloſſoms green.* Common Ovata

Ophrys bifolia. *Gerard.* 402. *Bauh. pin.* 87.
Bifolium ſylveſtre vulgare. *Park.* 504. *Ray's Syn.* 385.
Woods and moiſt rough grounds. P. May—June.
Cows and Goats eat it.

Least Cordata TWAYBLADE. Bulb fibrous. Stem with two leaves. Leaves heart-ſhaped—*The cloven lip of the honey cup hath a little tooth on each ſide, which doth not exiſt in the preceding ſpecies.* Bloſſoms *whitiſh.*

Ophrys minima. *Bauh. pin.* 87.
Bifolium minimum. *Ray's Syn.* 385.
Moiſt Heaths and Turf Bogs. P. July.

* * *Bulbs round.*

Dwarf Lilifolia TWAYBLADE. Bulb roundiſh; ſtalk naked; leaves ſpear-ſhaped; lip of the honey-cup entire. Petals on the back of the flower ſtrap-ſhaped—*Bulbs pear-ſhaped; encompaſſed with a leafy ſheath.* Stalk *naked; three cornered.* Root-leaves *ſtringy.* Bloſſoms *red; but the three narrow petals ſtrap-ſhaped and greeniſh.*

Pſeudo-orchis bifolia paluſtris. *Ray's Syn.* 382.
Chamæorchis lilifolia. *Bauh. pin.* 84.
Chamæorchis latifolia Zelandica. *Park.* 1354.
Dwarf Orchis.
Marſhes. P. July.

Marſh Paludoſa TWAYBLADE. Bulb roundiſh; ſtalk nearly naked; with five edges. Leaves rough at the end. Lip of the honey-cup entire—*Bulb egg-ſhaped; crooked; with a root at the bottom. Root-Leaves three or four; Spatula ſhaped; alternate.* Flowers *greeniſh yellow; ſeveral in a bunch. The two lateral petals bent back; upright; oblong egg-ſhaped. The two inner petals lateral; narrow; crooked. The upper petal ſtraight.*

Bifolium paluſtre. *Park.* 505. *Ray's Syn.* 385.
Orchis minima bulboſa. *Ray's Syn.* 378.
Ophrys Paluſtris. *Hudſon.* 339.
Leaſt Orchis.
Turf Bogs. P. July—Auguſt.

Yellow Monorchis TWAYBLADE. Bulb globular; ſtalk naked. Lip of the honey-cup cloven into three ſegments; croſs-ſhaped.—*Root-leaves two or three.* Bloſſoms *greeniſh yellow; without a ſpur.*

Orchis odorata moſchata, ſive Monorchis. *Bauh. pin.* 84. *Ray's Syn.* 378.
Orchis puſilla odorata. *Park.* 1354.
Yellow Orchis. Muſk. Orchis.
Barren paſtures. P. July.

TWAY-

TWAYBLADE. Bulbs roundifh. Stalk leafy. Lip of the honey-cup ftrap-fhaped; divieed into three fegments; the longeft in the middle cloven.—*Bloffoms greenifh; without a fpur.* Green Anthropophora

Orchis anthropophora orcades. *Ray's Syn.* 379. fæmina. *Park.* 1348.

Orchis flore nudi hominis effigiam repræfentans, fæmina. *Bauh pin.* 82.

Green Man Orchis.

Chalky and gravelly foils. P. June.

TWAYBLADE. Bulbs roundifh. Stalk leafy. Lip of the honey-cup divided into four or five lobes— Infect Infectifera

1. Refembling a fly; the lip of the honey cup cloven into four fegments; helmet and wings greenifh. *Fly.*

Orchis myodes galea et alis herbidis. *Ray's Syn.* 379.

Orchis mufcæ corpus referens minor, vel galea et alis herbidis. *Bauh. pin.* 83.

Orchis myodes minor. *Park.* 1352.

Orchis mufcifera. *Hudfon.* 340.

Common Fly Orchis.

In meadows and on chalky hills. P. May.

2. Orchis myodes major. *Park.* 1351. *Ray's Syn.* 379. *Greater Fly.*

Greater Fly Orchis.

In paftures and chalk-pits. P. May.

1. Refembling a bee; the lip of the honey-cup divided into five lobes, bent inwards underneath. (Hudfon)—*Outermoft petals of the bloffom large; expanding; of a purplifh colour; the two innermoft green. The* lower lip of the honey-cup *large, but fhorter than the petals, of a brownifh purple, mixed with yellow: divided into three lobes; the innermoft the largeft, and divided again into three lobes, bent downwards.* Upper lip of the Honey-cup *longer than the lower; narrowing upwards to a point; of a green colour.* Threads *long;* tips *very large.* Seedbud *longer than the petals, but fhorter than the floral leaves.* Martyn *Cat. Cant.* 171. *Bee.*

Orchis fuciflora galea, et alis purpurafcentibus. *Ray's Syn.* 379.

Orchis fphegodes feu fucum referens. *Park.* 1350.

Orchis, five Tefticulus vulpinus duo, fphegodes. *Gerard.* 212.

Orchis fucum referens major foliolis fuperioribus candidis et pupurafcentibus. *Bauh. pin.* 83.

Orchis apifera. *Hudfon* 340.

Common Humble Bee. Satyrion, or Bee-flower.

In dry paftures. P. June.

Drone — 2. Helmet and wings green; lip of the honey-cup hairy; without any mixture of yellow,—the whole bloſſom of a darker colour, and more reſembling a drone than a bee.

Orchis, five Teſticulus ſphegodes hirſuto flore. *Ray's Syn.* 380.

Orchis fucum referens colore rubiginoſo. *Bauh. pin.* 83.

Orchis ſphegodes altera. *Park.* 1351.

Teſticulus vulpinus major ſphegodes. *Gerard.* 212.

Humble-Bee Satyrion with green wings, or Waſp Orchis.

In chalk and gravel. P. April.

353 HELLEBORINE. 1012 Serapias.

EMPAL. *Sheaths* ſcattered. *Fruit-ſtalk* ſimple. *Cup* none.

BLOSS. *Petals* five; oblong egg-ſhaped; open but rather upright; approaching upwards.

Honey-cup as long as the petals; hollowed at the baſe; filled with honey; egg-ſhaped; hunched beneath; cloven into three ſegments; ſharp: the middle ſegment heart-ſhaped; blunt; cloven at the ſeam of the baſe; with three teeth.

CHIVES. *Threads* two; very ſhort; fixed to the pointal. *Tips* upright; placed under the upper lip of the honey-cup.

POINT. *Seedbud* beneath: oblong; twiſted. *Shaft* growing to the upper lip of the honey-cup. *Summit* indiſtinct.

S. VESS. *Capſule* inverſely egg-ſhaped; with three blunt edges; three keels, growing to it; three valves; opening under the keels, and one cell.

SEEDS. Numerous; like ſaw-duſt. *Receptacle* ſtrap-ſhaped; growing to each valve of the ſeed-veſſel.

Broad-leaved Latifolia. HELLEBORINE. Bulbs fibrous; leaves eggſhaped; embracing the ſtem. Flowers pendant.—*Bloſſoms browniſh green, with a purple ſpot at the bottom of the honey-cup.* Lip *of the honey-cup blunt, ſcolloped, equal to the petals.* Petals *egg-ſhaped.* Flowers *forming a long ſpike.* Fruit-ſtalks *long.*

Helleborine latifolia montana. *Bauh. pin.* 186. *Ray's Syn.* 383.

Helleborine. *Gerard.* 442.

Elleborine, flore viridante. *Park.* 218.

1. Flowers dark red. *Bauh. pin.* 186. *Ray's Syn.* 383.
2. Flowers purple on the outſide; white within.

Helleborine paluſtris noſtras. *Ray's Syn.* 384.

Broad-leaved baſtard Hellebore.

Woods and moiſt hedges. 2 in marſhes. P. July—Auguſt.

HELLE-

HELLEBORINE. Bulbs fibrous. Leaves ſword-ſhaped; ſitting. Flowers pendant —*Stem taller than the preceding ſpecies; Flowers twice as large.* Honey-cup *adhering to the other petals; marked with purple lines.* Petals *white on the outſide, with yellowiſh ſpots and lines within; ſpear-ſhaped.* Floral leaves *longer than the flower.* Long-leaved Longifolia

Helleborine flore albo. *Gerard.* 442. *Ray's Syn.* 383.

Helleborine flore albo, vel Damaſonium montanum latifolium. *Bauh. pin.* 187.

Elleborine minor flore albo. *Park.* 218.

1. With broader leaves and white bloſſoms. *Ray's Syn.* 384.
2. With very long, narrow, pointed leaves. *Ray's Syn.* 384.
3. With narrow leaves and purpliſh bloſſoms. *Bauh. pin.* 187.

White flowered baſtard Hellebore.

Woods and rough grounds. P. Auguſt.

354 SLIPPER. 1015 Cypripedium.

EMPAL. *Sheath* ſcattered. *Fruit-ſtalk* ſimple. *Cup* none.

BLOSS. *Petals* four or five; narrow and ſpear-ſhaped; very long; expanding; upright.

Honey-cup within the lower petal; ſhaped like a ſlipper; blown up; blunt; hollow; ſhorter and broader than the petals; the *upper Lip* ſmall; egg-ſhaped; flat; bent inwards.

CHIVES. *Threads* two; very ſhort; fixed to the pointal. *Tips* upright; covered by the upper lip of the honey-cup.

POINT. *Seedbud* beneath: long; twiſted. *Shaft* very ſhort; growing to the upper lip of the honey-cup. *Summit* indiſtinct.

S.VESS. *Capſule* inverſely egg-ſhaped; with three blunt edges; and three ſeams, under which it opens at the angles; *Valves* three; *Cell* one.

SEEDS. Numerous; very ſmall. *Receptacle* ſtrap-ſhaped; growing lengthways to each valve of the ſeed-veſſel.

Ladies Calceolus SLIPPER. Roots fibrous; leaves on the ſtem betwixt egg and ſpear-ſhaped—*Petals purple: honey-cups pale yellow.*

By cultivation the colour of the bloſſoms are changed to

1, Pale purple.
2. Golden yellow.
3 L rge yellow.
4. Red.
5. Various coloured.

Calceolus Mariæ. *Gerard.* 443. *Ray's Syn.* 385.
Helleborine flore rotundo ſeu Calceolus. *Bauh pin.* 187.
Elleborine major ſeu Calceolus Mariæ. *Park.* 217.
Woods and rough grounds. P. July.

Goats eat it.

Order VIII. Many Chives.

355 CUCKOWPINT. 1028 Arum.

EMPAL. *Sheath* one leaf; very large; oblong; lapped round at the baſe; approaching at the top; compreſſed in the middle; coloured on the inſide.

Sheathed Fruitſtalk club-ſhaped; undivided; a little ſhorter than the ſheath; coloured; ſet round with ſeedbuds on the lower part; above the ſeedbud ſhrivelling.

BLOSS. None.

CHIVES. *Threads* none; unleſs the honey-cups, which are thick at the baſe, and terminated in thread-ſhaped tendrils, growing in two rows round the middle of the fruit-ſtalk. *Tips* many; ſitting; four-edged; fixed to the fruit-ſtalk, and diſpoſed between the two rows of tendrils.

POINT. *Seedbuds* many; inverſely egg-ſhaped; covering the baſe of the fruit-ſtalk beneath the chives. *Shafts* none. *Summits* bearded with ſoft hairs.

S.VESS. *Berries* numerous; globular; with one cell.

SEEDS. Many; roundiſh.

OBS. *The wonderful and unparallelled ſtructure of this flower hath given riſe to many diſputes amongſt the moſt eminent Botaniſts.*

The Receptacle *is long; in ſhape of a club; with the ſeedbuds ſurrounding its baſe. The* Chives *are fixed to the receptacle amongſt the ſeedbuds, ſo that there is no occaſion for the tips to be ſupported upon threads. For what uſe are thoſe tendrils mentioned above?*

CUCK-

CUCKOWPINT. Without a ſtem. Leaves halberd-ſhaped; very entire. Fruit-ſtalk club-ſhaped.—*Leaves generally ſpotted.* Sheath *conical*; *pale green*. Berries *red when ripe*; *growing in a naked cluſter*. Spotted Maculatum

Arum. *Ray's Syn.* 266. vulgare. *Gerard.* 834.

Arum vulgare maculatum et non maculatum. *Park.* 373.

Arum vulgare maculatum. *Bauh. pin.* 195.

Wake-Robin. Cuckowpint. Lords and Ladies.

Shady places, ditch-banks and rough grounds. P. May.

It is ſuſceptible of the following variations.

1. Spotted with white.
2. Spotted with black.
3. Leaves circular and ſpotted.
4. Leaves with white ſtreaks and black ſpots.
5. Leaves not ſpotted.

The root and the leaves when recent, are ſo extremely acrid that is it highly diſagreeable to taſte them. The root has been employed in medicine as a ſtimulant, but when reduced to powder it loſes much of its acrimony, and there is reaſon to ſuppoſe that the compound powder which takes its name from that plant, owes its virtues chiefly to the other ingredients. There is no doubt but this acrid quality may be turned to very uſeful purpoſes, but we muſt firſt learn how to aſcertain its doſe.—The root dried and powdered, is uſed by the French to waſh their ſkin with. It is ſold at a high price under the name of Cypreſs Powder. It is undoubtedly a good, and an innocent coſmetic.—When the acrimony of the roots is extracted either by boiling or baking, they certainly will afford a very mild and wholeſome nouriſhment. Many nations prepare the only bread they have, from plants as acrimonious as this; firſt diſſipating the noxious qualities by the force of heat.—Starch may be made ftom the roots.

356 GRASSWRACK. 1032 Zostera.

EMPAL. *Leaves* forming a sheath at the base; approaching length-ways; notched at the upper part on each side; inclosing the fruit-stalk. *Sheathed fruit-stalk* strap-shaped; flat; furnished with chives on one side upon the upper, and pointals on the lower part. *Cup* none.

BLOSS. None,

CHIVES. *Threads* alternate; many; very short; fixed upon the fruit-stalk above the seed-buds. *Tips* oblong egg-shaped; nodding; blunt; awl-shaped upwards and backwards; crooked.

POINT. *Seedbuds* not so numerous as the chives; egg-shaped; compressed; two-edged; standing upon little foot-stalks fixed to the end; nodding; alternate. *Shafts* none. *Summits* hair-like; simple.

S.VESS. Membranaceous; unchanged; opening lengthways at the side-edge.

SEED. Single; egg-shaped.

Sea Marina

GRASSWRACK. Seed-vessels sitting.—*Leaves growing under water and floating with the tide.*

Alga. *Gerard.* 1569. *Ray's Syn.* 52.

Alga angustifolia Vitriariorum. *Baub. pin.* 364. *Ray's Syn.* 53.

Fucus marinus, sive Alga marina graminea. *Park.* 1242.

Fucus sive Alga marina graminea angustifolia seminifera ramosior. *Ray's Syn.* 58.

1. *Stem* branching, about two ells or more in length, *Leaves* about half a foot long, and half an inch broad.

Branched grass-leaved, Seeding Seawrack.

Fucus marinus sive Alga marina graminea minor. *Ray's Syn.* 52.

2. Smaller in every respect than the common species—Scarce a foot high.

Lesser Grass-wrack.

Potamogeton marinum inutriculis epiphyllospermon minus. *Ray's Syn.* 53.

3. Very much resembling variety 1, but much smaller seldom above a foot high.

This plant is thrown upon the sea-shore by the tide, in great plenty. Exposure to the weather bleaches it white. It is of great use as a manure. Buildings are thatched with it, and it endures a long time.—Horses and Swine eat it; Cows are not fond of it.

CLASS

CLASS. XXI.

ALL the Claſſes hitherto mentioned, are compoſed of Plants, whoſe Flowers contain the *Chives* and the *Pointals* within the *ſame* empalement ; but the plants of this Claſs have the *Chives* in one flower, or in one empalement, and the *Pointals* in another. The flowers that contain the Chives are BARREN ; thoſe that contain the Pointals are FERTILE. The *Barren* and the *Fertile* flowers of this Claſs, are always found upon the *ſame* plant.

From this account it is evident, that in giving the Generic characters, it is neceſſary to deſcribe both the Barren, and the Fertile Flowers. Therefore
B ſignifies the *Barren Flower* or that with Chives only; and
F ſignifies the *Fertile Flower*, or that with only Pointals.

OBS. *Florets incloſed in a* Common Empalement, *though agreeing in the diſpoſition of the Chives and Pointals ; are not to be referred to this Claſs. This caution is neceſſary in order to exclude ſome of the* Rundle bearing Plants *of the* fifth, *and ſome of the* Compound Flowers *of the* nineteenth *Claſs.*

CLASS XXI.

Chives and Pointals Separate.

Order I. One Chive.

357 STONEWORT. B. *Cup* none. *Bloſſ.* none.
F *Cup* four leaves. *Bloſſ.* none. *Summit* with five clefts. *Seed* one.

358 LAKEWEED. B *Cup* none. *Bloſſ.* none.
F *Cup* one leaf. *Bloſſ.* none. *Point.* four. *Seeds* four.

† *Vernal Stargraſs.*

Order II. Two Chives.

359 DUCKMEAT. B *Cup* one leaf. *Bloſſ.* none.
F *Cup* one leaf. *Bloſſ.* none. *Shaft* one. *Capſule* one cell.

Order III. Three Chives.

360 SEDGE. B *Catkin* with one flower in each ſcale. *Bloſſ.* none.
F *Catkin* with one flower in each ſcale. *Bloſſ.* one. *Shaft* one *Seed* one ; coated.

361 BURREED.

361 BURREED. B *Cup* three leaves. *Bloss.* none.
F *Cup* three leaves. *Bloss.* none. *Summits* two ; seed one.

362 REEDMACE. B *Cup* three leaves. *Bloss.* none.
F *Cup* hair-like. *Bloss.* none. *Shaft* one. *Seed* one; with a feather.

† *Everlasting Blite.*

Order IV. Four Chives.

363 NETTLE. B *Cup* four leaves. *Bloss.* none. *Honey-cup* glass-shaped.
F *Cup* two valves. *Bloss.* none. *Summit* hairy. *Seed* one ; egg-shaped.

364 BOX. - B *Cup* three leaves. *Bloss.* two petals.
F *Cup* four leaves. *Bloss.* three petals. *Summits* three. *Capsule* three cells.

365 BIRCH. - B *Catkin* with three flowers in each scale. *Bloss.* with four divisions.
F *Catkin* with two flowers in each scale. *Bloss.* none. *Shafts* two. *Seed* one ; egg-shaped.

† *Tree Holly.* † *Grass-leaved Plaintain.*

Order V. Five Chives.

366 DITCHBUR. B *Cup* common ; with many leaves. *Bloss.* five clefts. *Threads* connected.
F *Cup* none. *Bloss :* none. *Shafts* two. *Nut* with two cells.

367 EVERLASTING. B *Cup* individual, five leaves. *Bloss.* none. *Chives* three or five.
F *Cup* individual, five leaves. *Bloss.* none. *Shafts* three. *Cups* cut round.

Order VIII. Many Chives, (more than Seven.)

368 Arrowhead. B Cup three leaves. *Bloſſ.* three petals. *Chives* about twenty-four.

F *Cup* three leaves. *Bloſſ.* three petals. *Pointals* one hundred. *Seeds* numerous.

369 Milfoil. B *Cup* four leaves. *Bloſſ.* none. *Chives* eight.

F *Cup* four leaves. *Bloſſ.* none. *Pointals* four. *Seeds* four.

370 Hornweed. B. *Cup* with about ſeven diviſions. *Bloſſ.* none. *Chives* about eighteen.

F *Cup* with about ſeven diviſions. *Bloſſ.* none. *Pointal* one. Seed one.

371 Ironwort. B *Cup* three or four leaves. *Bloſſ.* with four diviſions. *Chives* about thirty two.

F *Cup* three or four leaves. *Bloſſ.* four petals. *Pointals* two. *Seeds* two; incloſed.

372 Beech. B *Cup* with five clefts. *Bloſſ.* none *Chives* about twelve.

F *Cup* with four clefts. *Bloſſ.* none. *Shafts* three. *Capſule* with two ſeeds.

373 Oak. - B *Cup* with five clefts. *Bloſſ.* none. *Chives* about ten.

F *Cup* entire. *Bloſſ.* none. *Shafts* five. *Seed* an *Acorn.*

374 Hazle. B *Catkin* tiled. *Bloſſ.* none. *Chives* eight.

F *Cup* two leaves. *Bloſſ.* none. *Shafts* two. *Fruit* a Nut.

375 Hornbeam. B *Catkin* tiled. *Bloſſ.* none. *Chives* ten.

F *Cup* with ſix clefts. *Bloſſ.* none. *Pointals* two. *Fruit* a Nut.

Order

Order IX. Threads United.

376 FIR. - - B *Cup* four leaves. *Bloss.* none. *Chives* many.
F *Catkin* like a cone. *Bloss.* none. *Point.* two. *Nuts* two ; winged.

Order X. Tips United.

377 BRYONY. B *Cup* with five teeth. *Bloss.* with five divisions. *Chives* three.
F *Cup* with five teeth. *Bloss.* with five divisions. *Shaft* with three clefts ; fruit a *Berry*.

357 STONEWORT. 1203 Chara.

B. *Fertile Flowers.*

EMPAL. *Cup* four leaves; awl-ſhaped; upright; permanent. The *two outer* ones oppoſite, and longer than the other two.

BLOSS. None.

POINT. *Seedbud* turban-ſhaped. *Shaft* none. *Summit* with five clefts; oblong; deciduous.

S. VESS. The *Coat* egg-ſhaped; of one cell; adhering to the ſeed.

SEED. Single; egg-ſhaped; marked with ſpiral lines.

F. *Barren Flower* at the baſe of the ſeedbud, on the outward ſide of the empalement.

EMPAL. *Cup* as above.

BLOSS. None

CHIVE. *Threads* none. *Tip* globular, placed before and below the ſeedbud, without the empalement.

Brittle
Tomentoſa

STONEWORT. Prickles on the ſtem egg-ſhaped—*The root fibrous. The whole plant is brittle and gritty in the mouth. Whilſt growing it is of a fleſh colour; but when dry it turns of an aſh colour.*

Chara major, ſub-cinerea fragilis. *Ray' Syn.* 132.
Equiſetum ſeu Hippuris coralloides. *Gerard.* 1115.
Brittle Chara. *Hudſon* 465.
Ditches and ſtagnant waters. P. ?

Common
Vulgaris.

STONEWORT. Stems ſmooth; leaves toothed on the inner ſide—*Flowers in the diviſions of the ſtem. The whole plant is yellowiſh or reddiſh green.*

Chara vulgaris fœtida. *Ray's Syn.* 132.
Equiſetum fœtidum ſub aqua repens. *Gerard.* 111 . *Bauh. pin.* 16. *Park.* 1201.
Stinking Water Horſe-tail.
Common Chara. *Hudſon* 465.
Ditches and ſtagnant waters. P. ?

Prickly
Hiſpida

STONEWORT. Prickles on the ſtem, hair-like and crowded together—*Whitiſh green.*

Chara maior caulibus ſpinoſis. *Ray's Syn.* 132.
Prickly Chara. *Hudſon* 465.
Ditches and turf bogs. P. ?

STONE-

STONEWORT. Stems without prickles at the joints; semi-transparent; broadest towards the top—*Leaves long; slender; entire. Flowers at the base of the leaves; generally eight together.* Smooth Flexilis

Chara translucens minor flexilis. *Ray's Syn.* 133

Smooth Chara. *Hudson* 465.

Ponds, Ditches and Bogs. P.

STONEWORT. Stems and leaves very fine and slender—*Longer than those of the* common STONEWORT, *of a deeper green, and more flexible.* Seeds *roundish, shining, of a reddish brown.—It appears to float under water.* Creeping Repens

Chara minor caulibus et foliis tenuissimis. *Ray's Syn.* 13.

Creeping Water Horse-tail.

Ponds and ditches. P.

This though evidently a different species does not appear to have been noticed by Linnæus or Hudson. The specific character is therefore formed from Ray's description.

358 LAKEWEED. 1034 Zannichellia.

B. *Barren Flower.*

EMPAL. None.

BLOSS. None.

CHIVE. Thread single; simple; long; upright. *Tip* egg-shaped; upright.

F. *Fertile Flower standing near the other.*

EMPAL. *Cup* one leaf; hardly perceptible; distended; with two teeth.

BLOSS. None.

POINT. *Seedbuds* four; like little horns; approaching. *Shafts* four; simple; rather expanding. *Summits* egg-shaped; flat; expanding outwards.

S. VESS. None.

SEEDS. Four; oblong; tapering at each end; hunched on one side; covered with a bark; crooked; reflected.

LAKEWEED. As there is only one species known Linnæus gives no description of it—*Stems jointed; branched; floating in the water.* Leaves *opposite.* Flowers *at the base of the leaves.* Horned Palustris

Aponogiton aquaticum gramini folium, staminibus singularibus. *Ray's Syn.* 135.

Horned Pondweed

Ditches and stagnant waters. A. July.

Order II. Two Chives.

359 DUCKMEAT. 1038 Lemna.

B.

EMPAL. One leaf; circular; opening at the ſide; obliquely dilated outwardly; blunt; expanding; depreſſed; large; entire.

BLOSS. None.

CHIVES. *Threads* two; awl-ſhaped: crooked; as long as the cup. *Tips* double; globular.

POINT. *Seedbud* egg-ſhaped. *Shaft* ſhort. *Summit* indiſtinct.

S.VESS. Barren.

F. *Fertile Flowers.*

EMPAL. *Cup* as above.

BLOSS. None.

POINT. *Seedbud* ſomewhat egg ſhaped. *Shaft* ſhort; permanent. *Summit* ſimple.

S.VESS. *Capſuſe* globular, with a point at the end; and with one cell.

SEEDS. Several; oblong; ſharp at each end; nearly as long as the capſule; ſcored on one ſide.

OBS. *If the pointals in the* B. *flower had been perfect and produced perfect ſeeds, this genus muſt have been referred to the twenty-third claſs.*

Ivy-leaved Trifulca — DUCKMEAT. Leaves ſpear-ſhaped; on leaf ſtalks.—*Stems flatted and proliferous; croſſing each other.*
Lenticula aquatica triſulca. *Bauh. pin.* 362. *Ray's Syn.* 129.
Hederula aquatica. *Gerard.* 830.
Ranunculus hederaceus aquaticus. *Park.* 1260.
Ditches and ſtagnant waters. P. May—June.

Leaſt Minor — DUCKMEAT. Leaves flattiſh on each ſide. Roots ſolitary—
Lens paluſtris. *Gerard.* 829. *Ray's Syn.* 129. tab. 4. fig. 1.
Lens paluſtris vulgaris. *Bauh. pin.* 362.
Lens paluſtris ſeu aquatica vulgaris. *Park.* 1262.
On ſtagnant waters. P. June.

Greater Polyrhiza — DUCKMEAT. Leaves ſitting; roots crowded together.—*Leaves green on the upper, purple on the under ſurface.*
Lenticula paluſtris major. *Ray's Syn.* 129. tab. 4. fig. 2.
Ditches. A. June—July.
All the ſpecies are acceptable food for Ducks and Geeſe.

Order

Order III. Three Chives.

360 SEDGE. 1046 Carex.

B. *Barren Flowers forming a Spike.*

EMPAL. *Catkin* oblong; tiled; conſiſting of *Scales*, each including a ſingle flower; ſpear-ſhaped; ſharp; concave; permanent.

BLOSS. None.

CHIVES. *Threads* three; briſtle-ſhaped; upright; longer than the incloſing ſcale. *Tips* upright; long; ſtrap-ſhaped.

F. *Fertile Flowers the ſame, but ſometimes on diſtinct Plants.*

EMPAL. *Catkin* as above.

BLOSS. *Petals* none.

Honey-cup blown up; oblong egg-ſhaped; with two or three teeth at the end; contracted towards the upper part; mouth open; permanent.

POINT. *Seedbud* three-cornered; within the honey-cup. *Shaft* very ſhort. *Summits* three or two; awl-ſhaped; bent inwards; long; tapering; downy.

S. VESS. None. The *Honey-cup* enlarging contains the ſeed.

SEED. Single; egg-ſhaped but ſharp; three-cornered; one of the angles ſmaller than the other two.

OBS. *In ſome ſpecies the two kinds of flowers are found upon the ſame ſpike; but in others they are upon different ſpikes* (See the Plate of *Graſſes*)

* *Spikes ſingle; ſimple; with fertile and barren flowers on diſtinct plants.*

SEDGE. Spike ſimple. The flowers with only chives and the flowers with only pointals on diſtinct plants.—*Leaves and ſtraw three-cornered.* Small Dioica

Gramen cyperoides minus ranunculi capitulo longiore. *Ray's Syn.* 425.

Turf bogs. P. June.

* * *Spike ſimple; compoſed of fertile and barren flowers.*

Round-leaved Capitata SEDGE. Spike ſimple; egg-ſhaped; the flowers with only chives forming the upper part of the ſpike. Capſules tiled, but not lying cloſe.—*Spike exactly egg-ſhaped.*

Gramen cyperoides minimum ranunculi capitulo rotundo. *Ray's Syn.* 425.

Turfy bogs. P. June.

Flea Pulicaris SEDGE. Spike ſimple: the flowers with chives forming the upper part of the ſpike. Capſules ſtraddling; reflected backward—*Straw cylindrical, but flattiſh on one ſide. When out of bloſſom the barren flowers fall off. The* Seeds *ſharp and hooked at the end.*

Gramen cyperoides minimum, ſeminibus deorſum reflexis puliciformibus. *Ray's Syn.* 424.

Muddy Marſhes P. June.

* * * *Spike compoſed of both fertile and barren flowers.*

Sea Arenaria SEDGE. Spike compound. Little ſpikes compoſed of fertile and barren flowers; the lower ones more remote and furniſhed with a leaf longer than the ſpike. Straw three cornered—*Root creeping; thread-ſhaped; ſimple; horizontal; jointed; about four inches beneath the ſurface of the ground. joints rather diſtant, and from each joint riſes a jointed* Stem, *which beneath the ſurface is covered with a brown ſheath; even with the ſurface it ſends out many* Leaves, *the outer ones ſhort; the inner ones narrow and as tall as the* Straw *which is naked, without knots; ſupporting the ſpike at the end.* Spike *of a ruſty iron colour.*

Gramen cyperoides ex monte Ballon ſimile humilius; in maritimis et arenoſis naſcens. *Ray's Syn.* 423.

Gramen cyperoides elegans multifera ſpica. *Park.* 1172.

On ſandy ſea ſhores. P. June—July.

Naked Leporina SEDGE. Spike compound. Little ſpikes ſitting; egg-ſhaped: growing near each other; alternate; compoſed of barren and fertile flowers not furniſhed with a leaf—*Spikes compoſed of five or ſix little ſpikes.* Florets *ſeparated by grey chaff, longer than the ſeeds.* Shafts *crooked.* Stalks *twelve or ſixteen inches high; hollow.*

Gramen cyperoides ſpica e pluribus ſpicis brevibus mollibus compoſita. *Ray's Syn.* 422.

Gramen cyperoides paluſtre majus ſpica diviſa. *Bauh. pin.* 6.

In marſhes and watery places. P. June—July.

SEDGE. Spike compound; little ſpikes round; crowded; compoſed of barrel and fertile flowers; each furniſhed with a little leaf longer than the ſpike; ſtraw nearly cylindrical—*Taller than the preceding; leaves longer and narrower; ſtraw not ſo evidently hollow;* little Spikes *ſmaller; each ſupported by a leaf which riſing above the ſpike gives it a reſemblance to the ſtraw of the* RUSH. Marſh Diviſa

Gramen cyperoides ex monte Ballon ſpica divulſa. *Ray's Syn.* 423.

Gramen cyperoides paluſtre majus ſpica divulſa. *Bauh. pin.* 6.

Gramen cyperoides paluſtre minus. *Park.* 1287.

Gramen cyperoides parvum. *Gerard.* 21.

In marſhes and low meadows. P. Auguſt.

SEDGE. Spike compound; nearly two-rowed; little ſpikes egg-ſhaped, tiled, compoſed of barren and fertile flowers, each furniſhed with a leaf longer than the ſpike; ſtraw three cornered—*Taller than either of the preceding; hollow, but leſs ſo than the former ſpecies;* Spike *of a ruſty iron colour, and at firſt ſoft to the touch.* Soft Diſticha

Gramini cyperoidi ex monte Ballon ſimile, ſpica totali e pluribus ſpicis compoſita. *Ray's Syn.* 423.

Gramen cyperoides elegans multifera ſpica. *Park.* 1172.

1. Gramen cyperoides elegans ſpica compoſita molli. *Plukenet ph.* 134. *f.* 4.

In marſhes and watery places. P. May—June.

SEDGE. Spike more than doubly compound; leaſt compact in the lower part. Little ſpikes egg-ſhaped; congregated; compoſed of fertile and barren flowers; the barren flowers placed above—*Straw thick; firm.* Spike *thick; rough on every ſide.* Sheath *of the leaves terminated on the inner ſide by a ſharp tongue, as in the graſſes.* Great Vulpina

Gramen cyperoides paluſtre majus, ſpica compacta. *Bauh. pin.* 6. *Ray's Syn.* 423.

Gramen paluſtre cyperoides. *Gerard.* 21.

Gramen cyperoides paluſtre majus. *Park.* 1266.

Banks of rivers. P. July.

Horſes and goats eat it; Swine refuſe it.

Spiked Spicata

SEDGE. Little ſpikes roundiſh; near together: compoſed of barren and fertile flowers; ſeed veſſels egg-ſhaped; ſharp. (Hudſon)—*Smaller than the preceding, which in many reſpects it very much reſembles.* Spikes *ſmaller and leſs compact.* Straws *leſs upright.* Leaves *narrow, and ſhorter than the ſtraw.*

Gramen cyperoides ſpicatum minus. *Ray's Syn.* 424.

Gramen cyperoides ſpicis minus compactis. *Park.* 1268.

Gramen cyperoides parvum. *Gerard.* 18.

Gramen cyperoides ſpicis minoribus minuſque compactis. *Bauh. pin.* 6.

Leſſer ſpiked cyperus-graſs.

In watery places, and on the banks of rivers. P. May—June.

Rough Brizoides

SEDGE. Spike compound; pointing from two oppoſite lines; naked. Little ſpikes oblong; contiguous; compoſed of fertile and barren flowers. Straw without leaves—

Gramen cyperoides paluſtre elegans ſpica compoſita aſperiore. *Ray's Syn.* 423.

In pools and marſhes. A. June.

Prickly Muricata

SEDGE. Little ſpikes ſomewhat egg-ſhaped; ſitting; at a diſtance from each other; compoſed of fertile and barren flowers. Capſules ſharp; diverging; thorny.—*Straw naked; three cornered: ſix or eight inches high.* Leaves *narrow; harſh; as long as the ſtraw.*

Gramen cyperoides ſpicatum minimum, ſpica divulſa aculeata. *Ray's Syn.* 424.

Gramen nemoroſum ſpicis parvis aſperis. *Bauh. pin.* 7.

Gramen cyperoides echinatum minimum. *Park.* 1272.

Marſhes and moiſt woods. P. June.

Horſes and Goats eat it.

Long-leaved Remota

SEDGE. Spikes egg-ſhaped; almoſt ſitting; at a diſtance from each other; compoſed of fertile and barren flowers. Floral leaves as tall as the ſtraw.—*Leaves narrow;* Straw *three cornered.* Scales *of the ſpike white, with a greeniſh line.*

Gramen cyperoides anguſtifolium, ſpicis parvis ſeſſilibus, in foliorum alis.

Ditch banks and moiſt woods. P. June—Auguſt.

Grey Caneſcens

SEDGE. Little ſpikes ſomewhat round; at a diſtance from each other; ſitting; blunt; compoſod of fertile and barren flowers. Capſules egg-ſhaped; rather blunt.

Gramen cyperoides ſpicatum minus, ſpica longa divulſa, ſeu interrupta. *Ray's Syn..* 424.

Hedges and moiſt woods, P. May—Auguſt.

SEDGE.

SEDGE. Flowers in a compound bunch, little ſpikes compoſed of fertile and barren flowers.—*Straw three cornered; harſh; three feet high.* Leaves *harſh; narrow.* Panicle *brown.* Panicled Paniculata

Gramen cyperoides paluſtre elatius, ſpica longiore laxa. *Ray's Syn.* 422.

Putrid bogs and marſhes. P. June—July.

**** *Barren and fertile Flowers on different ſpikes; the fertile ſpikes ſitting.*

SEDGE. Spikes crowded together; nearly ſitting; ſomewhat roundiſh. The barren ſpike narrow; capſules ſharp; bent back— Yellow Flava

Gramen cyperoides aculeatum Germanicum, vel minus. *Bauh. pin.* 7.

Gramen paluſtre echinatum. *Gerard.* 17. *Ray's Syn.* 421.

Gramen aculeatum Germanicum. *Park.* 1187.

1. There is a variety that is larger.

Gramen cyperoides echinatum majus. *Ray's Syn.* 421.

Marſh Hedge-hog Graſs.

Marſhes and wet meadows. P. June.

SEDGE. Spikes with fertile flowers ſitting; oblong. The lower ſpike at the baſe of the leaves. Leaves nearly thread-ſhaped.—*Straw naked, ſix inches long: about ſeven florets ine ach fertile ſpike.* Fertile *ſpikes two; alternate; dark ruſty brown.* Barren *ſpike ſingle; terminating: paler.* Birds-foot Pedata

Gramen caryophyllatum nemoroſum, ſpica multiplici.

Gramen caryophylleum anguſtiſſimis foliis. *Bauh. pin.* 4.

Spicis ſeſſilibus brevioribus erectis non compactis. *Ray's Syn.* 418.

In woods and moiſt paſtures. P.

SEDGE. Spikes terminating; crowded; roundiſh. Barren ſpike oblong.—*Leaves very ſlender, ſhort and ſoft;* fertile Spikes *from two to four.* Straw *about a foot high.* Globular Pilulifera

Gramen cyperoides tenuifolium ſpicis ad ſummum caulum ſeſſilibus globulorum æmulis. *Ray's Syn.* 422.

1. *Fertile Spikes* moſtly two, each ſupported by a leaf.

Gramen cyperoides anguſtifolium majus ſpicis ſeſſilibus in foliorum alis. *Ray's Syn.* 422

Dillenius makes this a variety of the preceding—It is ſaid to be larger, and with ſhort thick ſpikes, one ſmaller than the reſt, ſeated at the baſe of one of the leaves. *Ray's Syn.* 422.

Gramen cyperoides ſpicis brevibus congeſtis, folio molli. *Ray's Syn.* 421.

2. This variety ſeems to anſwer more exactly to Linnæus's character than the former. The *fertile Spikes*, which are three or four in number, being crowded nearer together.—They do not appear to be ſupported by a leaf as in the preceding variety, which ſeems to form one of the principal diſtinctions between them.

In marſhes and wet paſtures. P. June—July.

Vernal Saxatilis

SEDGE. Spikes three; egg-ſhaped; ſitting; alternate. Barren ſpike oblong—*Terminating; roundiſh; yellow.* Tips *yellow.* Fertile *ſpikes black; florets ſeparated by ſcales, as long as the pointals.* Seedbuds *three-cornered; black.* Shafts *black; long; cloven. Under the loweſt ſpike there is a ſlender leaf, ſhorter than the ſpike.* Mr. Ray *ſays the ſpikes are three or four, and ſometimes one lower than all, riſing from the baſe of one of the leaves.*

Gramen cyperoides vernum minimum. *Ray's Syn.* 421.
Gramen cyperoides ſpicatum. *Gerard.* 22.
Gramen caryophyllatæ foliis, ſpica divulſa. *Bauh. pin.* 3.
Gramen ſpicatum, foliis caryophylleis. *Park.* 1160.
Mountainous and barren paſtures. P. April—May.

* * * * * *Barren and fertile flowers on different ſpikes. The fertile ſpikes on fruit-ſtalks.*

Pale Palleſcens

SEDGE. Spikes pendant. Barren ſpike upright. Fertile ſpikes egg-ſhaped: tiled. Capſules crowded: blunt—*Leaves though hardly ſheathing the ſtraw yet riſe above it.* Fertile ſpikes *whitiſh or yellowiſh: tiled on every ſide, all growing on footſtalks.* Floral leaves *permanent.* Barren ſpike *greeniſh white.*

Gramen Cyperoides Polyſtachion flavicans, ſpicis brevibus prope ſummitatum Caulis. *Ray's Syn.* 419.

Moiſt paſtures. P. May.

Pink Panicea

SEDGE. Spikes on fruit-ſtalks; upright; at a diſtance from each other; fertile ſpikes ſtrap-ſhaped; capſules rather blunt; bladder-ſhaped—

Gramen cyperoides foliis Caryophylleis, ſpicis e rarioribus et tumidioribus granis compoſitis. *Ray's Syn.* 418.

Turfy bogs. P. June—July.
Cows, Sheep and Goats eat.

SEDGE. Spikes pendant. Fruit-ſtalks in pairs—*Capſules when ripe, awl-ſhaped; ſcored; tapering and ending in a double awn, which is bent back at the end.* Floral leaves *briſtle ſhaped. Spikes yellowiſh.* Baſtard Pſeudo cyperus

Gramen cyperoides ſpica pendula breviore. *Bauh. pin.* 6. *Ray's Syn.* 419.

Cyperus, ſeu Pſeudo-Cyperus ſpica pendula breviore. *Park.* 1266.

Pſeudo-cyperus. *Gerard.* 29.

Banks of wet ditches. P. July.

Cows, Sheep and Goats eat it.

SEDGE. Spikes pendant; barren ſpike upright. Fertile cylindrical; longer than the fruit-ſtalks. Capſules tapering to a point. (*Hudſon*)—*Spikes yellowiſh*, fertile ſpikes *moſtly five, or more.* Pendulous Pendula

Gramen cyperoides, ſpica pendula longiore. *Park.* 1267. *Ray's Syn.* 420.

Gramen ſpica pendula longiore et anguſtiore. *Bauh. pin.* 6.

Gramen cyperoides polyſtachion majuſculum latifolium, ſpicis multis, longis, ſtrigoſis. *Ray's Syn.* 419.

Woods and moiſt hedges. P. June—July.

SEDGE. Spikes upright; cylindrical, growing by threes; nearly ſitting; barren ſpike terminating. Straw three-cornered —*Leaves narrow; long; of a fine green. The two fertile ſpikes almoſt black; tiled with yellowiſh; egg-ſhaped; blunt; upright capſules.* Straw *naked; with three equal, flat ſides. Under the lower ſpike is a little leaf longer than the ſpike. The leaves under the other ſpikes are ſhorter.* Turfy Coeſpitoſa

Gramen cyperoides foliis caryophylleis, ſpicis erectis ſeſſilibus, e ſeminibus confertis compoſitis. *Ray's Syn.* 418.

Gramen cyperoides anguſtifolium, ſpica ſpadiceo-viridi minus. *Bauh. pin.* 6.

Meadows and moiſt woods. P. May—June.

This plant as well as the *Dwarf* BULLRUSH, begins ſpontaneouſly to grow upon boggy moſſes, and forming compact hillocks, affords a ſupport for men to walk upon; ſo that in the courſe of time, the moſt dangerous bogs are changed into firm and uſeful land.—Cows, Sheep, Goats and Horſes eat it.

SEDGE. Spikes on ſhort fruit-ſtalks; at a great diſtance from each other; ſheathed by a floral leaf. Capſules angular; ſharp-pointed—*Leaves ſoft. Spikes not hanging down.* Scales *of the flower-cup ſharp.* Looſe Diſtans

Gramen cyperoides ſpicis parvis, longiſſime diſtantibus. *Ray's Syn.* 420.

Turfy bogs. P. June.

SEDGE.

Wood Sylvatica

SEDGE. Spikes pendant; barren ſpike upright. Fertile ſpikes thread-ſhaped; ſhorter than the fruit-ſtalks. Capſules ſharp; at a diſtance from each other—Hudſon. 353. *Spikes of a yellowiſh colour, as is the whole plant.*

Gramen cyperoides ſylvarum tenuius ſpicatum. *Park.* 1171. *Ray's Syn.* 419.

In woods. P. May—June.

Linnæus conſiders it only as a variety of the *Bladder* SEDGE.

* * * * * * *Fertile and barren flowers on different ſpikes; the barren ſpikes moſt in number.*

Brown Acata

SEDGE. Barren ſpikes moſt numerous. Fertile ſpikes on very ſhort fruit-ſtalks; capſules rather blunt—*In dry ſituations blackiſh; in wet ones, reddiſh brown.* Shaft *cloven; hairy; white; not permanent.* Floral leaves *egg-ſhaped; brown, with a green ſtripe running lengthways.*

Gramen cyperoides. *Gerard.* 12. latifolium ſpica rufa, ſeu caule triangula. *Bauh. pin.* 6.

Gramen cyperoides cum paniculis nigris. *Ray's Syn.* 417.

Gramen cyperoides majus latifolium. *Park* 1265.

Gramen cyperoides majus anguſtifolium. *Gerard.* 1265.

1. Large; with narrow leaves. *Park.* 1265. *Ray's Syn.* 417. In this variety there are ſometimes more fertile than barren ſpikes. It is ſmaller than the preceding variety in every reſpect, except in height, in which it is equal.

2. Small, with narrow leaves.

Gramen cyperoides minus anguſtifolium. *Park.* 1266. fig. 3, a ſmaller variety of the preceding variety.

In watery places.

1. On the banks of brooks and rivers; ſometimes growing in the water. 2. In ponds and ditches.

Horſes, Cows, Sheep and Goats eat it.

Bladder Veſicaria

SEDGE. Barren ſpikes moſt numerous. Fertile ſpikes on fruit-ſtalks. Capſules bladder-ſhaped; tapering to a point—*The ſhaft in this ſpecies is very ſhort; permanent; cloven; but not very evidently ſo.* Floral leaves *ſpear-ſhaped; tapering; browniſh at the edges.* Fertile ſpikes *yellow.* Barren ſpikes *brown.*

Gramen cyperoides majus præcox, ſpicis turgidis, teretibus, flaveſcentibus. *Ray' Syn.* 420.

Greater Bladder Carex.

Marſhes and wet places. A. June.

Cows, Goats and Sheep eat it.

The Laplanders fill their ſhoes with it, to protect their feet from the cold in winter and from the heat in ſummer. For this purpoſe they pick, dry and card it.

SEDGE.

SEDGE. Barren ſpikes moſt numerous; fertile ſpikes ſit- Inflated
ting. Seedveſſels egg-ſhaped; diſtended; tapering to a point. Inflata
Floral leaves twice as long as the ſpike. *Hudſon.*

Gramen cyperoides paluſtre, ſpicis tribus ſubrotundis, vix aculeatis, ſpatio diſtantibus. *Ray's Syn.* 420.

Leſſer Bladder Carex. Hudſon. 354.

In Marſhes. P. June—July.

SEDGE. Spikes at a diſtance from each other. Barren ſpikes Hairy
moſt numerous. Fertile ſpikes on ſhort fruit ſtalks; upright. Hirta
Capſules hairy—*Leaves ſtanding out of their ſheathes; white and downy.*

Gramen cyperoides polyſtachion lanuginoſum. *Ray's Syn.* 418.

Gramen cyperoides Norvegicum pariumlanoſum. *Park.* 1172.

Meadows and wet paſtures. P. June.

361 BUR-REED. 1041 Sparganium.

B. *Barren Flowers numerous; collected into a little head.*

EMPAL. *Common Catkin* roundiſh; tiled very cloſely on every ſide; conſiſting of *proper Cups* with three leaves; narrow; deciduous.

BLOSS. *None.*

CHIVES. *Threads* three; hair-like; as long as the cup. *Tips* oblong.

F. *Fertile Flowers.*

EMPAL. As above. *Common Receptacle* roundiſh.

BLOSS. None.

POINT. *Seedbud* egg-ſhaped; ending in a ſhort awl-ſhaped *Shaft. Summits* two; ſharp; permanent.

S. VESS. Not very pulpy; turban-ſhaped, but terminated by a point; angular beneath.

SEEDS. *Nuts* two; hard as bone; oblong egg-ſhaped; angular.

OBS. *Sometimes there is only one, at other times two cells in the ſeed veſſel.*

BUR-REED. Leaves upright; three cornered.— Great
Sparganium ramoſum. *Park.* 1205. *Gerard.* 45. *Bauh.* Erectum
pin. 15. *Ray's Syn.* 437.

Sparganium non ramoſum. *Bauh. pin.* 15.

Wet ditches and banks of rivers. P. July.

Horſes and Swine eat it; Cows are not fond of it; Sheep and Goats refuſe it.

BUR-

Lesser
Natans

BUR-REED. Leaves drooping; flat.—
Sparganium minimum. *Ray's Syn.* 437.
Least Bur-reed.
Ponds and slow streams. P. July.
Cows eat it.

362 REEDMACE. 1040 Typha.

B. *Barren flowers numerous; forming a catkin at the end of the straw.*

EMPAL. *Catkin common,* cylindrical; very closely set; consisting of *individual Cups* with three leaves; *bristle-shaped.*

BLOSS. None.

CHIVES. *Threads* three; hair-like; as long as the cup. *Tips* oblong; pendant.

F. *Fertile flowers numerous; set exceedingly close; forming a catkin, which surrounds the staw.*

EMPAL. *Hairs* feathered.

BLOSS. None,

POINT. *Seedbud* sitting upon a bristle; egg shaped. *Shaft* awl-shaped. *Summit* hair-like; permanent.

S. VESS. None. The *Fruit* very numerous and forming a cylinder.

SEED. Single; egg-shaped; furnished with a shaft, and sitting upon a bristle. *Feather* hair like; fixed to the bristle supporting the seed, and and as long as the pointal.

Cat's-tail
Latifolia

REEDMACE. Leaves somewhat sword-shaped. The spikes of fertile and barren flowers placed near together—*Sheaths two; deciduous: one placed at the base and the other at the middle of the barren spike.* Spikes brown.
Typha, *Gerard.* 46. palustris major. *Bauh. pin.* 20. *Ray's Syn.* 436.
Typha palustris maxima. *Park.* 1204.
Great Cats-tail.
Banks of rivers and fish-ponds,

Smaller
Angustifolia

REEDMACE. Leaves semi-cylindrical; spikes of fertile and barren flowers placed at a distance from each other—
Typha palustris media. *Ray's Syn.* 436.
Typha palustr s clava gracili. *Bauh. pin.* 20.
Typha minor. *Park.* 1204.
Narrow-leaved Cats-tail.
Ditches and ponds. P, July.
Cows eat it; Swine refuse it.

Order IV.

Order IV. Four Chives.

363 NETTLE. 1054 Urtica.

B. *Barren flowers.*

EMPAL. *Cup* four leaves. *Little Leaves* circular; concave; blunt.

BLOSS. *Petals* none.

Honey-cup in the center of the flowers; glafs fhaped; entire; narrower at bottom; very fmall.

CHIVES. *Threads* four; awl-fhaped: as long as the cup; expanding; one placed within each leaf of the cup. *Tips* with two cells.

F. *Fertile flowers upon the fame, or upon diftinct plants.*

EMPAL. *Cup* with two valves; egg-fhaped; concave; upright; permanent.

BLOSS. None.

POINT. *Seedbud* egg-fhaped. *Shaft* none. *Summit* woolly.

S. VESS. None. *Cup* clofing.

SEED. Single; egg-fhaped; compreffed and blunt; fhining.

OBS. *In the* Common Nettle *the cup of the barren flower confifts of one leaf divided into four fegments; and the cup of the fertile flower confifts of four valves; the two innermoft and largeft fupplying the place of a feed veffel.*

NETTLE. Leaves oppofite; egg-fhaped; ferrated. Fruit bearing Catkins globular. Roman Pilulifera

Urtica urens pilulas ferens. *Bauh. pin.* 232.

Urtica Romana. *Gerard.* 784. *Park.* 440.

Urtica pilulifera, folio profundius urticæ majoris in modum ferrato, femine magno Lini. *Ray's Syn.* 140.

Amongft rubbifh. A. July.

NETTLE. Leaves oppofite; oval—*Deeply ferrated.* Stinging Urens

Urtica minor. *Gerard.* 704. *Ray's Syn.* 140.

Urtica urens minor. *Bauh. pin.* 232.

Leffer Nettle.

Gardens and amongft rubbifh. A. Auguft.

The ftings are very curious microfcopic objects: They confift of an exceedingly fine pointed, tapering, hollow fubftance, with a perforation at the point, and a bag at the bafe. When the fting is preffed upon, it readily punctures the fkin, and the fame preffure forces up an acrimonious fluid from the bag, which inftantly fquirts into the wound, and produces an effect that every one has experienced.

The

The leaves are gathered and cut to pieces, to mix with the food of young Turkeys.—Cows, Horſes, Sheep, Goats and Swine refuſe it.

Common Dioica

NETTLE. Leaves oppoſite; heart-ſhaped. Flowers in bunches, which grow in pairs—*Leaves ſerrated; ſtinging. Catkins brown. Barren and fertile flowers on diſtinct plants.*

Urtica urens *Gerard.* 404. maxima. *Bauh. pin.* 232.

Urtica racemifera major perennis. *Ray's Syn.* 139.

Ditch-banks and amongſt rubbiſh. P. July.

The chives afford a very pleaſing appearance: they are rolled inwards under the ſegments of the cup, which fold over them 'till they are ripe; when the ſegments gradually expanding, the chives releaſed from their confinement roll backwards with an elaſtic force, and diſcharge the contents of their tranſparent tips, which floats in the air like a cloud of duſt, that it may arrive at the ſummits of the fertile flowers, (which grow on diſtinct plants,) and fertilize the ſeedbud.—The ſtalks may be dreſſed like Flax or Hemp, for making Cloth, or Paper—The plant has formerly been uſed as an aſtringent but is now diſregarded.—A leaf put upon the tongue, and then preſſed againſt the roof of the mouth, is pretty efficacious in ſtopping a bleeding at the noſe.—Paralytic limbs have been recovered by ſtinging them with Nettles.—The young ſhoots are gathered early in the ſpring to boil in broth or gruel.—Cows eat the leaves when they are a little withered.—The leaves are cut to pieces to mix with the food of young Turkeys, and other Poultry.—The different ſpecies of NETTLE afford nouriſhment to the following Inſects:

Admiral, or Admirable Butter-fly, *Papilio Atalanta.*

Comma Butterfly, *Papilio C. album.*

Painted Lady Butterfly, *Papilio Cardui.*

Peacock Butterfly, *Papilio Io.*

Small Tortoiſe-ſhell Butterfly, *Papilio Urticæ.*

Great Fritillary Butterfly, *Papilio Paphiæ.*

Crimſon and Gold Moth. Dot Moth. Likeneſs egger Moth. Snout likeneſs egger Moth. Small Magpie Moth, *Phalæna Urticata.*

Dingy China mark likeneſs Moth. White plumed Moth, *Phalæna Didactyla.*

Angleſhades Moth, *Phalæna Meticuloſa.*

Large Tyger Moth, *Phalæna Caja.*

Scarlet Tyger Moth, *Phalæna Dominula.*

Wood Tyger Moth, *Phalæna Plantaginis.*

Spotted Buff Moth, *Phalæna Lubricipeda.*

Cream ſpot Tyger Moth, *Phalæna Villica.*

Mother of Pearl Moth, *Phalæna Verticalis. Phalæna Roſtralis.*

Nettle

Nettle Top Moth, Bloſſom Underwing Moth. Yellow Moth, *Phalæna Interrogationis.*

The Gothic Moth on the roots. Nettle Weevil, *Curculio Scaber.*

Nettle Louſe, *Aphis Urticæ. Chermes Urticæ.*

364 BOX. 1053 Buxus.

B. *Barren flowers projecting from the buds of the tree.*

EMPAL. *Cup* three leaves; circular; blunt; concave; expanding.

BLOSS. *Petals* two; circular; concave; reſembling the cup, but larger.

CHIVES. *Threads* four; awl-ſhaped; upright but expanding, generally longer than the cup. *Tips* upright; double.

POINT. *Seedbud* only a rudiment, without ſhaft or ſummit.

F. *Fertile Flowers in the ſame bud with the others.*

EMPAL. *Cup* four leaves. *Little Leaves* circular; blunt; concave; expanding.

BLOSS. *Petals* three; circular; concave; reſembling the cup but larger.

POINT. *Seedbud* roundiſh; with three blunt edges; ending in three very ſhort permanent *Shafts. Summits* blunt; rough with hair.

S.VESS. *Capſule* roundiſh; with three bills and three cells; opening elaſtically in three directions.

SEEDS. Two; oblong; roundiſh on one ſide; flat on the other.

BOX. As there is only one ſpecies known Linnæus gives no deſcription of it—*Leaves oval; thick; gloſſy.* Bloſſoms *greeniſh.* *white.* Tree Sempervirens

Buxus. *Gerard.* 1410. *Ray's Syn.* 445.

Buxus arboreſcens. *Bauh. pin.* 471.

Buxus arbor vulgaris. *Park.* 1428.

Buxus anguſtifolia. *Ray's Syn.* 445.

Box Tree.

Woods and hedges. P. April.

The

The varieties are,

1. Leaves ſtriped with yellow.
2. Leaves edged with yellow.
3. Leaves edged with white.
4. Leaves broad and ſtriped with white.
5. Narrow leaved.
6. Small leaved with yellow edges.
7. Leaves yellow at the ends.
8. Dwarf round-leaved.
9. Dwarf with ſtriped leaves.

The wood is very hard and ſmooth; and therefore well adapted for the uſe of the Turner. Combs, Mathematical Inſtruments, Knife-handles, and button-moulds are made of it.— An empyreumatic oil diſtilled from the ſhavings is often uſed as a topical application for the piles, and ſeldom fails to procure eaſe.

365 BIRCH. 1052 Betula.

B. *Barren flowers forming a cylindrical Catkin.*

EMPAL. *Catkin*, common; tiled on every ſide; looſe; cylindrical; conſiſting of *Scales*, with three flowers in each; to the ſides of each of which are fixed two exceedingly ſmall ſcales.

BLOSS. *Compound*, with three florets; equal; fixed to the center of each ſcale of the cup.

Individuals formed of one petal; with four diviſions; expanding; very ſmall. *Segments* egg-ſhaped; blunt.

CHIVES. *Threads* four; very ſmall. *Tips* double.

F. *Fertile flowers forming a Catkin on the ſame tree.*

EMPAL. *Catkin* common: tiled; *Scales* placed by threes; oppoſite; fixed to the ſpike-ſtalk; two flowers in each; heart-ſhaped; concave; ſhort; ſlightly notched at the end, with a ſharp point in the middle.

BLOSS. None that is perceptible.

POINT. *Seedbud* egg-ſhaped; very ſmall. *Shafts* two; briſtle-ſhaped; as long as the ſcales of the cup. *Summits* ſimple.

S. VESS. None. The ſeeds lie under each triple ſcale of the catkin.

SEEDS. Solitary; egg-ſhaped; bordered.

OBS. *In ſome ſpecies the flowers form a cylindrical catkin; in others a roundiſh cone.*

BIRCH. Leaves egg-ſhaped; tapering to a point; ſerrated —*Flowers forming a cylindrical catkin.* Bark *white; ſmooth; gloſſy.* White Alba

Betula. *Bauh. pin.* 427. *Ray's Syn.* 443. *Gerard.* 1478.

Birch Tree. Birk.

Woods and moiſt hedges. S. July.

It grows in all kinds of ſoil, but beſt in ſhady places. It bears cropping. It is hurtful to paſturage. The wood is firm, tough and white: Womens ſhoe-heels, and packing-boxes are made of it. It is planted along with hazle to make charcoal for forges. In the northern parts of Lancaſhire, they form the ſlender twigs into beſoms for exportation. (*Penn. Tour.*) The bark is extremely uſeful to the people in the north of Europe; they make hats and drinking cups of it in Kamſchatka. The Swediſh fiſhermen make ſhoes of it. The Norwegians cover their houſes with it, and upon this cover, they lay turf three or four inches thick. Torches are made of the bark ſliced and twiſted together. It abounds with a reſinous matter that is highly inflammable. If a hole is bored into the tree when the ſap riſes in the ſpring, a ſweet liquor diſtils from it, which properly fermented with the addition of ſugar, makes a pleaſant wine.—Horſes, Cows, Goats and Sheep eat it. Swine refuſe it.

It furniſhes food for the following inſects.

Willow Butterfly. *Papilio Antiopa.*
Brown hair-ſtreak Butterfly. *Papilio Betulæ.*
Emperor Moth. *Phalæna pavonia.*
Large Emerald Moth. *Phalæna Papilionaria.*
Great Egger Moth. *Phalæna Quercus.*
Chocolate-tip Moth. *Phalæna Curcula.*
Spotted Elm Moth. *Phalæna Beticlaria.*
Pebble Moth. *Phalæna Ziczac.*
- - - - - - *Tenthredo lutea.*
Birch Louſe. *Aphis Betulæ.*
- - - - - - *Coccus Betulæ.*
Birch Bug. *Cimex Betulæ.*
Birch Weevil. *Curculio Betulæ.*
- - - - - - *Chryſomela Betulæ.*

BIRCH. Fruit-ſtalks branched.—*Leaves nearly circular; clammy; ſerrated.* Catkins *brown.* Alder Alnus

Alnus. *Gerard.* 1249 *Ray's Syn.* 442. vulgaris. *Park.* 1408.

Alnus rotundifolia glutinoſa viridis. *Bauh. pin.* 428.

Common Alder. Owler.

Alnus vulgaris, ſub-conis ligulis membranaceis rubris donata. *Ray's Syn.* 442.

1. Scarlet Alder; there are likewiſe the *Black*, the *hoary-leaved*, the *indented* or *Elm-leaved*; the *curled leaved*; the *broad curled leaved* and the *white* ALDERS.

Wet ſituations. S. July.

 It

It flouriſhes beſt in low marſhy ſituations, in which it is frequently planted to make hedges. It will not live in a chalky ſoil. It is eaſily propagated by ſeeds, but not by ſlips or cuttings. Graſs grows well beneath its ſhade.—The wood is ſoft and brittle. It endures a long time under water and therefore is uſed for pipes; and to lay under the foundations of buildings ſituated upon bogs. Womens ſhoe heels, ploughmens clogs and various articles of the Turner are made of it.—The bark gives a red colour, and with the addition of copperas a black. It is principally uſed by fiſhermen to ſtain their nets.—In the Highlands of Scotland near Dundonnel, Mr. Pennant ſays, the boughs cut in the ſummer, ſpread over the fields, and left during the winter to rot, are found to anſwer as a manure. In March the ground is cleared of the undecayed parts, and then ploughed.—The freſh-gathered leaves are covered with a glutinous liquor; and ſome people ſtrew them upon their floors to deſtroy fleas; the fleas are ſaid to be entangled in the tenacious liquor like birds are by bird-lime.—The berries dye green.—The whole plant is aſtringent.—Horſes, Cows, Goats and Sheep eat it; Swine refuſe it.

The following inſects live upon it.

Puſs Moth. *Phalæna Vinula.*
Buff-tip Moth. *Phalæna Bucephala.*
Yellow Tail Moth. *Phalæna Chryſorrhœa.*
White ſpot tuſſock Moth. *Phalæna antiqua.*
Dagger Moth. *Phalæna Pſi.*
- - - - - - *Tenthredo Lutea.*
- - - - - - *Chermes Alni.*
Birch Weevil. *Curculio Betulæ.*
Alder Weevil. *Curculio Alni.*
Two ſpotted Lady Cow. *Coccinella 2 punctata.*
- - - - - - - - - *Chryſomela Alni.*

Order

Order V. Five Chives.

366 DITCHBUR. 1056 Xanthium.

B. *Barren flowers compound.*

EMPAL; *Cup*, common to many florets; formed of many leaves; tiled with ſlender ſcales, as long as the florets; equal.

BLOSS. *Compound*, uniform; tubular; equal; formed into an hemiſphere.

Individual, one petal; tubular; funnel-ſhaped; upright; with five clefts.

CHIVES. *Threads* five; forming a hollow cylinder. *Tips* upright: parallel; not united.

RECEPT. *Common*, next to none; the florets being ſeparated by chaff.

F. *Fertile flowers beneath the others on the ſame plant; two together.*

EMPAL. *Fence* containing two flowers; formed of two *Leaves*; oppoſite; each divided into three ſharp lobes; the middle lobe projecting fartheſt; ſet round with hooked prickles; ſurrounding and entirely covering the ſeedbuds to which they are fixed. *Little Segments* looſe.

BLOSS. None.

POINT. *Seedbud* oval; rough with hair. *Shafts* two; ſimilar; hair-like. *Summits* ſimple.

S. VESS. *Berry* dry; oblong egg-ſhaped; cloven at the end; beſet cloſely with hooked prickles.

SEED. *Nut*; with two cells.

DITCHBUR. Stem without thorns. Leaves heart-ſhaped; three fibred.—*Stem generally ſpotted.* Flowers *at the baſe of the leaves.* Burdock Strumarium

Xanthium, ſeu Lappa minor. *Park.* 1222. *Ray's Syn.* 140.
Lappa minor; Xanthium Dioſcoridis. *Bauh. pin.* 198.
Bardana minor. *Gerard.* 809.
Leſſer Burdock.
On dunghills. A. Auguſt—September.

The leaves are bitter and aſtringent.—A decoction of the whole plant affords a ſhowy yellow colour; but it is better if only the flowers are uſed. Horſes and Goats eat it; Cows, Sheep and Swine refuſe it.

367 EVERLASTING 1060 Amaranthus

B. *Barren flowers on the ſame plant with the fertile ones.*

EMPAL. *Cup* five or three leaves; upright; coloured, permanent; *little Leaves* ſpear-ſhaped; ſharp.

BLOSS. None. Unleſs you conſider the empalement as ſuch.

CHIVES. *Threads* five or three; hair-like; upright, but ſtanding rather open; as long as the cup. *Tips* oblong; vane-like.

F. *Fertile flowers in the ſame bunch with the others.*

EMPAL. *Cup* the ſame as the other.

BLOSS. None.

POINT. *Seedbud* egg-ſhaped. *Shafts* three; ſhort; awl-ſhaped. *Summits* ſimple; permanent.

S.VESS. *Capſule* egg-ſhaped; ſomewhat compreſſed; the ſize of the cup which contains it, and coloured like that. Bills three; cell one; cut round.

SEED. Single; globular; compreſſed; large.

Blite
Blitum

EVERLASTING. Flowers in lateral cluſters; each cup with three clefts. Leaves egg-ſhaped, dented. Stem ſpreading.—*Sometimes the leaves are of a ſilvery white in the middle, and with or without a brown ſpot.*

Blitum rubrum minus. *Bauh. pin.* 118. *Ray's Syn.* 157.

1. There is a variety that is white.

Leaſt Blite.

Amongſt rubbiſh. A. Auguſt.

Order

Order VIII. Many Chives.

368 ARROWHEAD. 1067 Sagittaria.

B. *Barren flowers many.*

EMPAL. *Cup* three leaves; *little Leaves* egg-ſhaped; concave; permanent.

BLOSS. *Petals* three; circular; blunt; flat; expanding; three times as large as the cup.

CHIVES. *Threads* many, (generally twenty-four,) awl-ſhaped; collected into a little head. *Tips* upright; as long as the cup.

F. *Fertile flowers few, and ſtanding beneath the others.*

EMPAL. *Cup* as above.

BLOSS. *Petals* three as above.

POINT. *Seedbuds* numerous; compreſſed; forming a little head; hunched on the outer-ſide, ending in very ſhort *Shafts*. *Summits* ſharp; permanent.

S. VESS. None. *Receptacle* globular; and ſet round with the ſeeds ſo as to form a globe.

SEEDS. Numerous; oblong; compreſſed; encompaſſed lengthways by a broad membranaceous border, which is hunched on one ſide, and tapering towards each end.

ARROWHEAD. Laaves arrow-ſhaped; ſharp—*On Leaf-ſtalks*. Bloſſoms *white*. Common

Sagitta. *Ray's Syn.* 258. minor latifolia. *Park.* 1247. Sagittifolia

Sagitta aquatica minor latifolia. *Bauh. pin.* 194.

Sagitta aquatica omnium minima. *Ray's Syn.* 258.

Wet Ditches and Banks of rivers. P. June.

There is always a bulb at the lower part of the root, growing in the ſolid earth, beneath the mud. This bulb conſtitutes a conſiderable part of the food of the Chineſe, and upon that account they cultivate it. Horſes, Goats and Swine eat it; Cows are not fond of it.

369 MILFOIL. 1066 Myriophyllum.

B. *Barren flowers.*

EMPAL. *Cup* four leaves; oblong; upright; the *outermost* larger and the *innermost* ſmaller than the reſt.

BLOSS. None.

CHIVES. *Threads* eight; hair-like; longer than the cup; limber. *Tips* oblong.

F. *Fertile flowers placed under the others.*

EMPAL. *Cup* as above.

BLOSS. None.

POINT. *Seedbuds* four; oblong. *Shafts* none. *Summits* downy.

S. VESS. None.

SEEDS. Four; oblong.

OBS. *In the ſecond ſpecies the chives and pointals are frequently found in the ſame empalement.*

Spiked Spicatum MILFOIL. Barren flowers in interrupted ſpikes.—*Stems hollow; ſcored; jointed; branched.* Fruit-bearing ſtalks *without leaves.*

Potamogiton foliis pennatis. *Ray's Syn.* 150.

Millefolium aquaticum pennatum ſpicatum. *Bauh. pin.* 141. *Park.* 1257.

Spiked Water Milfoil.

Ponds and ſlow ſtreams. P. June—July.

Sheep and Goats refuſe it.

Whorled Verticillatum MILFOIL. All the flowers growing in whorls; frequently the chives and pointals are found in the ſame flower.—

Myriophyllum aquaticum minus. *Gerard.* 828.

Pentapterophyllon aquaticum floſculis ad foliorum nodos. *Ray's Syn.* 316.

Millefolium aquaticum minus. *Park.* 1257.

Millefolium aquaticum floſculis ad foliorum nodos. *Bauh. pin.* 141.

Verticillated Water Milfoil.

Ponds and ſlow ſtreams. P. July.

370 HORNWEED. 1065 Ceratophyllum.

B. *Barren Flowers.*

EMPAL. *Cup* with many divisions. *Segments* awl-shaped; equal.

BLOSS. None.

CHIVES. *Threads* twice as many as there are segments in the cup (from sixteen to twenty;) hardly discernible. *Tips* oblong; upright; longer than the cup.

F. *Fertile flowers on the same plant with the others.*

EMPAL. *Cup* as above.

BLOSS. None.

POINT. *Seedbud* egg-shaped; compressed. *Shaft* none. *Summit* blunt; oblique.

S. VESS. None.

SEED. *Nut*, egg-shaped; of one cell; tapering to a point.

HORNWEED. Leaves twice forked; in double pairs. Fruit with three thorns—*Leaves in whorls.* Horsetail Demersum

Hydroceratophyllon folio aspero, quatuor cornibus armato. *Ray's Syn.* 135.

1. Leaves smooth with eight horns. *Ray's Syn.* 135.

Horned Pondweed.

Slow streams and wet ditches. P. July.

The uncommon breadth of the summits is the most remarkable peculiarity in this plant; and whether they float upon the surface of the water, or are sunk beneath it, they are equally calculated to receive the dust from the chives which stand above them.

371 IRONWORT. 1069 Poterium.

B. *Barren flowers forming a ſpike.*

EMPAL. *Cup* three leaves; *little Leaves* egg-ſhaped; coloured; ſhedding.

BLOSS. With four diviſions. *Petals* egg-ſhaped; concave; expanding; united at the baſe; permanent,

CHIVES. *Threads* many, (from twenty to fifty;) hair-like; very long; limber. *Tips* roundiſh; double.

F. *Fertile flowers in the ſame ſpike, but placed above the others.*

EMPAL. *Cup* as above.

BLOSS. *Petal* one; wheel-ſhaped. *Tube* ſhort; roundiſh; cloſing at the mouth. *Border* with four diviſions. *Segments* egg-ſhaped; flat; reflected; permanent.

POINT. *Seedbuds* two; oblong egg-ſhaped; within the tube of the bloſſom. *Shafts* two; hair-like; coloured; limber; as long as the bloſſom. *Summit* pencil-ſhaped; coloured.

S. VESS. *Berry* formed of the tube of the bloſſom, which grows thick; hard; and cloſes upon the ſeeds.

SEEDS. Two.

OBS. *There is one foreign ſpecies in which the Berry is fleſhy and globular; the ſeeds three; oblong; cylindrical. But in the Britiſh ſpecies the Berry is dry and angular, and the ſeeds four cornered; tapering at each end. There are two feeble pointals in each barren flower in the* Burnet IRONWORT.

Burnet
Sanguiſorba

IRONWORT. Without thorns; ſtem ſomewhat angular. —*Leaves winged, ſerrated. Spikes of flowers purple.*

Sanguiſorba minor. *Ray's Syn.* 203.

Pimpinella ſanguiſorba minor hirſuta. *Bauh. pin.* 160.

Pimpinella vulgaris minor. *Park.* 582.

Pimpinella ſylveſtris. *Gerard.* 1045.

Burnet.

High chalky paſtures and moiſt meadows. P. May—Auguſt.

The young leaves are ſometimes uſed in ſallads, and in cool tankards.

372 BEECH. 1072 Fagus.

B. *Barren flowers fixed to a common receptacle, somewhat like a catkin.*

EMPAL. *Cup* one leaf; bell-ſhaped; with five clefts.

BLOSS. None.

CHIVES. *Threads* many, (about twelve) as long as the cup; briſtle-ſhaped. *Tips* oblong.

F. *Fertile flowers in a bud, on the ſame tree.*

EMPAL. *Cup* one leaf; with four teeth; upright and ſharp.

BLOSS. None.

POINT. *Seedbud* incloſed by the cup. *Shafts* three; awl-ſhaped. *Summits* ſimple; reflected.

S. VESS. *Capſule* roundiſh; (formed of the cup;) large; beſet with ſoft thorns; with one cell, and four valves.

SEEDS. *Nuts* two; egg-ſhaped; three cornered: with three valves; tapering.

OBS. *The barren flowers ſometimes form a cylinder, at others a globe.*

BEECH. Leaves ſpear-ſhaped; tapering to a point; ſerrated; naked underneath.— Cheſnut Caſtanea

Caſtanea. *Gerard.* 1442. *Ray's Syn.* 442. vulgaris. *Park.* 1400.

Caſtanea ſylveſtris. *Bauh. pin.* 419.

Cheſnut Tree.

Woods and Hedges, in Kent. S. May.

1. The leaves are ſomething ſtriped with yellow,

Nothing will thrive under its ſhade—The wood is applicable to the ſame uſes that Oak is—If the bark is not taken off, it makes poles for eſpaliers, dead fences and hop yards, and pipes to convey water under ground, which will laſt longer than Elm or Oak—Some of the oldeſt buildings in London are ſaid to be conſtructed with this wood—At Tortworth in Glouceſterſhire there is a tree fifty-two feet round. It is proved to have ſtood there ever ſince the year 1150, and was then ſo remarkable as to be called the great Cheſnut of Tortworth. It fixes the boundary of the manor, and is probably near 1000 years old. See Gent. Mag. 1766 p. 321, where there is a drawing of it. Mr. Collinſon, the author of this account, makes no doubt of the Cheſnut being a native of England, and aſſigns the great profit that ariſes from them when cut for hop-poles, as the reaſon why it is ſo rare to ſee large trees in the woods. The nuts are uſed for whitening linen cloth, and for making ſtarch.

Common Sylvatica

BEECH. Leaves egg-ſhaped; indiſtinctly ſerrated.—*Bark ſmooth; white.* Catkin *globular*.

Fagus. *Gerard*.. 1444. *Bauh. pin.* 419. *Park.* 1403. *Ray's Syn* 439.

Woods and hedges. S. May.

1. The leaves are ſometimes variegated with white or yellow ſtripes.

The tree is large and beautiful, but no verdure will flouriſh under its ſhade. It loves a fertile Soil. Whilſt young it is apt to ſuffer from expoſure. It is difficult to tranſplant. It retains its old leaves through the winter. It bears lopping well, and may be trained to form very lofty hedges—the wood is brittle; ſoon decays in the air, but endures under water. It is formed into tool handles, planes, mallets, chairs and bedſteads. Split into thin layers it is uſed to make ſcabbards for ſwords. It is excellent fuel, and when burnt it affords a large quantity of Pot-aſh—The leaves gathered in Autumn before they are much injured by the froſts, make infinitely better mattraſſes than ſtraw or chaff, and endure for ſeven or eight years—The nuts, or maſts as they are called, when eaten occaſion giddineſs and head ache; but when well dried and powdered they make wholeſome bread. They are ſometimes roaſted and ſubſtituted for coffee. They fatten Swine. The poor people in Sileſia uſe the expreſſed oil inſtead of butter.—

Sheep and Goats eat the leaves.—

The following inſects feed upon both Species.

December Moth. *Phalæna Populi.*
Nut-tree Tuſſock Moth. *Phalæna Coryli.*
Yellow Tuſſock Moth. *Phalæna Pudebunda.*
Beech Weevil. *Curculio Fagi.*
Cockchaffer Beetle. *Scarabæus Melolontha.*
Beech Louſe. *Aphis Fagi.*

373 OAK. 1070 Quercus.

B. *Barren flowers forming a loose catkin.*

EMPAL. *Cup* one leaf; with four or five clefts; *Segments* sharp; often cloven.

BLOSS. None.

CHIVES. *Threads* many, (five, eight or ten;) very short. *Tips* large; double.

F. *Fertile Flowers seated in a bud on the same tree.*

EMPAL. *Cup* one leaf; like leather; hemispherical; rough; very entire; hardly discernible whilst the flower continues.

BLOSS. None.

POINT. *Seedbud* egg-shaped; very small. *Shafts* from two to five; longer than the cup. *Summit* simple; permanent.

S. VESS, None.

SEED. *Nut* oval; formed of a leather-like coat of one valve, which appears as if rasped at the base; fixed to a short cup; cylindrical; smooth.

OBS. *The number of segments in the cup generally corresponds with the number of chives.*

OAK. Leaves deciduous; oblong; broadest towards the end; with rather sharp indentations but blunt angles.—*Blossoms pale green. The* Cup *is composed of about twenty little, spear-shaped; membranaceous; coloured leaves.* English Robur

Quercus vulgaris. *Gerard.* 1339.

Quercus latifolia. *Park.* 1385. *Ray's Syn.* 440.

Quercus cum longo pediculo. *Bauh. pin.* 420.

1. There is one variety with shorter fruit stalks. *Bauh. pin.* 419

2. And and another with striped leaves.

Common Oak.

Woods and hedges. S. April.

It loves hilly better than boggy ground, and thrives best, while young, in large plantations. Its roots descend deep into the earth, and therefore will not bear to be transplanted. Much lopping destroys it. Grass will hardly grow beneath it—The wood is hard; tough; tolerably flexible; not easily splintering; and therefore is preferred before all other timber for building ships of war. It it is well adapted to almost every purpose of the carpenter; but an attempt to enumerate all the uses of this well known wood, would be equally superfluous and difficult.—Oak saw-dust is the principal indigenous vegetable used in dying fustian. All the varieties of drabs and different shades of brown are made with Oak saw-dust, variously managed and compounded

pounded—The balls, or Oak apples are likewise used in dying as a substitute for galls: the black got from them by the addition of copperas is more beautiful than that from galls, but not so durable—The bark is universally used to tan leather. An infusion of it with a small quantity of Copperas is used by the common people to dye woollen of a purplish blue: The colour, tho' not very bright is durable—The balls, or galls upon the leaves are occasioned by a small insect with four wings, called *Cynips querci folii*, which deposits an egg in the substance of the leaf, by making a small perforation on the under surface. The Ball presently begins to grow, and the egg in the center of it changes to a worm; the worm again changes to a nymph, and the nymph to the flying insect with four wings.—

Horses, Cows, Sheep and Goats eat the leaves; Swine and Deer fatten on the acorns.

The following insects feed upon it.

Emperor of the Woods. *Papilio Iris.*
Purple Hair-streak Butterfly. *Papilio Quercus.*
Puss Moth. *Phalæna Vinula.*
Great Egger Moth. *Phalæna Quercus.*
Black Arches Moth. *Phalæna Monacha.*
Gipsey Moth. *Phalæna Dispar.*
Yellow Tail Moth. *Phalæna Chrysorrhœa.*
Yellow Tussock Moth. *Phalæna Pudibunda.*
Scollop winged Oak Moth. *Phalena Oo.*
Spotted Buff Moth. *Phalæna Lubricipeda.*
Yellow July Oak Moth. *Phalæna Quadra.*
Willow Red-under wing Moth. *Phalæna Pacta.*
Dagger Moth. *Phalæna Psi.*
Buff Tip Moth. *Phalæna Bucephala.*
Wild Rose Moth. *Phalæna Lacertinaria.*
Small Oak Moth *Phalæna Viridana.*
Green Silver Lined Moth. *Phalæna Prasinana.*
Buff Argus Moth. *Phalæna Amataria,*

Red arches Moth. Broad bar Moth. Oak bar Moth. Scollop broad wing bar Moth. Triple bar Moth. Dun bar Moth. Oak beauty Moth. Pale Oak beauty Moth. Maiden blush Moth. Clouded border Moth. Laced border Moth. Small brindle beauty Moth. Brindled Moth. Hornsey carpet Moth. Marvel de jour Moth. July high flyer Moth. Gold fringe Moth. Pea green Moth. Heart Moth. Half mourner Moth. Japan long horn Moth. Golden long horn Moth. Maid of honour Moth. Green golden horn Moth. Large Japan Moth. Brussels lace Moth. Triple lines Moth. November Moth. Orange band Moth. Orange Moth. O Moth. Orange companion Moth. Owl Moth. October Moth. Panther Moth. Dark prominent Moth. Red neck Moth. Red shell Moth. Scolloped shell Moth. Snout Moth.

Moth. Grey ſparkling Moth. Orange ſparkling Moth. Clouded ſtraw Moth. Spider Moth. Brown tail Moth. Grey tuſſock Moth. Tiſſue Moth. Blood veined Moth. Waved umber Moth. Mottled Umber Moth. Copper underwing Moth. Orange upper wing Moth. Spring uſher Moth. March cloſe wing Moth. Crimſon under wing Moth, *Phalæna Nupta.*

- - - - - *Phalæna Vindata.*
- - - - - *Cynips Quercus Baccæ.*
- - - - - *Cynips Quercus Folii.*
- - - - - *Cynips Quercus Petioli.*
- - - - - *Cynips Quercus Gemmæ.*
- - - - - *Chermes Quercus.*

Oak Louſe. *Aphis Quercus.*
Oak Weevil. *Curculio Quercus.*

374 HAZLE. 1074 Corylus.

B. *Barren flowers forming a long catkin,*

EMPAL. *Catkin common,* tiled on every ſide; cylindrical; conſiſting of *Scales,* each incloſing a ſingle flower; narrower at the baſe, broader and more blunt at at the end; bent inwards with three clefts. The *Middle Segment* as long, but twice as broad as the others, and covering them.

BLOSS. None.

CHIVES. *Threads* eight; very ſhort; fixed to the inner ſide of the ſcale of the cup. *Tips* egg-ſhaped; oblong; ſhorter than the cup; upright.

F. *Fertile flowers at a diſtance from the others, on the ſame plant; ſitting; incloſed in the bud.*

EMPAL. *Cup* two leaves; like leather; jagged at the edge; upright; as long as the fruit; ſo ſmall as to be hardly diſcernable during thethe time of flowering.

BLOSS. None.

POINT. *Seedbud* roundiſh; very ſmall. *Shafts* two; briſtle-ſhaped; much longer than the cup; coloured. *Summits* ſimple.

S. VESS. None.

SEED. *Nut* ſomewhat egg ſhaped; appearing as if raſped at the baſe; point rather flatted; and a little tapering toward the end.

OBS. *This genus is nearly allied to the* HORNBEAM.

590 Nut Avellana

HAZLE. Props egg-ſhaped; blunt.—*Leaves oval; pointed; ſerrated; wrinkled. Catkins green; afterwards brown.*

Corylus ſylveſtris. *Gerard.* 1479. *Bauh. pin.* 418. *Ray's Syn.* 439.

Corylus, ſive Nux avellana ſylveſtris. *Park.* 1416.

Haſel-nut Tree.

Woods and hedges. S. March.

1. There are ſome varieties in the form of the fruit.

It is frequently planted in hedges, and in coppices to make charcoal for forges. The owners cut them down in equal portions in the rotation of ſixteen years, and raiſe regular revenues out of them; often more than the rent of the land, for freeholders of fifteen or twenty-five pounds per annum are known to make conſtantly ſixty-pounds a year from their woods. *Pennants Tour.* 1772. p. 29. The wood is uſed for fiſhing rods, walking ſticks, crates, hoops for barrels, &c. the ſhoots for ſpringles to faſten down thatch. The roots are preferred where beautiful wood is required for inlaying or ſtaining. It is a practice in Italy to put the chips of hazle into turbid wine to clear it, which it does in twenty-four hours: and in countries where yeaſt is ſcarce, they take the twigs of hazle, and twiſt them together ſo as to be full of chinks; theſe they ſteep in ale during its fermentation; then hang them up to dry, and when they brew again they put them into the the wort inſtead of yeaſt—Painters and engravers prepare coals for delineating their deſigns thus. They take pieces of Hazle about as thick as a man's arm and four or five inches long, dry and then cleave them into pieces about as thick as ones finger. Theſe they put into a large pot full of ſand, and then cover the top of pot with clay. This is expoſed in a potter's oven, or any other ſufficient degree of heat, and when cooled again, the ſticks are found converted into charcoal which draws freely and eaſily rubs out again—The nuts are agreeable to moſt people. Squirrels live chiefly upon them. An expreſſed oil is obtained from them for the uſe of painters—Goats and Horſes eat the leaves; Sheep and Swine refuſe them. The following inſects are found upon the hazle: Brindle Moth. Spider Moth. December Moth, *Phalæna Populi.* Emperor Moth, *Phalæna Pavonia.* Nut tree tuſſock Moth, *Phalæna Coryli.* Dagger Moth, *Phalæna Pſi.* Orange tuſſock Moth, *Phalæna Gonoſtigma. Attelabus Coryli* and *Curculio Nucum.*

375 HORNBEAM. 1073 Carpinus.

B. *Barren flowers formed into a cylindrical catkin.*

EMPAL. *Common Catkin,* loofely tiled on every fide; confifting of *Scales*, with a fingle flower in each; egg-fhaped; concave; fharp; fringed.

BLOSS. None.

CHIVES. *Threads* generally ten; very fmall. *Tips* double; compreffed; woolly at the end; with two valves.

F. *Fertile flowers forming an oblong catkin on the fame tree.*

EMPAL. *Common Catkin,* loofely tiled; confifting of *Scales*, inclofing a fingle flower; fpear-fhaped; woolly; reflected at the end.

BLOSS. Shaped like the cup; of one leaf, with fix clefts. Two of the *Segments* larger than the others.

POINT. *Seedbuds* two; very fhort; each furnifhed with two *Shafts*; hair-like; coloured; long. *Summits* fimple.

S. VESS. None. The *Catkin* growing very large contains the feed at the bafe of of the fcales.

SEED. *Nut* egg-fhaped; angular.

OBS. Linnæus *in one place fays there are ten, and in another twenty chives. I have not had an opportunity to determine which of thefe is true. Probably the number of chives varies.*

HORNBEAM. Scales of the cones flat.—*Bark fmooth; white.* Leaves *oval; pointed; fharply ferrated.* Smooth Betulus

Oftrya ulmo fimilis, fructu in umbilicis foliaceis. *Bauh. pin.* 427. *Ray's Syn.* 451.

Oftrys, five Oftrya. *Park.* 1406.

Betulus five Carpinus. *Gerard.* 1479.

1. There is a variety with ftriped yellow leaves.

Horn-beam Tree. Hard-beam Tree. Horfe or Horn-beech Tree.

In woods. S. May.

This tree loves a poor ftiff foil, on the fides of hills. It is eafily tranfplanted, and bears lopping. Cattle eat the leaves, but pafturage will not flourifh in its fhade.—The wood burns like a candle; it is very white, very tough, harder than hawthorn, and capable of fupporting a great weight. It is ufeful in turning, and for many implements of hufbandry. It makes cogs for mill wheels even fuperior to yew—The inner bark is much ufed in Scandinavia to dye yellow

The *Phalæna Brumata* and *Roftralis* feed upon it.

Order

Order IX. Threads United.

376 F I R. 1077 Pinus.

B. *Barren flowers in bunches.*

EMPAL. None, but the gaping ſcales of the buds.
BLOSS. None.
CHIVES. *Threads* many; united below into an upright pillar, divided at the top. *Tips* upright.

F. *Fertile Flowers on the ſame tree.*

EMPAL. *Common Cone*, ſomewhat egg-ſhaped; compoſed of *Scales*, with two flowers in each; oblong; tiled; permanent, inflexible.
BLOSS. None.
POINT. *Seedbud* very ſmall. *Shaft* awl-ſhaped. *Summit* ſimple.
S. VESS. None. The *Scales* of the cone which before ſtood open cloſing upon the ſeed.
SEED. *Nut*, enlarged by a membranaceous wing, larger than the ſeed, but ſmaller than the ſcales of the cone; oblong; on one ſide ſtraight but hunched on the other.

Scotch Sylveſtris FIR. Leaves in pairs, in their firſt growth ſolitary; ſmooth —*Cones pendant*; *whitiſh*.
Pinus ſylveſtris. *Baub. pin.* 491.
Pinus ſylveſtris, foliis brevibus glaucis, conis parvis alventibus. *Ray's Syn.* 442.
Mountains in Scotland. S. May.
It flouriſhes beſt in a poor ſandy ſoil. In a grove, the trunk becomes tall and naked; in ſunny open places, branched. On rocks or bogs, it ſeldom attains a large ſize, In black ſoil it becomes diſeaſed, and in chalky land it dies. Sometimes it will thrive near running, but never near ſtagnant waters. None but the terminating buds ſend forth branches, therefore it will not bear the leaſt clipping. The roots ſpread very near to the ſurface of the earth, all but the central root which grows perpendicularly downwards; and if this is broken off, or interrupted in its paſſage by rocks, the ſtem ceaſes to ſhoot upwards and the tree for ever remains a dwarf. Upon this account it is apt to ſuffer by tranſplanting—This tree furniſhes us with the beſt red or yellow deal. It is ſmooth, light, and eaſily cloven. The bark will tan leather—The inhabitants of the North of Europe make bread from this tree in the following manner. They chooſe

choose a tree whose trunk is even, for these contain the least resin, and strip off the bark in the spring when it separates most readily, This they first dry gently in the shade; then in a greater heat; and reduce it to powder. With this powder they mix a small quantity of corn-meal, and with water knead it into bread. This they eat, not only in years of scarcity, but at other times, from an apprehension that long disuse might render it disagreeable to them. Their children are very fond of the fresh bark in the spring time, either shaved with a knife or grated with a rasp—the young shoots distilled afford a flagrant essential oil.—Sheep and Goats are not fond it; Horses refuse it.

It affords nourishment to the following insects.

Pine Lappit Moth. *Phalæna Pini.*
Yellow July Oak Moth. *Phalæna Quadra.*
Fir Weevil. *Curculio Pini.*
Fir Bug. *Cimex Abietis.*

FIR. Leaves solitary; notched at the end.—*Grey on the under surface.* Cones *upright.* Yew-leaved Picea

Abies. *Gerard.* 1363. *Park.* 1539. *Ray's Syn.* 441.
Abies conis sursum spectantibus, seu mas. *Bauh. pin.* 505.
Mountains in Scotland.

The thirty-six fine trees of this species, mentioned by Mr. Ray as growing at Wareton near Newport in Shropshire, are now no more. Pitch, tar, and turpentine may be got from all the species of Fir.

FIR. Leaves solitary; awl-shaped, sharp pointed; smooth; pointing in two opposite directions—*Leaves compressed; shining on the upper surface.* Bark *reddish brown.* Cones *long; pendant.* Pitch Abies

Abies mas Theophrasti. *Ray's Syn.* 441.
Picea. *Park.* 1538. major. *Gerard.* 1354.
Picea major prima, sive Abies rubra. *Bauh. pin.* 493.
Common Fir. Pitch Tree.
Mountains in Scotland.

It will grow in various situations; either in thick woods or sunny exposures; on barren commons or in rich pastures; but in very dry or chalky soils it dies. It is difficult to transplant, for its roots spread very wide near the surface of the earth: but if it survives the first removal, it may afterwards be transplanted at pleasure; for the roots that have been cut off, send out numbers of little fibres, sufficient to supply it with nourishment; but it must be an invariable rule always to plant it at the same depth it stood at before, and to let the side which formerly faced the south, be placed again in the same direction—The wood is very light; white; rots in the air and crackles in the fire. It is uſed for making muſical inſtruments, packing boxes, &c. the Lapanders make ropes of the roots and employ them for faſtening together the thin planks of their portable canoes—The inhabitants of Canada prepare a pleaſant and wholeſome liquor from the leaves.

Goats eat it; Sheep refuſe it.

The Fir Bug, *Cimex Abietis*, and the *Chermes Abietis* live upon it.

Order X. Tips United.

377 BRYONY. 1093 Bryonia.

B. *Barren Flowers.*

EMPAL. *Cup* one leaf; bell-ſhaped; with five awl-ſhaped teeth.

BLOSS. With five diviſions; bell-ſhaped; fixed to the cup; *Segments* egg-ſhaped.

CHIVES. *Threads* three; very ſhort. *Tips* five; two upon each thread, and only one upon the third thread.

F. *Fertile Flowers upon the ſame plant.*

EMPAL. *Cup* as above; ſuperiour; permanent.

BLOSS. As above.

POINT. *Seedbud* beneath. *Shaft* with three clefts; as long as the bloſſom when open. *Summits* notched at the end; open.

S. VESS. *Berry* oval; ſmooth.

SEEDS. Several; fixed to the outer coat of the berry; ſomewhat egg-ſhaped.

White Alba — BRYONY. Leaves hand-ſhaped; rough with callous points on both ſurfaces—*The flowers bearing only chives, and thoſe having only pointals are ſometimes, though very rarely, found on diſtinct plants.* Root *very large.* Berry *red.* *Bloſſoms pale green.*

OBS. *In many parts of England the barren and fertile flowers are always on diſtinct plants.*

Bryonia alba. *Gerard.* 869. *Ray's Syn.* 261, vulgaris. *Park.* 178.

Bryonia aſpera ſive alba Baccis rubris. *Bauh. pin.* 297.

Wild Vine.

Hedges and roughs. P. May.

1. There is a variety with black berries.

The root is purgative and acrid; a dram of it in ſubſtance, or half an ounce of it infuſed in wine is a full doſe. A cold infuſion

ſion of the root in water is uſed externally in ſciatic pains. A cataplaſm of it is a moſt powerful diſcutient—A decoction made with one pound of the freſh root is the beſt purge for horned cattle—The active virtues of this plant ſeem to claim more attention than is now beſtowed upon it—Some people have a method of forming the roots into human figures, and ſelling them for mandrakes.

Goats eat it; Horſes, Cows, Sheep and Swine refuſe it.

CLASS XXII.

IN the preceding Claſs the Flowers with only Chives, and the Flowers with only Pointals were found upon the *ſame* plant; but in this they are upon *different* plants.

All the plants therefore of this Claſs are neceſſarily either BARREN or FERTILE; the Flowers of the former containing *Chives* only; thoſe of the latter, only *Pointals*.

Both ſorts are propagated from ſeeds, which are the product of the fertile plants.

B. ſignifies the barren, and

F. the fertile flowers.

CLASS XXII.

Chives and Pointals diſtinct.

Order II. Two Chives.

378 WILLOW. - B. *Catkin* ſcaly. *Bloſſ.* none. *Chives* two; rarely more.
F. *Catkin* ſcaly. *Bloſſ.* none. *Summits* two. *Capſ.* two valves. *Seeds* downy.

Order III. Three Chives.

379 CROWBERRY. B. *Cup* with three diviſions. *Bloſſ.* three petals.
F. *Cup* with three diviſions. *Bloſſ.* three petals. *Shafts* nine. *Berry* nine ſeeds.

† *Small Sedge.* † *White Bryony,* † *Marſh Valerian.*

Order IV. Four Chives.

380 SALLOWTHORN. B. *Cup* with two diviſions. *Bloſſ.* none.
F. *Cup* with two clefts. *Bloſſ.* none. *Point.* one. *Berry* one ſeed, with a lopped ſeed-coat.

381 MISLETOE. B. *Cup* with four divisions. *Bloss.* none.
F. *Cup* four leaves. *Bloss.* none. *Summit* blunt. *Berry* one seed: beneath.

382 GALE. - - B. *Catkin* scaly. *Bloss.* none.
F. Catkin scaly. *Bloss.* none. *Shafts* two. *Berry* one seed.

† *Purging Buckthorn.* † *Common Nettle.* † *Tree Holly.*

Order V. Five Chives.

383 HOP. - - - B. *Cup* five leaves. *Bloss.* none.
F. *Cup* one leaf. *Bloss.* none. *Shafts* two. *Seeds* winged by the cup.

† *Sweet Willow.*

Order VI. Six Chives..

384 LADYSEAL. B. *Cup* six leaves. *Bloss.* none.
F. *Cup* six leaves. *Bloss.* none. *Shaft* with three clefts. *Berry* with three cells; beneath.

Sorrel Dock. † *Little Dock.*

Order VII. Eight Chives.

385 POPLAR. - - B. *Catkin* ragged. *Bloss.* none. *Honey-cup* egg-shaped. *Chives* from eight to sixteen.
F. *Catkin* ragged. *Bloss.* none. *Summit* with four clefts. *Capsule* with two valves. *Seeds* feathered.

386 ROSEWORT. B. *Cup* with four divisions. *Bloss.* four petals.
F. *Cup* with four divisions. *Bloss.* none. *Pointals* four. *Capsules* four. *Seeds* many.

Order

Order VIII. Nine Chives.

387 MERCURY. - B. *Cup* three leaves. *Bloss.* none. *Chives* from nine to twelve.
F. *Cup* three leaves. *Bloss.* none. *Shafts* two. *Capsule* two berries.

388 FROGBIT. - B. *Cup* three leaves. *Bloss.* three petals.
F. *Cup* three leaves. *Bloss.* three petals. *Shafts* six. *Capsule* beneath; with six cells.

Order IX. Ten Chives.

† *Campion Cuckow-flower.* † *Catch-fly Campion.*

Order XI. Twenty Chives.

† *Cloudberry Bramble.*

Order XII. Many Chives.

† *Mercury.*

Order XIII. Threads United.

389 JUNIPER. - - B. *Catkin. Bloss.* none. *Chives* three.
F. *Cup* with three divisions. *Bloss.* three petals. *Shafts* three. *Berry* beneath; with three seeds, and three tubercles formed by the cup.

390 YEW. - - B. *Cup* with four leaves. *Bloss,* none. *Tips* with eight clefts.
F. *Cup* four leaves. *Bloss.* none. *Summit* one. *Berry* one seed; not covered at the end with the pulp of the berry.

Order XIV. Tips United.

391 PETTIGREE. - B. *Cup* six leaves. *Bloss.* none. *Chives* five.
F. *Cup* six leaves. *Bloss.* none. *Pointal* one. *Berry* with three cells and two seeds.

† *Mountain Cats-foot.* † *White Bryony.*

378 WILLOW. 1098 Salix.

B *Barren Flowers.*

EMPAL. *Common Catkin.* oblong; tiled on every side; inclosed by a fence formed of a bud which is composed of

Scales, inclosing a single flower; oblong; flat; expanding.

BLOSS. *Petals* none.

Honey-cup a cylindrical gland; very small; lopped; containing honey; placed in the center of the flower.

CHIVES. *Threads* two; straight; thread-shaped; longer than the cup. *Tips* double; with four cells.

F *Fertile Flowers.*

EMPAL. *Catkin* as above.

Scales as above.

BLOSS. None.

POINT. *Seedbud* egg-shaped; tapering into a *Shaft* hardly distinct from the seedbud but rather longer than the scale of the empalement. *Summits* two; cloven; upright.

S. VESS. *Capsule* betwixt egg and awl-shaped; with one cell, and two valves. The *Valves* rolling back.

SEEDS. Numerous; egg-shaped; very small; crowned with a simple hairy *Feather.*

OBS. *In some species there are three or five chives; unequal in length. In the first* species *the chives and pointals are inclosed within the same empalement—There are frequently large excrescences upon the branches, leaves, and leaf-stalks of Willows, which are the habitations of different species of* Cynipes—*Whoever desires to shade a walk with Willows, should set those which bear only chives in the catkins, or else they will soon multiply so as to form a thicket instead of a walk. The same observation holds good of the* Poplar—*The flowers of all the species are delightful to Bees.*

* *Leaves smooth; serrated.*

Shining Hermaphroditica WILLOW. Leaves serrated, smooth. Chives two; in the same empalement with the pointal—*Generally six leaves grow out of each bud.* Catkins *woolly. Branches palish red; veins of the leaves hollow. The whole plant is covered frequently with little red grains, but so small that they are hardly visible to the naked eye.*

Salix latifolia, folio splendente. *Ray's Syn.* 450.

Wet hedge-rows. S. March.

WILLOW.

WILLOW. Leaves ſerrated; ſmooth. Barren flowers with five chives—*Leaves yellowiſh green; the teeth at the edges pour out a yellow gum, ſo that put freſh into a book and compreſſed, there remains as many yellow dots on the paper as there are teeth in the leaf.* Catkins *very yellow. It grows about five or ſix feet high and hath purpliſh or yellowiſh branches. The* Buds *conſiſt of two oppoſite valves. The ſame buds ſend out both catkins and leaves.* Sweet Pentandra

Salix folio Laureo, ſeu lato glabro odorato. *Ray's Syn.* 449.

Bay-leaved Willow.

On hills. S. April.

The wood crackles greatly in the fire—The branches are cut to make ſpringles—The dried leaves give out a yellow colour—Sheep and Goats eat it.

WILLOW. Leaves ſerrated; egg-ſhaped; ſharp: ſmooth; griſtly at the edges; leaf-ſtalks dotted with little callous points —*Perhaps if this was neither cultivated nor cut, it might degenerate into the* White Willow. Yellow Vitellina

Salix ſativa lutea, folio cænato. *Bauh. pin.* 473.

Cultivated in plantations. S. April—May.

The ſhoots are uſed by crate and baſket makers.

WILLOW. Leaves ſerrated; ſmooth; ſpear-ſhaped; on leaf-ſtalks. Props in ſhape of an irregular ſquare—*Branches firſt green; afterwards deep purple; brittle.* Almondleave. Amygdalina

Salix folio auriculato ſplendente flexilis. *Ray's Syn.* 448.

Salix folio Amygdalino utrinque virente aurito. *Bauh. pin.* 473.

Salix viminalis nigra. *Park.* 1431.

Banks of rivers. S. May.

Horſes and Goats eat it.

WILLOW. Leaves ſerrated; ſmooth; betwixt egg and ſpear ſhaped; leaf-ſtalks toothed and glandular—*The branches when ſtruck with the finger break off at the joint of the laſt years ſhoot.* Crack Fragilis

Salix folio longo latoque ſplendente fragilis. *Ray's Syn.* 448.

Salix fragilis. *Bauh. pin.* 474. *Park.* 1431.

Wet marſhy places. S. May.

It will thrive in moſt kinds of ſoil if they are ſufficiently moiſt. It is a quick grower, and bears cropping. The white ſattin moth ſometimes eats all its leaves.

1. Salix folio Amygdalino utrinque aurito corticem abjiciens. *Ray's Syn.* 448. Leaves very much resembling those of the *almond-leaved*, but longer and narrower, and furnished at the base with a kind of appendages, besides the props which grow at the base of the leaf-stalks. These appendages one should imagine might form a specific distinction, but HUDSON notwithstanding arranges it only as a variety of the *Crack* WILLOW, though RAY it is evident considered it as a distinct species, though most nearly allied to the *almond-leaved.* Its casting its bark forms another characteristic.

Purple
Purpurea

WILLOW. Leaves serrated; smooth; spear-shaped. The lower leaves standing opposite—*Branches often as red as coral; extremely tough. Each bud produces three leaves. The flowering buds beneath the ends of the branches.* Empalements *hairy and brown.*

Salix folio longo sub-luteo non auriculata, viminibus rubris. *Ray's Syn.* 450.

Banks of rivers. P. May.

Baskets, cradles, and all sorts of twig-work, are made of the long, slender and flexible shoots of this plant—The inner bark is of a full yellow colour.

Rose
Helix

WILLOW. Leaves serrated; smooth; betwixt strap and spear-shaped; the upper leaves opposite, but oblique—*Branches angular.* Leaves *when full grown bluish green on the under surface.* Bark *yellowish.* Twigs *purple; not cylindrical.* Catkin *compact; downy; with black scales.*

Wet marshy places. S. May.

Herbaceous
Herbacea

WILLOW. Leaves serrated; smooth; round—*This is the smallest of all trees. The branches hardly a fingers length, and seldom more than three leaves upon each branch.* Fruit-stalks *arise from the same buds with the leaves.* Pointals *smooth.* Flowers *yellow, or purplish.*

Salix alpina, alni rotundo folio repens. *Ray's Syn.* 448.

Salix saxatilis minima. *Bauh. pin.* 474.

On high mountains. P. June.

* * *Leaves ſmooth; very entire.*

WILLOW. Leaves very entire; ſmooth; egg-ſhaped; blunt —*Green and wrinkled on the upper, but bluiſh and with a network of veins on the under ſurface. The veins at firſt red, afterwards green. The flowers and the leaves ſpring out of the ſame bud. Somewhat larger than the preceding ſpecies.* Catkins *oblong.* Empalements *blunt; conave; hairy; fleſh-coloured.* Chives *two; longer than the empalement; fleſh-coloured.* Seedbuds *hairy.* Network Reticulata

Salix pumila folio rotundo. *Ray's Syn.* 449.

Round-leaved Willow.

On mountains. P. June.

WILLOW. Leaves entire; ſmooth; betwixt ſtrap and ſpear-ſhaped; ſharp—(Hudſon 364). *Buds reddiſh.* Catkins *red at firſt, but afterwards of a greeniſh yellow colour.* Leaves *like thoſe of the Oſier* WILLOW, *but ſmooth on both ſides.* Red Rubra

In marſhes. S. April—May.

* * * *Leaves very entire, woolly.*

WILLOW. Leaves very entire; woolly on both ſides; inverſely egg-ſhaped; furniſhed with appendages—*This is a ſlender flexible tree, and ſometimes riſes to a conſiderable height.* Round-leaved Aurita

Salix folio rotundo minore. *Ray's Syn.* 450.

1\. Variety with the leaves downy on the under ſurface, and ſeldom riſing to above a foot high. *Dwarf*

Salix caprea pumila folio ſubrotundo, ſubtus incano. *Ray's Syn.* 450.

Woods and hedges. S. May.

The ſhoots are ſlender and tolerably flexible.

WILLOW. Leaves very entire; egg-ſhaped; ſharp; a little woolly above; downy beneath— Sand Arenaria

Salix pumila, foliis utrinque candicantibus, et lanuginoſis. *Ray's Syn.* 447. Tab. 19. fig. 3.

Sandy paſtures. P. June.

Creeping Repens

WILLOW. Leaves very entire; spear-shaped; somewhat hairy on each surface. Stem creeping—*As thick as a man's finger.* Branches *in rod-shaped bundles; drooping. Little branches with a few soft white hairs.* Catkins *lateral. Fertile capsules reddish. But little larger than the* Herbaceous Willow.

Salix pumila angustifolia inferne lanuginosa. *Ray's Syn.* 447.
Salix pumila brevi angustoque folio incano. *Bauh. pin.* 474.
Salix humilis angustifolia repens. *Park.* 1434.
1. With leaves more circular, and greyish on the under surface. *Bauh pin.* 474. *Ray's Syn.* 448.
On turfy bogs. P. May.
The leaves are intensely bitter; but Horses eat them.

Brown Fusca

WILLOW. Leaves very entire; egg-shaped; woolly on the under surface: shining—*This is a small creeping shrub. The upper surface of the leaves and the catkins generally blackish.* Bark *brown.*

Salix pumila angustifolia prona parte cinerea. *Ray's Syn.* 447.
Salix pumila linifolia incana. *Bauh pin.* 474.
Salix pumila angustifolia recta. *Park.* 1434.
Chamœitea sive salix pumila. *Gerard.* 1391.
On moist heaths. P. May—June.

Rosemary Rosmarinifolia

WILLOW. Leaves very entire; betwixt spear and strap-shaped; stiff and straight; sitting; downy on the under surface—*The under surface of the leaves silky and shining.* Props *none.*

Salix pumila Rhamni secundi Clusii folio. *Ray's Syn.* 447.
Salix oblongo incano acuto folio. *Bauh. pin.* 159.
Rosemary-leaved Willow.
On boggy mountains. P. May.

* * * * *Leaves somewhat serrated; woolly.*

Sallow Caprea

WILLOW. Leaves egg-shaped; wrinkled; downy on the under surface; waved at the edge, with little teeth towards the end—*The lower buds send forth leaves; the upper buds catkins without leaves.* Bark *rough; grey.*

Salix caprea rotundifolia. *Gerard.* 1319.
Salix latifolia rotunda. *Bauh pin.* 474. *Park.* 1432. *Ray's Syn.* 419.
1. Leaves with a sharpish point at the end. *Bauh. pin.* 474.
Common Sallow. Broad-leaved Willow.
Woods and hedges. P. April—May.

This requires a dryer soil than any of the other species, for it will thrive upon the driest hills. It may be topped every second or third year—The wood is smooth, soft and flexible. It is converted into charcoal for making gunpowder and drawing pencils—The Laplanders make a sort of leather of the bark, which

which they manufacture into gloves. They give a decoction of the leaves for the heart burn—The flowers are particularly grateful to Bees, and the leaves are eaten by Horses, Cows, Goats and Sheep. The purple Emperor Butterfly, *Papilio Iris.* The high flyer Moth; the scarce Silver-line-Moth, and the Copper Under-wing Moth feed upon it.

WILLOW. Leaves almost entire; betwixt spear and strap-shaped; very long: sharp; silky on the under surface. Branches long and slender—*Some buds produce only leaves; others only catkins.* Osier Viminali

Salix folio longissimo. *Ray's Syn.* 450.

Salix folio longissimo angustissimo utrinque albido. *Bauh. pin.* 474.

Common Osier.

In plantations, and near rivulets. P. April.

The branches are much used for making hoops and crates. It is ofen planted to prevent the banks of rivers from being washed away by the force of the current—Horses, Cows, Sheep and Goats eat it.

WILLOW. Leaves spear-shaped; tapering to a point; serrated; downy on both surfaces; with little glands at the edges towards the base—*Bark smooth and whitish.* White Alba

Salix. *Gerard.* 1389. *Ray's Syn.* 447.

Salix arborea angustifolia, alba vulgaris. *Park.* 1430.

Salix vulgaris alba arborescens. *Bauh. pin.* 473.

Common Willow.

Woods and hedges. P. April.

It loves a moist and open situation; grows quick, and bears lopping—The Reverend Mr. Stone, in the *Philos. Trans. Vol.* 53. p. 195. gives us an account of the great efficacy of the bark of this tree in curing intermitting Fevers. He gathers the bark in summer, when it is full of sap; dries it by a gentle heat, and gives a dram of it powdered every four hours betwixt the fits. In a few obstinate cases he mixed it with one fifth part of Peruvian bark. It is remarkable that intermittents are most prevalent in wet countries; and this tree grows naturally in such situations. Whilst the Peruvian bark remained at its usual moderate price it was hardly worth while to seek for a substitute; but now its price is more than doubled, and the supply from South America hardly equal to the consumption, we may expect to find it dearer and more adulterated every year. The *White* Willow Bark is therefore likely to become an object worth the attention of physicians, and if its success upon a more enlarged scale of practice proves equal to Mr. Stone's experiments, the world will be much indebted to that gentleman for his communication—This bark will tan leather---Horses, Cows, Sheep and Goats eat it.

The

The different species of WILLOW support the following insects.

Willow Butterfly, *Papilio Antiopa.*
Great Tortoise-shell Butterfly, *Papilio Polychloros.*
Eyed Willow Hawk Moth, *Sphinx Ocellata.*
Poplar Hawk Moth, *Sphinx Populi.*
Lappit Moth, *Phalæna Quercifolia.*
Fox coloured Moth, *Phalæna Rubi.*
Small Egger Moth, *Phalæna Lanestris.*
Puss Moth, *Phalæna Vinula.*
Dismal Moth. White Sattin Moth, *Phalæna Salicis.*
Buff-tip Moth, *Phalæna Bucephala.*
Black Arches Moth, *Phalæna Monacha.*
Pebble Moth, *Phalæna Ziczac.*
Goat Moth, *Phalæna Cossus.*
Willow red-under-wing Moth, *Phalæna Pacta.*
Sallow Moth, *Phalæna Citrago.*
Cream bordered Pea-green Moth. Province Rose Moth, *Phalæna Salicella.*
Bushy Prominent Moth. Emperor Moth, *Phalæna Pavonia.*
Furbelow Moth, *Phalæna Libatrix.*
Grey Dagger Moth, *Phalæna Psi.*
Willow Beauty Moth, *Tenthredo Lutea.*
- - - - - - - - *Tenthredo Caprea.*
- - - - - - - - *Tenthredo Rustica.*
- - - - - - - - *Ichneumon Glomeratus.*
Willow Louse, *Aphis Salicis.*
- - - - - - - - *Chrysomela Polita.*
- - - - - - - - *Chrysomela Sericea.*
Willow Weevil, *Curculio Nebulosus.*
White spotted Willow Lady-cow, *Coccinella 14 Guttatus.*
Horned Frog-hopper, *Cicada Cornuta.*

Order

Order III. Three Chives.

379 CROWBERRY. 1100 Empetrum.

B. *Barren Flowers.*

EMPAL. *Cup* with three divisions. *Segments* egg-shaped; permanent.

BLOSS. *Petals* three; oblong egg-shaped; narrowest at the base; larger than the cup: shrivelling.

CHIVES. *Threads* three; hair-like; very long; hanging downwards. *Tips* upright; short; cloven.

F. *Fertile Flowers.*

EMPAL. *Cup* as above.

BLOSS. *Petals* as above.

POINT. *Seedbud* depressed. *Shaft* hardly any. *Summits* nine; reflected and expanding.

S. VESS. *Berry* round and flat; depressed; larger than the cup: with one cell.

SEEDS. Nine; placed in a jointed circle; hunched on one side; angular on the other.

OBS. *Sometimes though very rarely flowers have been found containing both chives and pointals.*

CROWBERRY. Stems trailing—*This is a small drooping shrub; the outer bark brown and deciduous; the inner bark yellow.* Branches *rough with the remains of leaf-stalks. The terminating* bud *produces five little leaves; membranaceous: hairy at the edges; standing on five little leaf-stalks; four of them placed in a whorl.* Leaves *somewhat three cornered, with a white line along the back rib; standing upon leaf-stalks, four together.* Flowers *at the base of the leaves; sitting; solitary; sustained by a floral leaf, cloven into three parts, and resembling an outer cup.* Cup *whitish.* Petals *purple.* Threads *very long; purple.* Tips *black. The plant with flowers bearing only pointals resembles the other but the* stem *is more of a red colour; the* leaves *a fuller green and growing by fives; the* pointal *black; the* berry *black.* Heath Nigrum

Empetrum montanum fructu nigro. *Ray's Syn.* 444.
Erica baccifera procumbens nigra. *Bauh. pin.* 486.
Erica baccifera procumbens. *Gerard.* 1383.
Erica baccifera nigra. *Park.* 1485.
Black-berried Heath. Crow-berries. Crake-berries.
On boggy heaths and mountains. P. April. May.

Greedy

Greedy children ſometimes eat the berries; but when taken in large quantities they occaſion head-ache—Grouſe feed upon them—Boiled with alum they afford a dark purple dye—Goats are not fond of it; Cows, Sheep and Horſes refuſe it.

Order IV. Four Chives.

380 SALLOWTHORN. 1100 Hippophae.

B *Barren Flowers.*

EMPAL. *Cup* one leaf; divided into two parts forming two valves. Baſe of the leaf undivided. *Segments* circular; blunt; concave; upright, but the points approaching; open at the ſides.

BLOSS. None.

CHIVES. *Threads* four; very ſhort. *Tips* oblong; angular; almoſt as long as the cup.

F. *Fertile Flowers.*

EMPAL. *Cup* one leaf; oblong egg-ſhaped; tubular; reſembling a club; cloven at the rim; deciduous.

BLOSS. None.

POINT. *Seedbud* roundiſh; ſmall. *Shaft* ſimple; very ſhort. *Summit* rather thick; oblong; upright; twice as long as the cup.

S.VESS. *Berry* globular; with one cell.

SEED. Single; roundiſh.

Sea Rhamnoides SALLOWTHORN. Leaves ſpear-ſhaped — *The* barren *flowers ſolitary; beneath the leaves, betwixt the branches and the buds. The* fertile *flowers ſolitary and ſitting at the baſe of the lower leaves.*

Rhamnoides fructifera, foliis Salicis, baccis leviter flaveſcentibus. *Ray's Syn.* 445.

Rhamnus. *Gerard.* 1334.

Rhamnus Salicis folio anguſto, fructu flaveſcente. *Bauh. pin.* 477.

Rhamnus primus Dioſcoridis Lobelio, ſive littoralis. *Park.* 1006.

Sea Buckthorn.

On the ſea-coaſt. S. April.

Cows refuſe it; Goats Sheep and Horſes eat it.

381 MISLETOE. 1105 Viſcum.

B. *Barrren flowers.*

EMPAL. *Cup* with four diviſions. *Little leaves* egg ſhaped; equal.

BLOSS. None.

CHIVES. Four. *Threads* none. *Tips* oblong; tapering; one fixed to each leaf of the cup.

F. *Fertile flowers moſtly growing oppoſite the others*

EMPAL. *Cup* four leaves; egg ſhaped; ſmall; ſitting on the ſeedbud; deciduous.

BLOSS. None.

POINT. *Seedbud* beneath; oblong; three edged; indiſtinctly crowned with a border with four clefts. *Shaft* none. *Summit* blunt; a little notched.

S.VESS. *Berry* globular; with one cell; ſmooth.

SEED. Single; inverſely heart-ſhaped; compreſſed; blunt; fleſhy.

MISLETOE. Leaves ſpear-ſhaped; blunt. Stem forked; flowers in ſpikes in the boſom of the leaves—*This is a paraſitical evergreen ſhrub.* Bloſſoms *greeniſh white.* Berries *whitiſh. The root inſinuates its fibres into the woody ſubſtance of the plant on which it grows.* White Album

Viſcum. *Gerard.* 1315. *Ray's Syn.* 464. vulgare. *Park.* 1392.

Viſcum baccis albis. *Bauh. pin.* 423.

White Miſſel.

It is found upon Willows, Oaks, Hazels, Peartrees, and Appletrees, but on Crabtrees principally. S. May.

Birdlime may be made from the berries and from the bark—The Miſletoe Bird, the Fieldfare, and the Thruſh eat the berries, the ſeeds of which paſs through them unchanged, and along with their excrements adhere to the branches of trees, where they vegetate. Some authors obſerving that the roots are always inſerted on the under ſide of the branches, deny this method of propagation; but they do not recollect that the rains will ſoon waſh them into that ſituation, though it is highly probable that they firſt fell upon the upper ſide of the branch--No art hath yet been invented to make theſe plants take root in the earth—Miſletoe was formerly in great repute as a remedy for Epileptic and other complaints; but it is now very much diſregarded; and indeed its ſenſible qualities promiſe but little. Some remains of Druidical ſuperſtitions probably gave birth to its medical fame, and an opinion which many people are fond to entertain, (viz.) that every climate produces remedies to counterballance the diſeaſes of its inhabitants, is one reaſon why in these more enlightened days such ideas still continue to exist. Without entering into the discussion of a question which would give rise to much altercation, and be productive of little real advantage, it may be sufficient to observe, that whilst the inhabitants of this island were contented with the products of their own country, it is probable that the remedies of this country were equal to its diseases: but when foreign productions and foreign manners were introduced; it then became necessary to introduce foreign remedies.

382 GALE. 1107 Myrica.

B. *Barren Flowers.*

EMPAL. *Catkin* oblong egg-shaped; tiled on every side; limber; consisting of *Scales* inclosing a single flower; crescent-shaped; tapering to a blunt point; concave.

Individual Cup none.

BLOSS. None.

CHIVES. *Threads* four, (sometimes but rarely six; thread-shaped; short; upright. *Tips* large; double; with the lobes cloven.

F. *Fertile Flowers.*

EMPAL. As above.

BLOSS. None.

POINT. *Seedbud* somewhat egg-shaped. *Shafts* two; thread-shaped; longer than the cup. *Summits* simple.

S. VESS. *Berry* one cell.

SEED. Single.

OBS. *In the* Sweet GALE *the fruit is a dry berry, or rather a leather like coat, compressed at the end, and formed of three lobes.*

Sweet Gale — GALE. Leaves spear-shaped; somewhat serrated. Stem somewhat woody—*Smooth; of the colour of rusty iron; with white dots. The* Buds *producing flowers are placed at the ends of the branches; those producing leaves, at the sides. Each bud is composed of nine shining leafy scales.* Leaves *on leaf-stalks, serrated towards the ends and beset with resinous dots.*

Gale, frutex odoratus septentrionalium; Elæagnus Cordi. *Ray's Syn.* 443.

Rhus Myrtifolio Belgica. *Bauh. pin.* 414.

Rhus sylvestris, sive Myrtus Brabantica vel Anglica. *Park.* 1451.

Myrtus Brabantica, seu Elæagnus Cordi. *Gerard.* 1414.

Goule. Sweet Willow. Dutch Myrtle.

In marshy barren ground. S. May.

The

The northern nations formerly used this plant instead of hops; but unless it is boiled a long time it is apt to occasion head-ach—The catkins boiled in water throw up a waxy scum, which gathered in sufficient quantity would make candles. From another species of this plant, which is a native of warmer climates, the myrtle candles are prepared—Gathered in the autumn it dyes wool yellow—It is used to tan calf-skins. Horses and Goats eat it; Sheep and Cows refuse it.

Order IV. Five Chives.

383 HOP. 1116 Humulus.

B. *Barren Flowers.*

EMPAL. *Cup* five leaves; oblong; concave; blunt.

BLOSS. None.

CHIVES. *Threads* five; hairlike; very short. *Tips* oblong.

F. *Fertile Flowers.*

EMPAL. *General Fence* with four clefts; sharp.

Partial Fence with four leaves; egg-shaped; inclosing eight florets, each of which is furnished with a

Cup of one leaf; egg-shaped; very large; flat on the outer side; approaching at the base; obliquely expanding; entire.

BLOSS. None.

POINT. *Seedbud* very small, *Shafts* two; awl shaped; reflected and standing wide. *Summits* sharp.

S. VESS. None. The *Cup* closing at the base contains the seed.

SEED. One. Roundish; covered by a coat.

HOP. As there is only one species known, Linnæus gives no description of it.—*Leaves divided into lobes*; *serrated.* Flowers *greenish yellow.* Stems *climbing.* Brewers Lupulus

Lupulus mas et fæmina. *Bauh. pin.* 298. *Ray's Syn.* 137.

Lupulus sylvestris. *Park.* 176.

Lupus salictarius. *Gerard.* 885.

Hops.

In hedges. P. June.

Soil and cultivation occasion some varieties, as the *Garlic Hop*: the *Long White Hop* and the *Oval Hop*; but for the common purposes of brewing they are distinguished as *Kentish* Hops or *Worcestershire* Hops.

 If

If the hop-yards were covered with stones, the plants would be less liable to suffer from the honey-dew or from the Otter-Moth; for the honey-dew is the excrement of a species of louse, (*Aphis*) but these insects seldom increase so as to endanger the plant, unless it is in a weak condition; and the larvæ of the Otter-moth at the roots, first occasion the plant to be sickly. Now when the hop grows wild in stony places and in fissures of rocks, where the moth cannot penetrate to deposit its eggs, the hop is never known to suffer from the honey dew.—The flowers of the fert le plants are very generally infused in wort or boiled along with it to prevent the ale growing sour.—The young shoots are eaten early in the spring as sparagus, and are sold under the name of Hop-tops—Strong cloth is made in Sweden from the stalks. For this purpose they must be gathered in autumn, soaked in water all winter; and in March, after being dried in a stove they are dressed like flax. Horses, Cows, Sheep, Goats and Swine eat it.—The Peacock Butterfly, *Papilio Jo*, the Common Butterfly, *Papilio C. album*, the Otter Moth, *Phalæna Humuli*, and the *Phalæna Rostralis* live upon it.

Order VI Six Chives.

384 LADYSEAL. 1119 Tamus.

B. *Barren flowers.*

EMPAL. *Cup* with six divisions. *Little Leaves* betwixt egg- and spear-shaped; expanding towards the top.

BLOSS. None.

CHIVES. *Threads* six; simple; shorter than the cup. *Tips* notched at the end.

F. *Fertile Flowers.*

EMPAL. *Cup* one leaf, with six divisions; bell shaped; expanding. *Segments* spear-shaped; superior; shrivelling.

BLOSS. *Petals* none.

Honey-cup, an oblong dot at the base of each segment of the cup, on the inner side.

POINT. *Seedbud* beneath; oblong egg-shaped; large; smooth. *Shaft* cylindrical, as long as the cup. *Summits* three; reflected; notched at the end; sharp.

S. VESS. *Berry* egg-shaped; with three cells.

SEEDS. Two; globular.

LADYSEAL. Leaves heart-ſhaped; undivided.—*Stems* *twining*. Bloſſoms *greeniſh*. Berries *red*. Root *very large*. Bryony Communis

Tamnus racemoſa, flore minore luteo palleſcente. *Ray's Syn.* 262.

Bryonia nigra. *Gerard.* 869.

Bryonia ſylveſtris nigra. *Park.* 178

Bryonia lævis ſeu nigra racemoſa. *Baub. pin.* 297.

Black Briony.

Hedges and rough places. P. June.

The young ſhoots are good eating when dreſſed like ſparagus—The root is acrid and ſtimulating—Horſes will not eat this plant.

Order VII. Eight Chives

385 POPLAR. 1123 Populus.

B. *Barren Flowers.*

EMPAL. *Common Catkin* oblong; looſely tiled; cylindrical; conſiſting of *Scales*, incloſing a ſingle flower; oblong; flat; ragged at the edge.

BLOSS. *Petals* none.

Honey-cup one leaf; turban-ſhaped beneath; tubular; ending at the top obliquely, in an egg-ſhaped border.

CHIVES. *Threads* eight; extremely ſhort. *Tips* four-edged; large.

F. *Fertile Flowers.*

EMPAL. *Catkin* and *Scales* as above.

BLOSS. *Petals* none.

Honey-cup as above.

POINT. *Seedbud* egg-ſhaped, but tapering. *Shaft* hardly diſcernible. *Summit* with four clefts.

S. VESS. *Capſule* egg-ſhaped; with two cells and two valves. *Valves* reflected.

SEEDS. Numerous; egg-ſhaped; furniſhed with a hair-like *Feather*.

Abele
Alba

POPLAR. Leaves circular; toothed and angular; downy on the under ſurface—*The flowers are altogether the ſame as thoſe in next ſpecies.* Leaves *entirely deſtitute of glands.*

Populus alba. *Gerard.* 1486. *Park.* 1400. *Ray's Syn.* 446.
Populus alba majoribus foliis. *Bauh. pin.* 429.
Abele Tree.
Hedges and near brooks. · S. March.

White

1. There is one variety with variegated, and another with ſmaller leaves.

It loves low ſituations, and flouriſhes beſt in clay. It grows quick and bears cropping, but it is unfavourable to paſturage—The wood is ſoft, white and ſtringy: it makes good wainſcoting, being but little ſubject to ſwell or ſhrink. Floors, laths, packing boxes and turners ware are made of it.

Horſes, Sheep and Goats eat it; Cows are not fond of it.

Trembling
Tremula

POPLAR. Leaves circular; toothed and angular; ſmooth on both ſides—*On long leaf ſtalks. The* Leaf Stalks *are compreſſed towards the top, from whence the leaves have a tremulous or vibrating motion. There is a double glandular ſubſtance on the inner ſide of the baſe of each leaf*

Populus lybica. *Gerard.* 1487. *Park.* 1411. *Ray's Syn.* 446.
Populus Tremula. *Bauh. pin.* 429.
Aſp. Aſpen Tree.
Moiſt woods. S. March.

It will grow in all ſituations, and in all ſoils, but worſt in clay. It impoveriſhes the land; its leaves deſtroy the graſs, and the numerous ſhoots of the roots ſpread ſo near the ſurface of the earth that they will not permit any thing elſe to grow. It is eaſily tranſplanted—The wood is extremely light, white, ſmooth, woolly, ſoft; durable in the air—The bark is the principal food of Beavers. The bark of the young trees is made into torches. —The leaves and leaf ſtalks ſometimes are ſet with red globular ſubſtances about as large as a Pea, which are the neſts of the *Tipula Juniperina.*—Sheep and Goats eat it; Horſes and Swine refuſe it.

Black
Nigra

POPLAR. Leaves triangularly ſpear ſhaped; broad; tapering; ſerrated—*The* Leaves *have no glands at the baſe; but the teeth at the edges are glandular.* Leaf Stalks *yellowiſh.*

Populus nigra. *Gerard.* 1485. *Bauh. pin.* 429. *Park.* 1410. *Ray's Syn.* 446.

Near rivers and wet ſhady places. S. March.

It loves a moiſt black ſoil; grows rapidly and bears cropping—The wood is not apt to ſplinter—The bark being light like cork, ſerves to ſupport the nets of fiſhermen—The red ſubſtances like berries upon the leaf-ſtalks, as large as a cherry, hunched on one ſide and gaping on the other, are occaſioned by an inſect called the *Aphis Burſaria.*—Horſes, Cows, Sheep and Goats eat it.

The

The ſeveral ſpecies ſupport the following Inſects.

Poplar hawk Moth. *Sphinx Populi.*
Puſs Moth. *Phalæna Vinula.*
December Moth. *Phalæna Populi.*
Black tuſſock Moth. *Phalæna Faſcelina.*
Orange under-wing Moth.
Poplar Louſe. *Aphis Populi.*
- - - - - - *Chryſomela Polita.*
- - - - - - *Chryſomela Populi.*
Poplar Weevil. *Curculio Tortrix.*
Poplar Bug. *Cimex Populi.*
Hornet Moth. *Sphinx Apiformis.*

386 ROSEWORT. 1124 Rhodiola.

B. *Barren Flowers.*

EMPAL. *Cup* with four diviſions; concave; upright; blunt; permanent.

BLOSS. *Petals* four; oblong; blunt; upright but expanding; twice as long as the cup; deciduous.

Honey-cups four; upright; notched at the end; ſhorter than the cup.

CHIVES. *Threads* eight; awl-ſhaped; longer than the bloſſom. *Tips* ſimple.

POINT. *Seedbuds* four; oblong; tapering. *Shafts* and *Summits* imperfect.

S. VESS. Barren.

F. *Fertile Flowers.*

EMPAL. *Cup* as above.

BLOSS. *Petals* four; rude; upright; blunt; equal in height to the cup; permanent.

Honey-cups as above.

POINT. *Seedbuds* four; oblong; tapering; ending in ſtraight ſimple *Shafts*. *Summits* blunt.

S. VESS. *Capſules* four; horned; opening inwards.

SEEDS. Many; roundiſh.

ROSEWORT. As there is only one ſpecies known, Linnæus gives no deſcription of it—*Root white; with the odour of a Roſe.* Stem *ſimple; upright; leafy.* Leaves *ſerrated.* Bloſſoms *terminating; yellow.* Yellow Roſea

Anacampſeros radice Roſam ſpirante major. *Ray's Syn.* 260.
Rhodia radix. *Bauh. pin.* 286. *Gerard.* 532. *Park.* 729.
On mountains. P. June—July.

The root has the fragrance of a Rofe, particularly when dried; but cultivated in a garden it lofes moft of its fweetnefs.—Goats and Sheep eat it; Cows and Swine refufe it.

Order VIII. Nine Chives.

387 MERCURY. 1125 Mercurialis.

B. *Barren Flowers.*

EMPAL. *Cup* with three divifions. *Segments* betwixt egg- and fpear-fhaped; concave; expanding.

BLOSS. None.

CHIVES, *Threads* nine, or twelve; hair-like; ftraight; as long as the cup. *Tips* globular; double.

F. *Fertile Flowers.*

EMPAL. *Cup* as above.

BLOSS. None.

Honey-cups, two awl-fhaped pointed fubftances; one placed on each fide the feedbud, and preffed into its furrows.

POINT. *Seedbud* roundifh; compreffed, with a hollow furrow on each fide; rough with hairs. *Shafts* two; reflected; horned; rough with hair. *Summit* fharp; reflected.

S. VESS. *Capfule* roundifh; purfe-fhaped; double; with two cells.

SEEDS. Solitary; roundifh.

Dog s Perennis MERCURY. Stem undivided; leaves rough—*Oppofite ferrated.*

Mercurialis perennius repens Cynocrambe dicta. *Ray's Syn.* 138. ⊙

Mercurialis montana fpicata, et Mercurialis montana fpicata. *Bauh. pin.* 122.

Mercurialis fylveftris Cynocrambe dicta vulgaris, mas et fæmina. *Park.* 295.

Cynocrambe mas et fæmina. *Gerard.* 333.

Woods and ditchbanks. P. April—May.

This plant dreffed like fpinach is very good eating *early* in the Spring, and is frequently gathered for that purpofe; but it is faid to be hurtful to Sheep; and Mr. Ray relates the cafe of a Man, his Wife and three Children, who experienced highly deleterious effects from eating it fried with bacon; but this was probably

probably when the ſpring was more advanced, and the plant was become acrimonious—Steeped in water it affords a fine deep blue colour.—Sheep and Goats eat it; Cows and Horſes refuſe it.

MERCURY. Stem branched; leaves ſmooth; flowers in ſpikes.— French Annua

Mercurialis annua glabra vulgaris. *Ray's Syn.* 139.
Mercurialis teſticulata ſeu Mas. *Bauh. pin* 121.
Mercurialis ſpicata ſeu fæmina. *Bauh. pin.* 121.
Mercurialis mas et fæmina. *Gerard.* 333.
Mercurialis vulgaris mas et fæmina. *Park.* 295.
Herb Mercury
Amongſt rubbiſh. A. September.

The whole plant is mucilaginous, and was formerly much employed as an emollient, but is now diſregarded.—

The ſmall old Gentlewoman Moth, and the Angleſhade Moth, *Phalæna Meticuloſa*, feed upon it.

388 FROGBIT. 1126 Hydrocharis.

B. *Barren Flowers.*

EMPAL. *Sheath* of two leaves; oblong; incloſing three flowers.
Cup proper, of three *Leaves*; oblong egg-ſhaped; concave; membranaceous at the edge.

BLOSS. *Petals* three; circular; flat; large.

CHIVES, *Threads* nine; awl-ſhaped; upright; diſpoſed in three rows; the middlemoſt row in the center ſends out an awl-ſhaped little pillar, reſembling a ſhaft, from the inner ſide of the baſe. The other two rows are connected at the baſe, ſo that the outer and inner thread adhere together. *Tips* ſimple.

POINT. *Seedbud* only a rudiment; placed in the center of the flower.

F. *Fertile Flowers.*

EMPAL. *Sheath* none. Flowers ſolitary.
Cup as above: ſuperior.

BLOSS. As above.

POINT. *Seedbud* beneath: roundiſh. *Shafts* ſix; as long as the cup; compreſſed; cloven and furrowed. *Summits* cloven; tapering.

S. VESS. *Capſule* like leather; roundiſh; with ſix cells.

SEEDS. Numerous; very ſmall; roundiſh.

Water Morfus ranæ FROGBIT. As there is only one fpecies known Linnæus gives no defcription of it.—*Leaves kidney-fhaped*; *thick*; *fmooth*; *brown green.* Bloffoms *white.*

Nymphæa alba minima. *Paub. pin.* 193, five Morfus Ranæ. *Park.* 1252.

Morfus Ranæ. *Gerard.* 818.

Stratiotes foliis Aloes, femine rotundo. *Ray's Syn.* 290.

1. Variety with double flowers, of a very fweet fmell.—Obferved by Mr. Ray in a ditch on the fide of Audrey Caufey in the Ifle of Ely.

Nymphæa alba minima, fore pleno odoratiffimo. *Ray's Syn.* 290.

Slow ftreams and wet ditches. P. June.

Order XIII. Threads United.

389 JUNIPER. 1134 Juniperus.

B. *Barren Flowers.*

EMPAL. *Catkin* conical; confifting of a common fpike-ftalk, in which three oppofite flowers are placed in a triple row, and a tenth flower at the end. At the bafe of each flower is a

Scale; broad; fhort; fixed fide-ways to a little pillar like a footftalk.

BLOSS. None.

CHIVES. *Threads*, (in the *terminating* flower) three; awl-fhaped; united at the bottom into one body. The threads in the *lateral* flowers are hardly perceptible. *Tips* three: diftinct in the *terminating* flower; but in the *lateral* flowers fixed to the fcale of the empalement.

F. *Fertile Flowers.*

EMPAL. *Cup* with three divifions; very fmall; growing to the feedbud; permanent.

BLOSS. *Petals* three; ftiff; fharp; permanent.

POINT. *Seedbud* beneath. *Shafts* three; fimple. *Summits* fimple.

S. VESS. *Berry* flefhy; roundifh; marked in the lower part with three oppofite tubercles which were formerly the cup; and marked at the top by three little teeth which were originally the petals.

SEEDS. Three; hard as bone; convex on one fide; angular on the other; oblong.

JUNIPER.

JUNIPER. Leaves growing by threes; expanding; sharp-pointed; longer than the berry.—*Bark reddish.* Berries *blue black.* Common Communis

Juniperus vulgaris, Baccis parvis purpureis. *Ray's Syn.* 444.
Juniperus. *Gerard.* 1372. vulgaris. *Park.* 1028.
Juniperus vulgaris fruticosa. *Bauh. pin.* 488.
Juniperus vulgaris arbor. *Bauh. pin.* 488.
On Heaths and in woods. S. May.

1. With broader leaves and egg-shaped berries.—
Juniperus alpina. *Park.* 1028. *Ray's Syn.* 444.
Juniperus alpina minor. *Gerard.* 1372.
Juniperus minor montana, folio latiore, fructuque longiore. *Bauh. pin.* 489.

Heaths and woods. 1. Mountains.

It grows in fertile or in barren soils; on hills or in valleys; in open sandy plains, or in moist and close woods. On the sides of hills its trunk grows long; but on the tops of rocky mountains and on bogs it is little better than a shrub. It is easily transplanted and it bears cropping. Grass will not grow beneath it, but the *Meadow* Oat destroys it—The wood is hard and durable. The bark may be made into ropes—The berries are two years in ripening. When bruised they afford a pleasant diuretic liquor, but it is not easy to prevent its growing sour. It is esteemed a good antiscorbutic. The Swedes prepare an extract from the berries which some people eat for breakfast, but it is fitter for a medicine than for food. The spirit impregnated with the essential oil of these berries is every where known by the name of Gin, or Juniper water—The berries sometimes appear in an uncommon form, the leaves of the cup grow double the usual size; approaching, but not closing; and the three petals fit exactly close, so as to keep the air from the *Tipulæ Juniperi* which inhabit them—Gum Sandarach, more commonly called Pounce, is the product of this tree.—Horses, Sheep and Goats eat it.

The Juniper Bug, *Cimex Juniperinus*, the *Thrips Juniperina* and the *Coccinella 9 Punctata* feed upon it.

390 Y E W. 1135 Taxus.

B. *Barren Flowers.*

EMPAL. None; except the *Bud* which resembles a cup with three or four leaves.

BLOSS. None.

CHIVES. *Threads* numerous, united below into a column longer than the bud. *Tips* depressed; blunt at the edge; with eight clefts; opening each way at the base; and having parted with the dust, flat, target-shaped, and the clefts in the edge become more remarkable.

F. *Fertile Flowers.*

EMPAL. As above.

BLOSS. None.

POINT. *Seedbud* egg shaped; but tapering. *Shaft* none. *Summit* blunt.

S. VESS. *Berry* an expansion of the receptacle; succulent and globular; open at the end; coloured. In course of time it grows dry; decays, and disappears.

SEED. Single; egg-shaped but oblong; standing out of the open end of the berry

OBS. *This species of berry is very singular, and strictly speaking can hardly be called a seed-vessel.*

Common Baccata — YEW. Leaves growing near together—*The berries come to perfection the second year.* Bark *reddish.* Berries *when ripe, red.* Taxus. *Gerard.* 1370. *Park.* 1412. *Bauh pin.* 505. *Ray's Syn.* 445.

Yew Tree.

On mountains. S. March—April.

Varieties.

1. Leaves broad and shining.
2. Leaves variegated

It grows best in a moist loamy soil. On bogs or dry mountains it languishes. It bears transplanting even when old. It is often planted to make hedges, and as these hedges admit of clipping, they form excellent skreens to keep off the cold winds from tender plants—The wood is hard, smooth, and beautifully veined with red. It is converted into bows, axle-trees, spoons, cups, cogs for mill wheels and flood-gates for fish-ponds which hardly ever decay—The berries are sweet, and viscid Children often eat them in large quantities without any inconvenience—The fresh leaves are fatal to the human species. Cattle are sometimes killed by browsing upon the branches that are cut off and are half withered, but when growing it is doubtful whether

whether ſuch an effect would follow. Indeed from the Upſal experiments it appears that neither Cows nor Horſes will eat it in a recent ſtate. Sheep and Goats eat it, but the former are ſaid to have been killed by browſing upon the bark.

Order XIV. Tips United.

391 PETTIGREE. 1139 Ruſcus.

B. *Barren Flowers.*

EMPAL. *Cup* with ſix leaves; upright, but expanding. *Leaves* egg ſhaped, convex; the edges at the ſide reflected.

BLOSS. None. Unleſs you conſider every other leaf of the cup as ſuch.
Honey-cup egg-ſhaped; as large as the cup; blown up; open at the rim; upright.

CHIVES. *Threads* none. *Tips* three; expanding; placed upon the end of the honey-cup; united at the baſe.

F. *Fertile Flowers.*

EMPAL. *Cup* as above.

BLOSS. *Petals* as above.
Honey-cup as above.

POINT. *Seedbud* oblong egg-ſhaped; hidden within the honey-cup. *Shaft* cylindrical; as long as the honey-cup. *Summit* blunt; projecting through the mouth of the honey-cup.

S. VESS. *Berry* globular; with three cells.

SEEDS. Two; globular.

OBS. *In this and other genera nearly related to it, it is ſeldom that all the ſeeds come to perfection; for the moſt part one ſeed takes to enlarge and by preſſure deſtroys the others.*

PETTIGREE. Leaves bearing the flowers on the upper ſurface; naked.—*Stem tough*; *woody*, *branched*; *ſcored*. Leaves *betwixt egg and ſpear-ſhaped*; *pointed*. Bloſſoms *yellowiſh green*. Berries *red*. Prickly Aculeatus

Ruſcus. *Bauh. pin.* 470. *Park.* 253. *Ray's Syn.* 262.
Ruſcus ſive Bruſcus. *Gerard.* 907.
Knee Holly. Butchers Broom.
Woods and roughs. S. March—April.

In Italy it is made into beſoms, and the butchers uſe them to ſweep their blocks. Huckſters place the boughs round their bacon and cheeſe to defend them from the Mice, for they cannot make their way through the prickly leaves.

CLASS

CLASS XXIII.

THIS Claſs conſiſts of plants in which ſome of the flowers in the ſame ſpecies have *both* CHIVES and POINTALS, and others have *only* CHIVES, or *only* POINTALS.

H. Signifies the flowers that have both *Chives* and *Pointals*.

B. Signifies the flowers that have only *Chives*.

F. Signifies the flowers that have only *Pointals*.

It is remarkable that in trees with broad leaves. where the Chives and Pointals are in different Empalements, or on diſtinct trees, is the HAZLE, the POPLAR and the ASH, the flowers come out before the leaves are fully expanded, that the paſſage of the duſt from the Chives to the Pointals might not be interrupted : but where the leaves are narrow, as in the FIR and the YEW Tree, no ſuch proviſion takes place.

CLASS. XXIII.

VARIOUS DISPOSITIONS.

Order I. Upon one Plant.

392 Softgrass. H. *Husk* of two valves, containing one flower. *Chives* three. *Shafts* vo. *Seed* one.
B. *Husk* of two valves. containing one flower. *Chives* three.

393 Hardgrass. H. *Husk* with three awns, containing three flowers. *Chives* three. *Shafts* two. *Seed* one.
B. *Husk* with three awns, containing three flowers. *Chives* three.

394 Crosswort. H. *Empal.* none. *Bloss.* with four divisions. *Chives* four. *Shaft* cloven. *Seed* one.
B. *Empal.* none. *Bloss.* with three or four divisions. *Chives* three or four.

395 Maple. - - H. *Cup* with five clefts. *Bloss.* five petals. *Chives* eight. *Shafts* two. *Caps.* a double berry; winged.
B. *Cup* with five clefts. *Bloss.* five petals. *Chives* eight.

396 Pel-

396 PELLITORY. H. *Cup* with four clefts. *Bloss.* none. *Chives* four. *Shaft* one. *Seed* one.
F. *Cup* with four clefts. *Bloss.* none. *Shaft* one. *Seed* one.

397 ORACHE. H. *Cup* with five leaves. *Bloss.* none. *Chives* five. *Shaft* cloven. Seed one.
F. *Cup* with two leaves. *Bloss.* none. *Shaft* cloven; seed one.

† *Portland Spurge.* † *Red Spurge.* † *Duckmeat.*

Order II. Upon Two Plants.

398 ASH. - - - H. *Empal.* none; or with four divisions. *Bloss.* none; or with four petals. *Chives* two. *Point.* one. *Seed* one.
F. *Empal.* none; or with four divisions. *Bloss.* none; or with four petals. *Point.* one. *Seed* one.

† *Tree Holly.*

392 SOFTGRASS. 1146 Holcus.

H. *Flowers with Chives and Pointals.*

EMPAL. *Husk* inclosing one or two florets; with two valves; stiff; without awns.

Outer Valve egg-shaped; concave; large: inclosing the *Inner Valve*, which is oblong; with its sides rolled inwards.

BLOSS. *Husk* of two valves; tender; beset with soft hair; smaller than the empalement. *Outer Valve* generally furnished with a stiff awn, which is longer than the empalement. *Inner Valve* without an awn; very small.

CHIVES. *Threads* three; hairlike. *Tips* oblong.

POINT *Seedbud* turban-shaped. *Shafts* two; hair-like. *Summits* pencil-shaped.

S.VESS. None. The blossom incloses, protects and adheres to the seed.

SEED. Solitary; egg-shaped; inclosed.

B. *Barren Flowers smaller than the other.*

EMPAL. *Husk* with two valves. *Valves* betwixt egg and spear-shaped; rolled inwards; without awns; sharp.

BLOSS. None.

CHIVES. *Threads* three; hair-like. *Tips* oblong.

OBS. *See the Plate of* GRASSES.

SOFTGRASS. Husks inclosing two florets; almost naked. Florets with chives and pointals, without awns. Florets with only chives with jointed awns.—*Empalements ending in a sharp point.* Creeping Mollis

Gramen miliaceum aristatum molle. *Ray's Syn.* 404

Gramen caninum longius radicatum, majus et minus. *Bauh. pin.* 1.

Pastures and ditch-banks. P. July.

SOFTGRASS. Husks inclosing two florets; woolly. Florets with chives and pointals, without awns. Florets with only chives with crooked awns—*The straws are not collected into a bundle but spread about.* Meadow Lanatus

Gramen pratense paniculatum molle. *Bauh. pin* 2. *Park.* 1155.

Gramen miliaceum paniculatum molle. *Ray's Syn.* 404.

Pastures. P. June.

393 HARDGRASS. 1150 Ægilops.

H. *Florets with Chives and Pointals two; lateral.*

EMPAL. *Huſk* very large, with two valves; incloſing three florets. *Valves* egg ſhaped; lopped; ſcored; with awns variouſly diſpoſed; griſtly.

BLOSS. *Huſk* with two valves. *Outer Valve* egg ſhaped; terminated by a double or triple awn. *Inner Valve* ſpear-ſhaped; upright without an awn; the edges bent inwards length-ways.

CHIVES *Threads* three; hair-like. *Tips* oblong.

POINT. *Seedbud* turban-ſhaped. *Shafts* two; reflected; *Summits* hairy

S. VESS. None. The *inner Valve* of the bloſſom adheres to the ſeed and doth not open.

SEED. Oblong.

B. *Placed betwixt the other two.*

EMPAL. The bloſſom incloſed within the huſk deſcribed above.

BLOSS. As above.

CHIVES. As above.

POINT. *Seedbud* as above, but generally barren.

OBS. *See the plate of* GRASSES.

Sea Incurvata HARDGRASS. Spike awl-ſhaped; without awns; ſmooth; crooked; one floret in each cup—*Spike but little thicker than the Straw; cylindrical.* Cups *with two valves placed outwardly.*

Gramen parvum marinum Spica loliacea. *Gerard.* 78. *Ray's Syn.* 395.

Gramen loliaceum minus, Spica ſimplici. *Bauh. pin.* 9.

Phænix acerofa aculeata. *Park.* 1145.

On the ſea-coaſt. A. July—Auguſt.

394 CROSSWORT. 1151 Valantia.

H. *Flowers with chives and pointals solitary.*

EMPAL. Hardly any; the seedbud occupying its place.
BLOSS. *Petal* one; flat; with four divisions. *Segments* egg-shaped; sharp.
CHIVES. *Threads* four; as long as the blossom. *Tips* small.
POINT. *Seedbud* beneath; large. *Shaft* as long as the chives; cloven half way down. *Summit* a knob.
S. VESS. Like leather; compressed; reflected.
SEED. Single; globular.

B. *Barren flowers solitary*; *one placed on each side the fertile one.*

EMPAL. As above.
BLOSS. As above; with three or four divisions.
CHIVES. As above.
POINT. *Seedbud* beneath: small. *Shaft* and *Summits* imperfect and hardly discernible.
S. VESS. Barren; but there is a slender oblong rudiment which adheres to the side of the fertile flower.

CROSSWORT. Barren flowers with four clefts. Fruit-stalks with two leaves—*When the seeds ripen the leaves bend inwards and cover them so effectually that birds cannot get at them so long as the plant is entire.* Blossoms *yellow*. Yellow Cruciata

Cruciata. *Gerard.* 1123. *Ray's Syn.* 223.
Cruciata vulgaris. *Park.* 566.
Cruciata hirsuta. *Bauh. pin.* 335.
Crosswort. Mugweed.
Roughs and ditch banks. P. May—June.

395 MAPLE. 1155 Acer.

H. *With chives and pointals.*

EMPAL. *Cup* one leaf; with five clefts; ſharp; coloured; flat and entire at the baſe; permanent.

BLOSS. *Petals* five; egg ſhaped; broadeſt towards the end; blunt; but little larger than the cup; expanding.

CHIVES. *Threads* eight; awl-ſhaped; ſhort. *Tips* ſimple. *Duſt* croſs-ſhaped.

POINT. *Seedbud* compreſſed; nearly buried in a large perforated convex receptacle. *Shaft* thread-ſhaped; daily growing longer. *Summits* two; tapering; ſlender; reflected.

S. VESS. *Capſules* two; united at the baſe; roundiſh; compreſſed; each terminated by a very large membranaceous wing.

SEEDS. Solitary; roundiſh.

B. *Barren Flowers.*

EMPAL. As above.

BLOSS. As above.

CHIVES. As above.

POINT. *Seedbud* none. *Shaft* none. *Summit* cloven.

OBS. *At the firſt opening of the flower the* SUMMIT *only makes its appearance and after ſome days the* SHAFT *ſhoots out.*

In the Sycamore MAPLE *the bloſſom is hardly ſeparated from the cup, and the chives are long.*

In ſome flowers in the ſame rundle the lower ones have tips which do not ſhed their duſt; but the pointal brings forth perfect fruit; and the upper ones have tips which ſhed their duſt, but the pointals fall off and periſh.

Sycamore Pſeudoplatanus

MAPLE. Leaves with five lobes; unequally ſerrated. Flowers in bunches—*yellowiſh green.*

Acer montanum candidum. *Bauh. pin.* 430.

Acer majus. *Gerard.* 1484. *Ray's Syn.* 470.

Acer majus latifolium, Sycomorus falſo dictum. *Park.* 1425.

Greater Maple. Sycamore Tree.

Hedges. S. May—June.

1. There is a variety with ſtriped leaves.

It flouriſhes beſt in open places, and ſandy ground; but will thrive very well in richer ſoil. It grows quick; is eaſily tranſplanted; bears cropping, and the graſs flouriſhes under its ſhade. It is ſaid to grow better near the ſea than in any other ſituation, and that a plantation of theſe trees at fifty feet aſunder, with three ſea SALLOWTHORNS betwixt every two of them, will make a fence ſufficient to defend the herbage of the country from the ſpray of the ſea. *Gent. Mag.* 1757. *p.* 252.

The wood is ſoft, and very white. The turners form it into bowls, trenchers, &c.

If a hole is bored into the body of the tree when the ſap riſes in the ſpring, it diſcharges a conſiderable quantity of a ſweetiſh watery liquor which is uſed in making wine, and if inſpiſſated it affords a fine white ſugar—The duſt of the *Tips* appears globular in the miſcroſcope, but if touched with any thing moiſt, theſe globules burſt open with four valves which then appear in form of a croſs-- The Cockchafer Beetle, *Scarabœus Melolontha*, feeds upon the leaves.

MAPLE. Leaves lobed; blunt; notched—*Bark rough; furrowed.* Bloſſoms *terminating; pale green.* Common Campeſtre

Acer minus. *Gerard.* 1484. *Ray's Syn.* 470.

Acer minus et vulgare. *Park.* 1426.

Acer campeſtre et minus. *Bauh. pin.* 431.

1. There is a variety with red fruit. *Ray's Syn.* 470.

Hedges and roughs. S. April—May.

The wood is much uſed for turning in the lathe, and veſſels may be turned ſo thin as to tranſmit light.—Horſes will eat the leaves—The Sycamore Tuſſock moth, *Phalæna Aceris*, and the Maple Louſe, *Aphis Aceris*, are nouriſhed by both ſpecies.

396 PELLITORY. 1152 Parietaria.

H. *Two flowers containing both chives and pointals are inclosed within one flat fence of six leaves: the two opposite and outer leaves the largest.*

EMPAL. *Cup* one leaf, with four clefts; flat; blunt; half the size of the fence.

BLOSS. None; without you consider the cup as such.

CHIVES. *Threads* four; awl-shaped; longer than the cup; bursting it open; permanent. *Tips* double.

POINT. *Seedbud* egg-shaped. *Shaft* thread-shaped; coloured. *Summit* pencil shaped; with a knob.

S. VESS. None. The *Cup* becoming longer, larger, and bell-shaped, and its segments approaching, closes upon the seed.

SEED. Single; egg-shaped.

F. *Flowers with only pointals. One placed betwixt the other two within the same fence.*

EMPAL. As above.

BLOSS. None.

POINT. As above.

S. VESS. None. *Cup* slender; inclosing the fruit.

SEED. As above.

Wall
Officinalis

PELLITORY. Leaves betwixt spear and egg-shaped. Fruit-stalks forked. Cups with two leaves—*Blossoms greenish white. Flowers with only pointals, pyramidal aud four edged.* Stems *reddish. If you touch the* Tips *when ripe, with the point of a needle, they burst and throw out their dust with considerable force.*

Pariataria. *Gerard.* 331. *Ray's Syn.* 158.

Pariataria vulgaris. *Park.* 437.

Pariataria officinarum et Dioscoridis. *Bauh. pin.* 121.

Pellitory of the Wall.

On old walls and amongst rubbish. P. May—September.

This plant was formerly in repute as a Medicine, but it does not seem to possess any remarkable qualities—It is asserted that the leaves strewed in granaries destroy the Corn Weevil.

397 ORACHE. 1153 Atriplex.

H. *Flowers with chives and pointals.*

EMPAL. *Cup* five leaves; concave; permanent. *Segments* egg-ſhaped; concave; membranaceous at the edges.
BLOSS. None.
CHIVES. *Threads* five; awl ſhaped; placed oppoſite to the leaves of the cup and longer than them. *Tips* roundiſh; double.
POINT. *Seedbud* round. *Shaft* deeply divided; ſhort. *Summits* reflected.
S.VESS. None. The *Cup* cloſing, hath five ſides and five angles; the angles compreſſed. Deciduous.
SEED. Single; roundiſh; flatted and depreſſed.

F. *Flowers with only pointals, on the ſame plant.*

EMPAL. *Cup* two leaves. *Leaves* flat; upright; egg-ſhaped; ſharp; large; compreſſed.
BLOSS. None.
POINT. *Seedbud* compreſſed. *Shaft* deeply divided. *Summits* reflected; ſharp.
S. VESS. None. The valves of the cup which are large and heart-ſhaped incloſe the ſeed betwixt them.
SEED. Single; roundiſh; compreſſed.

OBS. *There is a very great affinity betwixt the* ORACHE *and the* BLITE, *the preſence of the* Flowers with only pointals *in the* ORACHE *is the only mark of diſtinction; for if the* BLITE *had theſe flowers it would be* ORACHE *and the* ORACHE *without them would be* BLITE.

ORACHE. Stem ſomewhat woody. Leaves inverſely egg-ſhaped—*Whitiſh. Spikes greeniſh purple.* Sea Portulacoides
Atriplex maritima fruticoſa, Halimus et Portulaca marina dicta, anguſtifolia. *Ray's Syn.* 153.
Halimus, ſeu Portulaca marina. *Bauh. pin.* 120.
Halimus vulgaris, ſeu Portulaca marina. *Gerard.* 323.
Portulaca marina noſtras. *Park.* 724.
Sea Purſlane.
On the ſea-ſhore. S. Auguſt.
Cows, Sheep and Goats eat it.

Jagged Laciniata

ORACHE. Stem herbaceous; leaves triangularly spear-shaped; broad and toothed; of a silver white beneath—*The whole plant is covered with a skin which peels off.* Stem *upright; cylindrical; naked; branched like a rod.* Spikes *with chives and pointals terminating;* Tips *of a pleasant red. Flowers with only pointals in pairs. When the fruit ripens the* Cup *is compressed, and hath five teeth; the middle tooth the largest.*

Atriplex maritima. *Ray's Syn.* 152.
Atriplex maritima laciniata. *Bauh. pin.* 120.
Atriplex marina. *Gerard.* 326. repens. *Park.* 758.
Jagged sea Orache. Shrubby sea Orache.
On the sea-shore. A. August.
Cows eat it.

Wild Hastata

ORACHE. Stem herbaceous. In the flowers with only pointals the valves of the cup are large; triangularly spear-shaped and indented—*Spikes whitish.*

Atriplex erecta. *Hudson.* 376.
Atriplex sylvestris, folio hastato, seu deltoide. *Ray's Syn.* 151.
Amongst rubbish; on dung-hills and on the sea-coast. A. August—September.

Varieties.

1. Perenial: with leaves not so hoary—Grows on the sea-shore. Atriplex maritima perennis folio deltoides seu triangulo minus incano. *Hist. Oxon.* II. 607. 19.
2. With hoary angular leaves, very much indented—On the sea shore. Atriplex maritima nostras procerior. foliis angulosis incanis admodum sinuatis. *Plukenet Alm.* 60.
3. With a kind of appendages at the base of the leaves; stem trailing, and the leaves but slightly indented—On the sea-shore. Atriplex maritima ad foliorum basin auriculata, procumbens et ne vix sinuata. *Plukenet Alm.* 61.

It is sometimes gathered as a potherb—Cows, Goats, Sheep and Swine eat it.

Narrowleaved Patula

ORACHE. Stem herbaceous, spreading. Leaves triangularly spear-shaped. The cups containing the seeds toothed in the center---*Spikes whitish.*

Atriplex sylvestris angustifolio. *Gerard.* 326. *Park.* 748. *Ray's Syn.* 151.
Atriplex angusto oblongo folio. *Bauh. pin.* 119.
Ditch-banks and amongst rubbish. A. August.

Spear-leaved Erecta

ORACHE. Stem herbaceous; leaves halberd-shaped and toothed. *Hudson.* 376.

Atriplex angustifolio laciniata. *Ray's Syn.* 152.
Amongst rubbish. A. August.

ORACHE.

ORACHE. Stem herbaceous; upright; leaves all ſtrap-ſhaped; very entire.— Graſs-leaved Littoralis

Atriplex anguſtiſſimo et longiſſimo folio. *Ray's Syn.* 153.

Ray mentions the two following varieties of this plant, the ſecond of which has a blunter leaf; but it does not appear wherein the others differs.

1. Atriplex maritima, ſcoparia folio. *Ray's Syn.* 153.
2. Atriplex maritima anguſtifolia obtuſiore folio. *Ray's Syn.* 153.

On dunghills; ſea-coaſts, and amongſt rubbiſh. A. Auguſt. September.

ORACHE. Stem herbaceous, upright; leaves ſtrap-ſhaped ſerrated. *Hudſon.* 377. Indented Serrata

Atriplex anguſtifolia maritima dentata. *Ray's Syn.* 152.

Atriplex maritima anguſtifolia. *Bauh. pin.* 152.

Amongſt rubbiſh and on the ſea-ſhore. A. Auguſt.

ORACHE. Stem herbaceous; ſtraddling. Leaves ſpear-ſhaped; blunt; entire. Flowers with only pointals on fruit-ſtalks—*Flowers in bunches; terminating; and on lateral fruit-ſtalks, at the end of which is placed a cup reſembling the fruit of the* Purſe Mithridate, *compoſed of three lobes; the middle lobe the ſmalleſt.* Stalked Pedunculata

Atriplex marina ſemine lato. *Ray's Syn.* 153.

Stalked ſea Orache.

On the ſea-coaſt. A. Auguſt—September.

1. There is ſome variety in the form of the leaves, as in the Atriplex maritima noſtras, Ocymi minoris folio. *Ray's Syn.* 153.

The Wild Arrach Moth. *Phalæna Atriplicis.*

The July Arrach Moth. The Spotted Buff Moth. *Phalæna Lubricipeda.*

The Sword-graſs Moth. *Phalæna Exſoleta.*

The Ealings Glory. *Phalæna Oxyacanthæ*, and the Orache Louſe. *Aphis Atriplicis* are found upon the different ſpecies.

Order

Order II. Upon Two Plants.

398 A S H. 1160 Fraxinus.

H. *With chives and pointals.*

EMPAL. None : or a *Cup* of one leaf; with four divifions; fharp ; upright, fmall.

BLOSS. None : or *Petals* four, ftrap-fhaped ; long ; fharp; upright.

CHIVES. *Threads* two ; upright ; much fhorter than the bloffom. *Tips* upright ; oblong ; with four furrows.

POINT. *Seedbud* egg fhaped ; compreffed. *Shaft* cylindrical ; upright. *Summit* rather thick ; cloven.

S. VESS. Spear-fhaped ; compreffed ; membranaceous ; with one cell.

SEED. Single ; flat ; fpear-fhaped.

F. *Flowers with only pointals.*

EMPAL. As above.

BLOSS. As above.

POINT. As above.

S. VESS. As above.

SEED. As above.

OBS. *It often happens that the tree bearing flowers with both chives and pointals hath fome with only pointals intermixed ; and the reverfe.*

Common Excelfior

ASH. Little leaves ferrated. Flowers without petals—*The lateral buds fend forth flowers ; the terminating buds, leaves. The flowers with only pointals have neither petals nor empalement.*

Fraxinus. *Gerard.* 1472. *Ray's Syn.* 469.

Fraxinus excelfior. *Bauh. pin* 416.

Woods and hedge-rows. S. May—June.

1. The leaves are fometimes variegated with white or ftraw-colour.

It flourifhes beft in groves, but it grows very well in rich foil in open fields. It bears tranfplanting and lopping—In the north of Lancafhire they lop the tops of this tree to feed the cattle in autumn when the grafs is upon the decline ; the cattle peeling off the bark as food. In Queen Elizabeths time the inhabitants of *Colton* and *Hawkfheadfells* remonftrated againft the number of forges in the country becaufe they confumed all the loppings and croppings, the fole winter food for their cattle. *Pennants*

Pennants Tour 1772. p. 29. The wood hath the ſingular advantage of being nearly as good when young as when old. It is hard and tough, and is much uſed to make the tools employed in huſbandry. The aſhes of the wood afford very good potaſh—The bark is uſed for tanning calf-ſkin. A ſlight infuſion of it appears of a pale yellowiſh colour when viewed betwixt and the light, but when looked down upon or placed betwixt the eye and an opake object it is blue. This blueneſs is deſtroyed by the addition of an acid, and alkalies recover it again—The ſeeds are acrid and bitter—In the church-yard of Lochaber in Scotland, Dr. Walker meaſured the trunk of a dead Aſh tree which at 5 feet from the ſurface of the ground was 58 feet in circumference—Horſes, Cows, Sheep and Goats eat it; but it ſpoils the milk of Cows, ſo that it ſhould not be planted in dairy farms.

The following inſects are found upon it.

The Leopard Wood Moth. The Green Silver-lined Moth. *Phalæna Praſinana.*
Privet Hawk Moth. *Sphinx Liguſtri.*
Scarlet Tyger Moth. *Phalæna Dominula.*
Cliefden Nonpareil. *Phalæna Fraxini.*
- - - - - - - *Chermes Fraxini.*

CLASS XXIV.

UNDER this Claſs are arranged a number of Vegetables whoſe Flowers are either but little known, or whoſe Chives and Pointals are too minute to admit of that mode of inveſtigation which prevails through the twenty-three Claſſes preceding. The ſtructure too of theſe vegetables differs conſiderably from that of other plants. They are divided into four Natural Orders, (viz.) FERNS; MOSSES; THONGS; and FUNGUSSES.

FERNS.

The plants of this order ſometimes have their flowers in ſpikes, as in the HORSETAIL (Plate I. A.) but they are generally diſpoſed in ſpots or lines on the under ſurface of the leaves, as in the Harts-tongue, Plate I. B.

EMPAL. A ſcale ſpringing out of the leaf; opening on one ſide. Underneath this ſcale, ſupported upon little foot-ſtalks, are

GLOBES, encompaſſed by an elaſtic ring, which burſts with violence, and ſcatters a powder.

OBSERVATIONS.

The ſeed veſſels on the under ſurface of the leaves are covered by a very fine, thin, ſemi-tranſparent ſkin, which tears open before the ſeeds ripen. The ſeed-veſſels themſelves are compoſed of three parts. Firſt, a little fruit-ſtalk, by means of which they are connected with the ſurface of the leaf. Second, a globular Capſule ſtanding upon the fruit-ſtalk. Third, an elaſtic cord, fixed to the top of the fruit-ſtalk and ſurrounding the Capſule. When the Seeds are ripe, the cord endeavours

to

to become ſtraight, and by its elaſticity tears open the Capſule. The Capſule opens like that of the PIMPERNEL, as if it had been cut round with a knife, forming two hollow hemiſpherical caps. The elaſtic force which tears it open, diſperſes the ſeeds abroad. Theſe are ſo minnte as hardly to be viſible to the naked eye. In the months of September and October this curious mechaniſm is very evident in the Common BRAKES or in the Harts-tongue SPLEENWORT, by the aſſiſtance of a good ſingle Microſcope with a reflecting Speculum. The ſudden jerk of the ſpringing cord frequently carries the object out of the field of view, ſo that it requires ſome patience to obſerve the whole of the proceſs.

As there are no certain diſtinctions in the Flowers themſelves ſufficient to eſtabliſh the Genera, we are guided by their diſpoſition under their covers.

Few of the FERNS are eſculent. They have a heavy diſagreeable ſmell. In larges doſes they deſtroy worms, and are purgative.

MOSSES.

This Order is ſubdivided according as the TIPS have a *Veil* or no VEIL: as they are upon the ſame plant with the POINTAL, or upon different plants; and as the Pointals are *ſolitary* or incorporated. The

SEEDS conſiſt only of a *Heart* without any *Coat* or *Seed-lobes*. The

TIPS according to Linnæus are rather to be conſidered as *Capſules*, and the *Duſt* which they contain as the real ſeeds; for within the cover of ſome of them he thinks he has obſerved real *Tips* hanging by threads, opening at the end, and letting fall a *Duſt* upon the hairs of the fringed ring, which he ſeems to conſider as ſo many Pointals.

OBSER-

CLASS XXIV.

OBSERVATIONS.

It has been doubted whether the powder in the heads of Mosses is the *Dust* or the *Seeds*. but as the dust of all plants explodes in water and flashes in the flame of a candle; and as this powder possesses these properties, there can no longer be a doubt what to call it. These Mosses then with heads or tips furnished with dust, are the barren plants; and except in the Genus Clubmoss, they have no *Threads*, but a *Tip* only; which in many of them is covered with a *Veil* or *Lid*. The *Seeds* of Mosses are therefore to be sought for in the Fertile Plants, and are sufficiently evident in the Fir Clubmoss.

The *Fringe* that surrounds the mouth of the tip in most of the Mosses, seems to answer the same, or at least a similar purpose to the down which crowns the seeds of the compound flowers in the nineteenth Class. Before the veil and the lid fall off, the little hairs which compose the fringe lie flat over the mouth of the tip, meeting in the center like the spokes of a wheel. If the lid is taken off by force when the tip is nearly ripe, these hairs immediately rise up and expand horizontally. Now in the natural progress of things as the tip becomes mature, is it not the expansion of this fringe which throws off the lid and opens a passage for the dust? I am indebted for this remark to a young gentleman who will probably one day clear up much of the darkness in which this numerous and neglected tribe of plants are still inveloped; and consequently reduce them to a system something more scientific than what we now can boast of.

The Fertile Flowers, or rather Seeds, in several Genera are totally undiscovered; but in others they are supposed to exist on the innerside at the base of the leaves, in certain starry or cone-like shoots. In plate I. C is a shoot of one of the Mosses. D is a flower.bearing stalk; (*a*) the veil; (*c*) the tip; (*b*) the lid.

Mosses thrive best in barren places. Most of them love cold and moisture. Trifling and insignificant as many people think them. their uses are by no means inconsiderable. They protect the more tender plants when they first begin to expand in the spring, as the experience of the Gardiner can testify, which teaches him to cover with Moss

Mofs the foil and pots that contain his tendereft plants; for it equally defends the roots again the fcorching fun-beams and the feverity of the froft. In the fpring, when the fun has confiderable power in the day-time and the frofts at night are fevere: the roots of young trees and fhrubs are liable to be thrown out of the ground, particularly in light fpongy foils. But if they are covered with Mofs, this accident never can happen. Thofe who are fond of raifing trees from feeds, will find their intereft in attending to this remark.

Moffes retain moifture a long time without being difpofed to putrefy. The angler takes advantage of this circumftance to preferve his worms, and the gardener to keep moift the roots of fuch plants as are to be tranfported to any confiderable diftance.

It is a vulgar error to fuppofe that Moffes impoverifh land. It is true they grow upon poor land that can fupport nothing elfe; but their roots penetrate very little, in general hardly a quarter of an inch into the earth. Take away the Mofs, and inftead of more grafs you will have lefs; but manure and drain the land; the grafs will increafe and the Mofs difappear.

The Grey Bogmoss, the Triangular Marshmoss, the dwarf and water Threadmoss, the hooked, fcorpion, floating and pointed Feathermoss, grow upon the fides and fhallower parts of pools and marfhes; and in procefs of time; occupying the fpace heretofore filled with water, are in their half decayed ftate dug up and ufed as fuel under the name of Peat. Thefe marfhes, drained partly by human induftry, and partly by the long continued operations of vegetables, are at length converted into fertile meadows.

Very few Moffes are eaten by cattle. The Bifhop Moth and the Bruffels Lace Moth feed upon fome of them. Their medicinal virtues are but little known, and lefs attended to. Some of the fpecies will probably be found very active, and therefore ufeful medicines.

THONGS.

The plants comprised under this division scarcely admit of a distinction of root, stem and leaf; much less are we enabled to describe the parts of the flowers. The Genera therefore are distinguished by the situation of what we suppose to be the flowers or seeds, or by the resemblance of the whole plant to some other substance we are well acquainted with.

Linnæus calls the plants of this Order SEA-WEEDS, but with no great propriety; for very few of the Genera have any thing to do with the sea. We rather choose to call them THONGS, because the substance of most of them is more or less like leather, and many of them are in the form of Thongs. In plate 1st. E. and F. are specimens of one of the Genera, and G. of another.

These plants, though generally looked upon as unworthy of notice, are of great consequence in the œconomy nature, and afford the first foundation for vegetation. Thus one species of the POWDERWORT, and several species of the CUPTHONG fix upon the barest rooks and are nourished by what slender supply the air and the rains afford them. When these die, they are converted into a very fine earth in which the tiled CUPTHONGS find nourishment, and when these putrify and fall to dust, various Mosses, as the THREADMOSS, FEATHERMOSS, &c. occupy their place; and in length of time when these perish in their turn, there is a sufficiency of soil in which trees and other plants take root. This process of nature is sufficiently apparent upon the smooth and barren rocks upon the sea-shore.

Some of the OARWEEDS are esculent.

Many of the CUPTHONGS are a grateful food to Goats; and the Rein-deer, which constitutes the whole œconomy of the Laplanders, and supports many thousand inhabitants, lives upon one of the species. Many of the species afford colours for dying. One of them brought from the Canary Islands, viz. the Orchel, or Argol, makes a very considerable article of traffic. It is not improbable that some of the species growing in our own island may afford very beauful and useful colours; but this matter hath not been sufficiently examined. Mr.

Hellot

Hellot gives us the following procefs for difcovering whether any of thefe plants will yield a red or purple colour. "Put about a quarter of an ounce of the plant in queftion "into a fmall glafs; moiften it well with equal parts of "ftrong Lime-water and fpirit of Sal Ammoniac; or the "Spirit of Sal Ammoniac made with quick-lime will do "without Lime-water. Tye a wet bladder clofe over "the top of the veffel and let it ftand three or four days. "If any colour is likely to be obtained, the fmall "quantity of liquor you will find in the glafs will be of "a deep crimfon red; and the plant will retain the fame "colour when the liquor is all dried up. If neither the "liquor nor the plant have taken any colour, it is need"lefs to make any further tryals with it."

FUNGUSSES.

We know very little about the Flowers or Seeds of the Funguffes; The generic characters are therefore taken from their external form. In plate 1. at H. a fpecies of MUSHROOM is reprefented to fhew (*a*) the *Cap*; (*b*) the *Pillar*; (*c*) the *Hat*.

All the Genera under this divifion, particularly the PUFFBALL and MOULD, abound with a black powder, which examined with a good microfcope is found to confift of globules which are fuppofed to be the feeds. But the Baron Otto MUNCKHAUSEN fays thefe globules are femi-tranfparent, containing a little black particle. He fays too that if this powder is mixed with water and kept in a warm place, the globules prefently fwell and are changed into egg-fhaped felf-moving animalcules. In about two days thefe animalcules unite and form a mafs of a pretty firm texture, or Fungus. When thefe Funguffes begin to grow, they appear like white veins, which are commonly fuppofed to be the roots; but in fact they are only tubes in which the animalcules move, and in a fhort time are transformed into a Fungus, which with plenty of moifture, and a proper degree of warmth grows to a very large fize. The black powder found betwix the gills of Mufhrooms, produces the fame phænomena.

A fact fo well attefted, and fo very fingular; could not fail to excite the attention of philofophers, and accordingly the accurate and ingenious Mr. Ellis, whofe discoveries in many abstruse parts of the animal and vegetable kingdoms, do him the highest honour, undertook the subject; and soon demonstrated that the motion of these globules is occasioned by a number of very minute animalcula feeding upon them; but the animalcula being much smaller than the globules are difficult to detect. See *Philos. Trans.* vol. 59. p. 138. See also *Gent. Mag.* for 1773. p. 316.

Fungusses have been sought for as food, upon account of their high flavour; but nobody supposes them to yield good nourishment; and many have been killed by them. Some of them have been found of considerable use in stopping hæmorrhages, and the acrimonious qualities of others will probably some day be turned to good account.

CLASS

CLASS XXIV.

Flowers Inconſpicuous.

FERNS.

* *Flowers in a Spike.*

399 Horsetail. - *Spike* ſcattered. *Flowers* target-ſhaped; with valves at the baſe.

400 Adderstongue. *Spike* jointed. *Flowers* cut round.

401 Moonwort. - *Spike* a ſort of bunch. *Flowers* two valves.

* * *Flowers on the under ſurface of the Leaves.*

402 Rustyback. Covering the whole ſurface of the leaf.

403 Polypody. In diſtinct ſpots upon the ſurface of the leaf.

404 Spleenwort. In ſeveral nearly parrallel lines, upon the ſurface of the leaf.

405 Brakes. - - In lines at the edge of the leaf.

406 Maidenhair. In ſpots, covered by the points of the leaves bent back.

407 Goldilocks. Flowers ſolitary; inſerted in the very edge of the leaf.

**** Fertile Flowers at the roots.*

408 PEPPERGRASS. Capsule with four cells.
409 QUILLWORT. - Capsule with two cells.

MOSSES.

** Without Veils.*

410 CLUBMOSS. - *Tip* with two valves; sitting.
411 BOGMOSS. - *Tip* with a smooth mouth.
412 EARTHMOSS. - *Tip* with a fringed mouth.

*** With Veils. Chives and Pointals distinct.*

413 BOTTLEMOSS. *Tip* with a large excrescence.

† *Bottle Earthmoss.*

414 HAIRMOSS. - *Tip* with a very small excrescence; bordered.

† *Hair Marshmoss.* † *Scored Threadmoss.*

415 MARSHMOSS. *Tip* without any excrescence.

**** With Veils. Chives and Pointals on the same plant.*

416 THREADMOSS. *Tip* on a fruit-stalk rising out of a tubercle at the end of the branch.

† *Oval Earth-moss.* † *Awl-shaped Earth-moss.*

417 FEATHERMOSS. *Tip* on a fruit-stalk rising out of a scaly bulb at the side of the branch.
418 WATERMOSS. *Tip* sitting; inclosed in a tiled scaly bulb.

THONGS.

** Growing on the Ground.*

419 STARTIP. - - Flowers with a simple empalement of four valves.
420 LIVERWORT. - Flowers on the under-side of a common target-shaped empalement.
421 VETCHCAP. - Flowers with an empalement of two valves.

422 HORN-

422 HORNFLOWER. Flowers with a tubular empalement. *Tip* awl-ſhaped, with two valves.
423 LEATHERCUP. Seeds cylindrical, tubular.
424 GRAINWORT. - Seeds little grains in the ſubſtance of the leaf.
425 CUPTHONG. - Seeds in a ſmooth, ſhining receptacle.
426 POWDERWORT. Subſtance woolly.

† *Star-jelly*

* * *Growing in the water.*

427 STARJELLY. - Subſtance like jelly.
428 LAVER. - - Subſtance membranaceous.
429 OARWEED. - Subſtance like leather.
430 RIVERWEED. Subſtance like hair.

FUNGUSSES.

* *Furniſhed with a Hat.*

431 MUSHROOM. Hat with gills on the underſide.
432 SPUNK. - - Hat with pores on the underſide.
433 PRICKLYCAP. - Hat with prickles on the underſide.
434 MORELL. - Hat ſmooth on the under-ſide.

** *Without a Hat.*

435 TURBANTOP. Shaped like a turban.
436 FUNNELTOP. Shaped like a bell.
437 CLUBTOP. - Shaped like a club.
438 PUFFBALL. - Globular.
439 MOULD. - - Little bladders ſupported upon pillars.

399 HORSETAIL. 1169 Equiſetum.

FLOWERS diſpoſed in an egg-ſhaped oblong ſpike.
Individuals round; opening at the baſe with many valves connected at the top, which is flat and target-ſhaped.—See Plate 1. Fig. A.

Wood
Sylvaticum
Common

HORSETAIL. Stem ſupporting a ſpike; leaves compound—*All the leaves diſpoſed in whorls; eight or ten in each whorl.*
1. Equiſetum ſylvaticum. *Gerard.* 1114. *Ray's Syn.* 130.
Equiſetum ſylvaticum tenuiſſimis ſetis.. *Bauh. pin.* 16.
Equiſetum omnium minus tenuifolium. *Park.* 1201.

Trailing. Hud.
2. Stem trailing, leaves pointing moſtly one way,
Equiſetum ſylvaticum procumbens, ſetis uno verſu diſpoſitis *Ray's Syn.* 131.

Marſh. Hud.
3. Leaves very long.
Equiſetum paluſtre tenuiſſimis et longiſſimis ſetis. *Bauh. pin.* 16.
Woods and moiſt ſhady places. P. April—May.
Horſes are fond of it, and in ſome parts of Sweden it is collected to ſerve them as winter food.

Corn
Arvenſe

HORSETAIL. Fruit bearing ſtalk naked; barren ſtalk leafy—*The fruit bearing ſtalks ſpring up firſt, but ſhrivel in a ſhort time; the leafy ſtalks continue much longer. The* Duſt *when ſhaken from the ſpike jumps about as if it was alive.*
Equiſetum arvenſe longioribus ſetis. *Bauh. pin.* 16. *Park.* 1202. *Ray's Syn.* 130.
Equiſetum ſegetale. *Gerard.* 1114.

Long-leaved. Hud.
1. With very long leaves.
Equiſetum pratenſe longiſſimis ſetis. *Bauh. pin.* 16.
Moiſt cornfields. P. March.
Cows will not eat it unleſs compelled by hunger.

Marſh
Paluſtre

HORSETAIL. Stem angular; leaves ſimple,—*Compoſed of ten or twelve joints.*
Equiſetum paluſtre brevioribus Setis. *Bauh. pin.* 16.
Equiſetum paluſtre. *Gerard.* 1114. *Ray's Syn.* 131. minus. *Park.*. 1200.

Many-headed.
1. Spikes numerous. Leaves with five or ſix joints. *Ray's Syn.* 311. tab. 5. fig. 3.
Equiſetum paluſtre minus polyſtachion. *Bauh. pin.* 16.
In watery places. P. June.

HORSE-

HORSETAIL. Stem ſcored; leaves generally ſimple.—*Twenty or more in each whorl.* River Fluviatile

Equiſetum majus. *Gerard.* 1115. *Ray's Syn.* 130.
Equiſetum majus paluſtre. *Park.* 1200.
Equiſetum paluſtre longioribus foliis. *Bauh. pin.* 15.
Banks of rivers and pools. P. May.
In ſome places they mix it with the food of Cows to increaſe the quantity of their milk. Horſes are not fond of it.

HORSETAIL. Stem almoſt naked; ſmooth.—*Leafy as the Summer advances.* Smooth Limoſum

Equiſetum nudum lævius noſtras, *Ray's Syn.* 131. tab. 5. fig. 2.

HORSETAIL. Stem naked, rough; ſometimes a little branched at the bottom—*The ſheaths of the joints white. Spike brown; terminating.* Shaving Hyemale

1. Equiſetum nudum. *Gerard.* 1113. *Ray's Syn.* 131. *Branched*
Equiſetum junceum ramoſum. *Park.* 1207.
Equiſetum foliis nudum ramoſum, *Bauh. pin.* 16.
2. Equiſetum nudum minus variegatum Baſilienſe. *Bauh. pin.* 16. *Variegated*
3. Stem undivided. *Undivided*
Equiſetum foliis nudum non ramoſum ſeu junceum. *Bauh. pin.* 16.
Equiſetum junceum ſeu nudum. *Park.* 1201.
4. Stem rather ſmooth. *Smoother.* Hud.
Equiſetum læve pæne nudum. *Petiver Conc. Gram.* 238.
Shave-graſs. Pewterwort.
In marſhy places. P. July—Auguſt.
The Turners and Cabinet makers uſe it to ſmooth their work.—It is wholeſome to Horſes, hurtful to Cows and hateful to Sheep.

400 ADDERSTONGUE. 1171 Ophioglossom.

CAPSULES. Pointing from two oppoſite lines; with numerous joints placed tranſverſely, and divided into as many cells as there are joints. When ripe every cell opens tranſverſely.

SEEDS. Numerous; very ſmall; ſomewhat egg-ſhaped.

ADDERSTONGUE. Leaf egg-ſhaped.—*Spike very ſlender; on a fruit-ſtalk.* Common Vulgatum

Ophioglossum. *Gerard.* 404. *Ray's Syn.* 128. ſeu Lingua ſerpentina. *Park.* 506.
Ophioglossum primum, ſive vulgatum. *Bauh. pin.* 354.
Moiſt meadows and damp walls. P. May.

401 MOONWORT. 1172 Osmunda.

CAPSULES. Globular, distinct disposed in a bunch; and opening horizontally.

SEEDS. Numerous; very small; egg-shaped.

Common Lunaria — MOONWORT. Stalk rising from the stem; solitary. Leaf winged, solitary.—*Within the base of the stem early in the spring, may be found a compleat rudiment of the next year's plant.* Capsules *yellowish.*

Lunaria minor. *Gerard.* 328. *Park.* 507. *Ray's Syn.* 128.
Lunaria racemosa minor. *Bauh. pin.* 354.

Jagged. Hud. 1. With jagged leaves.
Lunaria minor foliis dissectis. *Ray's Syn.* 129.
Hilly pastures. P. May.

Royal Regalis — MOONWORT. Leaf doubly winged; bearing bunches of flowers at the ends.—*Root thready; bundled; black.*

Osmunda regalis, seu Filix florida. *Park.* 1038.
Filix ramosa non dentata florida. *Bauh. pin.* 357.
Filix florida seu Osmunda regalis. *Gerard.* 1131.
Flowering Fern. Osmund Royal.
In putrid marshes. P. July—August.

The root boiled in water is very slimy and is used in the North to stiffen linen instead of starch. Only some of the leaves bear flowers.

Rough Spicant — MOONWORT. Leaves spear-shaped, with winged clefts. Segments very entire, parallel, running into each other.—*The flowering leaves much narrower than the barren leaves.*

Lonchitis aspera. *Gerard.* 1140. *Ray's Syn.* 118.
Lonchitis aspera minor. *Bauh. pin.* 359. *Park.* 1042.
Rough Spleenwort.
Groves and moist heaths. P. July.

Stone Crispa — MOONWORT. Leaves more than doubly compound: little leaves alternate: circular, but jagged—*With yellow lines on the under side, parallel to the middle rib.*

Adiantum album crispum alpinum. *Ray's Syn.* 126.
Stone Fern.
On Rocks. P. August.

402 RUSTYBACK. 1173 Acrostichum.

FLOWERS entirely covering the under surface of the leaf.

RUSTY-

RUSTYBACK. Leaf simply divided; naked; strap-shaped; jagged.— Forked Septentrionale
Filix saxatilis cordiculata. *Bauh. pin.* 358.
Filix saxatilis Tragi. *Park.* 1045. *Ray's Syn.* 120.
Horned Fern.
Old walls and clefts of rocks. P. August.

RUSTYBACK. Leaves almost doubly winged; little leaves opposite: united; blunt; hairy underneath; very entire at the base.—*About as long as ones finger.* Hairy Ilvense
Filix alpina; pedicularis rubræ foliis subtus villosis. *Ray's Syn.* 118.
Clefts of rocks. P.

RUSTYBACK. Leaves winged; little leaves with winged clefts; very entire—*Leaf-stalks smooth. Barren leaves broader and more blunt than the others. Circles of flowers ten pair or more.* Marsh Thelypteris
Filix minor palustris repens.. *Ray's Syn.* 122.
Dryopteris. *Gerard.* 1135. seu Filix querna repens. *Park.* 1041.
In turfy Bogs. P. August.

403 POLYPODY. 1179 Polypodium.

FLOWERS disposed in distinct circular dots on the under surface of the leaf.

* *Leaves with winged clefts. Lobes united.*

POLYPODY. Leaves with winged clefts; wings oblong; somewhat serrated; blunt. Root scaly,—*and hairy. Flowers yellowish brown; in rows, parallel to the rib of the little leaves.* Common Vulgare
Polypodium vulgare *Bauh. pin.* 359. *Park.* 1039.
Polypodium. *Gerard.* 1138. *Ray's Syn.* 117.
2. Little leaves serrated.— *Serrated.* Hud.
Polypodium murale pinnulis serratis. *Ray's Syn.* 117.
On old walls; shady places; and at the roots of trees. P.
The root is sweetish: by long boiling it becomes bitter. When fresh it is a gentle purgative. An infusion of six drams of it in half a pint of boiling water may be taken at twice.

POLYPODY. Leaves with winged clefts; wings spear-shaped: ragged with clefts, and serrated. Jagged Cambricum
Polypodium Cambro-britannicum pinnulis ad margines laciniatis. *Ray's Syn.* 117.
On rocks. P.

* * *Leaves.*

* * *Leaves winged.*

Spleenwort Lonchitis — POLYPODY. Leaves winged; wings crefcent-fhaped; fringed; ferrated, declining. Stalks befet with ftiff flathairs.—*Spots of flowers twelve pair or more.*

Lonchitis afpera major. *Gerard.* 1140. *Ray's Syn.* 118.
Lonchitis afpera. *Bauh. pin.* 359.
Lonchitis afpera major Matthiolo. *Park.* 1042.
Great Polypody. Spleenwort.
Clefts of rocks. P.

Rock Fontanum — POLYPODY. Leaves winged and jagged; wings circular, fharply and elegantly cut; ftalk fmooth—*It refembles the* Brittle POLYPODY *but the wings ftand clofer together and are not fo deeply fubdivided; the dots of flowers are larger, and proceed not from a roundifh fcale, but from an oblong white narrow valve or chink.*

Common 1. Filicula fontana minor. *Bauh. pin.* 358.
Elegant 2. Filix faxatilis omnium minima elegantiffima. *Plukenet Phyt.* tab. 89. fig. 3.
Old walls and clefts of rocks. P.

* * *Leaves almoft doubly winged.*

Wood Phegopteris — POLYPODY. Leaves almoft doubly winged; lower little leaves bent back in pairs; united by a four-cornered little wing.—

Filix minor Britannica, pediculo pallidiore, alis inferioibus deorfum fpectantibus. *Ray's Syn* 122.
Clefts of moift fhady rocks. P.

Sweet Fragrans — POLYPODY. Leaves almoft doubly winged; fpear-fhaped. Little leaves crowded together; lobes blunt, ferrated; ftalk chaffy—*Refembleng Fern* POLYPODY *but much fmaller.*

In the clefts of moift rocks. P.

Crefted Criftatum — POLYPODY. Leaves almoft doubly winged; little leaves egg-fhaped; lobes rather blunt, fharply ferrated towards the point—*The flowers in this fpecies are upon the upper, not upon the lower little leaves.*

Common 1. Filix mas ramofa, pinnulis dentatis. *Gerard.* 1129. *Ray's Syn.* 124.
Smaller. Hud. 2. Little leaves very elegantly ferrated.
Filix montana ramofa minor argute denticulata. *Ray's Syn.* 124.
Woods, heaths, and fhady places. P.

* * * *Leaves doubly winged.*

POLYPODY. Leaves doubly winged; wings blunt, a little ſcolloped; ſtalk chaffy—*Flowers kidney-ſhaped; ſix or ſeven on each wing.* Fern Filix mas.

Filix mas vulgaris. *Park.* 1036. *Ray's Syn.* 120.

Filix mas non ramoſa dentata. *Bauh. pin.* 258.

Filix mas non ramoſa, pinnulis latis denſis minutim dentatis *Gerard.* 1129.

Male Polypody. Male Fern.

Woods heaths and ſtony places. P.

The Siberians boil it in their ale, and admire the flavour of it.

POLYPODY. Leaf doubly winged; little wings ſpear-ſhaped, with winged clefts; ſharp—*Flowers egg-ſhaped, ſomewhat fringed, ſolitary.* Fringed Filix fœmina

Filix mas non ramoſa, pinnulis anguſtis raris profunde dentatis *Gerard.* 1130. *Ray's Syn.* 121.

Female Polypody.

POLYPODY. Leaves doubly winged; wings creſcent-ſhaped, fringed and toothed; ſtalk beſet with flat ſtiff hairs. Prickly Aculeatum

1. Filix mas aculeata major. *Bauh. pin.* 358. *Common*

Filix mas non ramoſa, pinnulis latis auriculatis ſpinoſis. *Gerard.* 1130. *Ray's Syn.* 121.

It ſometimes varies with and without ears. *Ray.*

2. With narrow leaves. *Narrow leaved* Hudſon.

Filix aculeata major, pinnulis auriculatis crebrioribus foliis integris anguſtioribus. *Ray's Syn.* 121.

3. With a kind of woollineſs on the leaf. *Woolly.* Hudſ.

Filix mas aculeata noſtras, alis expanſis muſcoſa lanugine aſperſa. *Plukenet Phyt.* tab. 180 fig. 1.

POLYPODY. Leaves doubly winged; wings creſcent-ſhaped, fringed and toothed, the upper ones divided into lobes at the baſe; ſtalk beſet with flat ſtiff hairs. *Hudſon* 39 . Lobed Lobatum

Filix lonchitidi affinis. *Ray's Syn.* 121.

Polypodium lobatum, *Hudſon* 390.

In ſhady places. P.

POLYPODY. Leaves doubly winged; little leaves and wings ſpear-ſhaped; diſtant from each other; teeth tapering to a point. Stone Rhæticum

Filix fontana major, ſeu Adiantum album filicis folio. *Bauh. pin.* 351.

Filix pumila ſaxatilis altera. *Ray's Syn.* 122.

Stony hills. P.

POLYPODY.

Brittle Fragile — POLYPODY. Leaves doubly winged; little leaves diſtant from each other; little wings circular; cut at the edges—*Flowers in large, black numerous ſpots.*

Common — 1. Stem ſlender and brittle.

Filix ſaxatilis caule tenui fragile. *Ray's Syn.* 125.

Maiden-hair Hudſ. — 2. Wings divided like *Cow-weed.*

Adiantum nigrum pinnulis cicutariæ diviſura. *Ray's Syn.* 126.

Dry ſtony places. P.

**** *Leaves more than doubly compound.*

Branched Dryopteris — POLYPODY. Leaves more than doubly compound; little leaves growing by threes; doubly winged.

Filix ſaxatilis ramoſa, maculis nigris punctata. *Bauh. pin.* 358.
Filix ramoſa minor. *Ray's Syn.* 125.
Filix pumila ſaxatilis prima Cluſii. *Park.* 1043.
Dryopteris Tragi. *Gerard.* 1135.
Dry ſtony places. P.
The Fern Moth feeds upon the different ſpecies.

404 SPLEENWORT. 1178 Aſplenium.

FLOWERS diſpoſed in ſtraight lines on the under ſurface of the leaf.

* *Leaf ſimple.*

Harts-tongue Scolopendrium — SPLEENWORT. Leaves ſimple; betwixt heart and tongue-ſhaped; very entire. Stalks hairy.

Phyllitis. *Gerard.* 1138. *Ray's Syn.* 116.
Phyllitis, ſeu Lingua Cervina vulgaris. *Park.* 1046
Lingua Cervina Officinarum. *Bauh. pin.* 353.
Moiſt ſhady rocks, and in the mouths of wells. P.

** *Leaf with winged clefts.*

Common Ceterach — SPLEENWORT. Leaves with winged clefts; lobes alternate, running into one another; blunt. The under ſurface of the leaf is covered with ſcales in ſuch a manner that the flowers are ſcarce diſcernible.

Aſplenium ſive Ceterach. *Gerard.* 1138. *Park.* 1046. *Ray's Syn.* 118.

Ceterach Officinarum. *Bauh. pin.* 354.
On old walls and clefts of moiſt rocks. P.

* * * *Leaves winged.*

SPLEENWORT. Leaves winged. Wings circular; ſcolloped.—*Stalk gloſſy, black.* Roots *black; fibrous.* Flowers *in three, four, or five rows,* Maiden-hair Trichomanoides

Trichomanes. *Park.* 1051. *Ray's Syn.* 119,
Trichomanes ſeu Polytrichum officinarum. *Bauh. pin.* 356.
Trichomanes mas. *Gerard.* 1146.
Common Maidenhair.
On old walls and ſhady ſtony places. P.

SPLEENWORT. Leaf winged. Wings inverſely egg-ſhaped; ſerrated; hunched towards the end, and blunt; wedge-ſhaped at the baſe.—*Stalks ſhining; almoſt black.* Sea Marinum

Filix marina anglica. *Park.* 1045.
Filicula petræa fæmina, ſeu Chamæfilix marina anglica. *Gerard.* 1143.
Chamæfilix marina anglica. *Ray's Syn.* 119.
Sea Maidenhair, Sea Fern.
On rocks near the Sea. P.

* * * * *Leaves doubly compound.*

SPLEENWORT. Leaves alternately doubly compound. Little leaves wedge-ſhaped and ſlightly ſcolloped—*Flowers in two or three rows.* White Rutamuraria

Ruta muraria. *Bauh. pin.* 356. *Gerard.* 1144. *Ray's Syn.* 122.
Ruta muraria ſeu Salvia Vitæ. *Park.* 1050.
White Maidenhair.
Clefts of rocks. P.

SPLEENWORT. Leaves almoſt triply winged; little leaves alternate; wings ſpear-ſhaped jagged and ſerrated—*The duſt is the colour of Saffron.* Stalks *black; gloſſy.* Flowers *in whitiſh rows; from three to ſeven together.* Black Adiantum nigrum

Adiantum nigrum Officinarum. *Ray's Syn.* 126.
Adiantum nigrum vulgare. *Park.* 1046.
Adiantum foliis longioribus pulverulentis, pediculo nigro. *Bauh. pin.* 356.
Onopteris mas. *Gerard.* 1137.
1. There are ſome varieties in the ſhape of the little leaves. Some being broader and others narrower. *Ray's Syn.* 127.
Black Maidenhair.
Old walls and moiſt ſhady rocks. P.

SPLEEN-

Green Trichomanes Ramoſum SPLEENWORT. Leaves doubly winged: lobes inverſely egg-ſhaped and ſcolloped; lower little leaves the ſmalleſt.—*Leaf winged; wings circular. lopped at the baſe. Hudſon* 385.
Trichomanes ramoſum. *Ray's Syn* 119.
Trichomanes fæmina. *Gerard.* 1146.
Trichomanes ramoſum majus et minus. *Bauh. pin.* 356.
Aſplenium viride. *Hudſon.* 385.
Green Maidenhair.
Moiſt rocks. P.

405 BRAKES. 1174 Pteris.

FLOWERS diſpoſed in a line, along the edge of the leaf, on the under ſide.

Common Aquilina BRAKES. Leaves more than doubly compound. Little leaves winged; wings ſpear-ſhaped; the lowermoſt with winged clefts; the upper ones ſmaller—*The roots extend ſo deep into the earth as not to be deſtroyed by burning the ſurface, or by the Plough. When cut tranſverſly there is a faint reſemblance of the Imperial Eagle; from which circumſtance Linnæus has named it the* P. Aquilina *or* Eagle Brakes.

Filix fæmina. *Gerard.* 1128. *Ray's Syn.* 124 vulgaris. *Park.* 1037.
Filix ramoſa major, pinnulis non dentatis. *Bauh. pin.* 357.

Small 1. Smaller than the above.
Filicula ſaxatilis ramoſa maritima noſtras. *Ray's Syn.* 125.
Female Fern.
Woods and heaths. P. Auguſt. (2) on rocks near the ſea, and on walls.

The root dried, powdered, and given in doſes of half an ounce, is a ſecret to kill the Tape-worm, and is ſuppoſed to be no leſs efficacious in killing other Worms—A tolerably pure alkaly is obtained from the aſhes. The common people in many parts of England mix the aſhes with water and form them into balls: theſe balls are afterwards made hot in the fire and then uſed to make lye for ſcouring linen—It makes a very durable thatch; and is an excellent litter for Horſes and Cows. Where coal is ſcarce, they uſe it to heat ovens and to burn limeſtone; for it affords a very violent heat. In the more inhoſpitable climates, bread is made of the roots.—The Fern Moth feeds upon it.

406 MAIDENHAIR. 1180 Adiantum.

LOWERS diſpoſed in oval ſpots, under the ends of the leaves, which are bent back upon them.

MAIDEN-

MAIDENHAIR. Leaves doubly compound; little leaves alternate. Wings wedge-ſhaped; divided into lobes and placed upon footſtalks.— True Capillus Veneris

Adiantum foliis Coriandri. *Bauh pin.* 355.
Adiantum verum, ſeu Capillus Veneris verus. *Park*, 1049.
Capillus Veneris verus. *Gerard.* 1144. *Ray's Syn.* 123.
On rocks. P.

MAIDENHAIR. Leaves more than doubly compound; little leavss alternate. Wings diamond-ſhaped; cut at the edges; with flowers upon both ſurfaces.— Shining Trapeziforme

Adiantum majus Coriandri folio, adianto vero affine, pediculo pallide rubente, *Ray's Syn.* 124.
Adianto vero affinis minor Scotica, folio obtuſo, ſaturate viridi. *Ray's Syn.* 124.
On rocks in Scotland.

407 GOLDILOCKS. 1181 Trichomanes.

Empal. Turban-ſhaped, ſingle, upright; riſing from the very edge of the leaf.
Shaft briſtle-ſhaped; terminating the Capſule.

GOLDILOCKS. Leaves almoſt doubly winged; wings alternate, crowded together, divided into lobes; ſtrap-ſhaped. Cup Pyxidiferum

Filix humilis repens, foliis pellucidis et ſplendentibus, caule alato. *Ray's Syn.* 127. tab.3. fig. 3-4.
On dripping rocks. P.

GOLDILOCKS Leaves winged. Wings oblong; forked; running along the ſtem; toothed.—*The ſeed like globules are produced betwixt two leafy valves.* Tunbridge Tunbrigenſe

Adiantum petræum perpuſillum anglicum, foliis bifidis, trifidiſque. *Ray's Syn.* 123.
In clefts of moiſt rocks. P. Auguſt.

408 PEPPERGRASS. 1183 Pilularia.

Barren Flowers like duſt, in a line on the under ſide of the leaf.
Fertile Flowers at the root.
Capsule globular; with four cells, containing many ſeeds.

Globular
Globulifera

PEPPERGRASS. Stem creeping. Leaves upright; slender.—

Pilularia paluſtris juncifolia. *Dillenius.* 538. tab. 79. fig. 1.

Graminifolia paluſtris repens, vaſculis granorum Piperis æmulis. *Ray's Syn.* 136.

Pillwort. Peppermoſs.

In grounds that have been overflowed, eſpecially in a ſandy ſoil. P. June—September.

409 QUILLWORT. 1184 Iſoetes.

Barren Flowers ſolitary; within the baſe of the inner leaves.

EMPAL. *Scale* heart-ſhaped; ſharp; ſitting.

BLOSS. None.

CHIVE. *Thread* none. *Tip* roundiſh, with one cell.

Fertile Flowers ſolitary, within the baſe of the outer leaves of the ſame plant.

EMPAL. As above.

BLOSS. None.

POINT. *Seedbud* egg-ſhaped, and together with the *Shaft* and *Summit* concealed within the leaf.

S. VESS. *Capſule* ſomewhat egg-ſhaped, with two cells; concealed within the baſe of the leaf.

SEEDS. Numerous; globular.

Mountain
Lacuſtris

QUILLWORT. Leaves jointed; awl-ſhaped; ſemi-cylindrical.—

Long leaved

1. Leaves long and ſlender.

Calamaria folio longiore et graciliore. *Dillenius.* 541. tab. 80. fig. 2.

Subularia lacuſtris ſeu calamiſtrum herba aquatica alpina. *Ray's Syn.* Ed. 1. p. 210. tab. 2.

Subularia fragilis folio longiore et tenuiore. *Ray's Syn.* 307.

Short-leaved. HUD.

2. Merlin's Graſs with ſhort thick leaves.

Calamaria folio breviore craſſiore. *Dillenius* 541. tab. 80. fig. 1.

Subularia vulgaris erecta folio rigidiſſimo. *Ray's Syn.* 306.

Flexible. HUD.

3. With more flexible leaves.

Subularia repens folio minus rigido. *Ray's Syn.* 306.

Leaves the ſame with the former variety but more flexible.

At the bottom of lakes upon mountains.

410 CLUB-

410 CLUBMOSS. 1185 Lycopodium.

Barren Flower at the baſe of the leaves; ſitting.
EMPAL. *Veil* none.
CHIVE. *Tip* kidney ſhaped; with two valves; ſitting.
Fertile Flower on the ſame plant.
EMPAL. *Cup* with four leaves.
POINT. None.
SEED. Without ſeed-lobes; growing to the empalement.

CLUBMOSS. Leaves ſcattered, terminating in threads. Spikes cylindrical, growing generally two on a footſtalk—*Though ſometimes three, or only one.* Common. Clavatum

Lycopodium vulgare piloſum anfragoſum et repens. *Dillenius* 441. tab. 58. fig. 1.

Lycopodium. *Ray's Syn.* 107.

Muſcus terreſtris clavatus. *Bauh. pin.* 360.

Muſcus clavatus, ſeu Lycopodium. *Gerard.* 1562. *Park.* 1307.

Wolf's claw.

On heaths and mountains. P. July.

In Sweden they form it into mats or baſſes, which lye at their doors to clean ſhoes upon.

CLUBMOSS. Leaves ſcattered; fringed; ſpear-ſhaped. Spikes ſingle; at the ends of the branches; leafy —*Capſules at the Baſe of the lower leaves, four together: each containing a large ſolid ſeed; but above theſe, upon the ſame ſtem are others containing four ſeeds.* Tips *round, of a yellowiſh colour.* Prickly Selaginoides

Selaginoides foliis ſpinoſis. *Ray's Syn.* 106. *Dillenius* 460. tab. 68. fig. 1.

In bogs, and in wet places on mountains. P. July—Auguſt—September.

CLUBMOSS. Leaves ſcattered; very entire; ſpikes at the ends of the branches; leafy—*Single; upright.* Stem *creeping.* Marſh Inundatum

Lycopodium paluſtre repens, Clava ſingulari. *Dillenius* 452. tab. 62. (61) fig. 7.

Muſcus terreſtris repens, clavis ſingularibus folioſis erectis. *Ray's Syn.* 108.

Moiſt heaths. P. July.

Fir
Selago

CLUBMOSS. Leaves ſcattered; pointing eight different ways. Stem forked; upright. Branches all of the ſame height. Flowers ſcattered—*The fertile flowers evidently ſhew us the nature of the flowers of Moſſes.*

Selago vulgaris, abietis rubræ facie. *Dillenius.* 435. tab. 56. fig. 1.

Selago foliis et facie abietis. *Ray's Syn.* 106.

Muſcus erectus ramoſus ſaturate v ridis. *Bauh. pin.* 360.

On heaths and mountains in the clefts of rocks. P. Auguſt.

The common people uſe a decoction of it to kill lice upon Cows and Swine. Taken inwardly it purges.

Welſh
Annotinum

CLUBMOSS. Leaves ſcattered, pointing five different ways; a little ſerrated. Stem jointed at every year's ſhoot. Spikes at the ends of the branches; ſmooth; upright—*Single; without foot-ſtalks; branches contracted at the joints.*

Lycopodium elatius Juniperinum, clavis ſingularibus ſine pediculis. *Dillenius* 455. tab. 63. fig. 9. *Ray's Syn.* 107.

On high mountains. P. July.

It is made into baſſes.

Mountain
Alpinum

CLUBMOSS. Leaves pointing four different ways; tiled; ſharp. Stems upright; cloven; ſpikes ſitting; cylindrical.—*When fully grown it becomes yellow.*

Lycopodium Sabinæ facie. *Dillenius.* 445. tab. 58. fig. 2. *Ray's Syn.* 108.

Muſcus clavatus cupreſſiformis. *Park.* 1309.

Muſcus clavatis foliis Cupreſſi. *Bauh pin.* 360. *Gerard.* 1562.

In wet places on mountains. P. Auguſt—September.

411 BOGMOSS. 1187 Sphagnum.

Barren Flowers on very ſhort foot-ſtalks.

EMPAL. *Veil* none.

BLOSS. None except a membrane of ſhort duration betwixt the receptacle and the tip.

CHIVE. *Tip* globular; mouth very entire; covered with a blunt *Lid.*

RECEPT. *Excreſence* bordered, imperfect; under the tip.

Fertile Flower hitherto undiſcovered.

BOGMOSS. Branches bent downwards.—*White, or reddish.* Grey
Tips *reddish brown, nearly globular.* Palustre
1. Leaves egg-shaped; with thicker branches. *Common*
Sphagnum palustre molle deflexum, squamis cymbiformibus. *Dillenius.* 240. tab. 32. fig. 1.
Sphagnum cauliferum et ramosum palustre molle candicans, reflexis ramulis, foliolis latioribus. *Ray's Syn.* 104.
2. Leaves awl-shaped with finer branches. *Finer*
Sphagnum palustre molle deflexum, squamis, capillaceis. *Dillenius* 243. tab. 32. fig. 2.
Sphagnum cauliferum et ramosum palustre molle candicans, reflexis ramulis foliolis angustioribus. *Ray's Syn.* 104.
In turfy bogs. P. July—August.

BOGMOSS. Somewhat branched; upright.—*Of a beautiful* Green
green. Tips *egg shaped.* Alpinum
Sphagnum subulatnm viridissimum, capsulis ovatis. *Dillenius.* 245. tab. 32. fig. 3.
Bogs on mountains. August—September—October.

BOGMOSS. Branched; creeping. Tips on the sides of the Creeping
branches pointing one way.—*Hairy.* Arboreum
Sphagnum heteromallum polycephalum. *Dillenius.* 248. tab. 32. fig. 6.
Sphagnum cauliferum et ramosum minus hirsutum, capitulis crebris pilosis per ramulorum longitudinem adnascentibus. *Ray's Syn.* 105.
Trunks of trees. Nov.—Dec.

412 EARTHMOSS. 1189 Phascum.

Barren flower nearly sitting, or on a very short fruit-stalk.

EMPAL. *Veil* none.

CHIVE. *Tip* oval; with a fringed mouth; covered with a *Lid* tapering to a point

RECEPT. *Excrescence* none.

Fertile flower hitherto undiscovered,

Oval 1. Acaulon — EARTHMOSS. Without a ftem; tips fitting; leaves egg-fhaped; fharp—*tiled.*

Greater — 1. Sphagnum acaulon bulbiforme, majus. *Dillenius* 251. tab. 32. fig. 11.

Sphagnum acaulon, foliis in bulbi formam congeftis, majus. *Ray's Syn.* 105.

Leffer — 2. Sphagnum acaulon bulbiforme, minus. *Dillenius* 252. tab. 32. fig. 12.

Sphagnum acaulon, foliis in bulbi formam congeftis, minus. *Ray's Syn.* 105.

Sandy ditch-banks and gardens. A. Jan.—Feb.

Awl fhaped 2. Subulatum — EARTHMOSS. Without a ftem; tip fitting; leaves briftly; awl-fhaped, open.—*Outer leaves flat, fpread upon the ground; inner leaves pale, terminating in brown hairs.*

Obs. *This and the preceding fpecies, have been fometimes obferved to have a veil, and might therefore with propriety be ranged with the* Threadmosses.

Leffer — 1. Sphagnum acaulon trichoïdes. *Ray's Syn.* 105. *Dillenius* 251. tab. 32. fig. 10.

Greater — 2. Confiderably larger than the preceding.

Sphagnum acaulon maximum, foliis in centro ciliaribus. *Dillenius* 253. tab. 32. fig. 13.

(1.) Moift fandy roads and heaths. (2.) Sandy meadows, and on rocks. A. Mar.—April.—(2.) Sept.—Oct. *Dill.*)

Upright 3. Pedunculatum *Hud.* — EARTHMOSS. Without a ftem; tip inverfely egg-fhaped; fupported by a fruit-ftalk.—*Leaves egg-fhaped, tranfparent.* Hudfon 397.

Bryum ampullaceum, foliis ferpylli pellucidis collo craffiore. *Dillenius* 344. tab. 44. fig. 4.

Bryum erectis gigartinis capitulis foliis ferpylli pellucidis obtufis. *Ray's Syn.* 93. tab. 3. fig. 2.

Moift heaths. May.

Bottle 4. Ampullaceum *Dill.* — EARTHMOSS? PHASCUM?

Without a ftem; tip oblong, fupported by a fruit-ftalk; leaves awl-fhaped,—*open.* Shoots *upright, moftly fimple, but fometimes divided towards the top.* Tip *upright, contracted at the mouth, fomewhat refembling that of the* Bottlemofs. *From* Dillenius.

Bryum ampullaceum, foliis et ampullis anguftioribus. *Dillenius* t. 41. f. 5.

On heaths, and in old rotten cow-dung. *Dill.*

EARTH-

EARTHMOSS. Stem creeping; tips on the ſides of the branches, ſitting.— Creeping 5. Repens

Sphagnum vermiculare repens, capſulis intra foliorum ſquamoſorum alas minimis. *Dillenius* 550. tab. 85. f. 16.

Trunks of trees. P. April.

413 BOTTLEMOSS. 1191 Splachnum.

Barren flower on a fruit-ſtalk.

EMPAL. *Veil* conical; ſmooth; ſhedding.

CHIVE. *Tip* cylindrical. *Mouth* opening with eight little reflected teeth; without a ring.

RECEPT. Membranaceous, coloured, very large; under the tip.

Fertile flower on a different plant.

EMPAL. *Common*, ſtar-like; at the ends of the ſhoots or branches; formed of ſeveral little awl-ſhaped leaves pointing in rays and tiled.

POINT. Many; in the centre; all of the ſame height; ſhort; coloured.

* *Bottle Earthmoſs.*

BOTTLEMOSS. Barren flower with its excreſcence; ſhaped like a glaſs bottle; excreſcence inverſely egg-ſhaped—*Yellow. leaves egg-ſhaped*; *ſharp*; *tranſparent*. Tips *upright*. Common 1. Ampullaceum

Bryum ampullaceum, foliis Thymi pellucidis, collo ſtrictiore. *Dillenius*. 343. t. 44. f. 3.

Bryum erectis gigartinis capitulis, foliis ſerpilli pellucidis acutis. *Ray's Syn.* 93.

Linnæus obſerves that it is never found but where the dung of animals has lain, and conjectures that it may be only a variety of the Fountain MARSHMOSS.

Turf bogs. March—April.

414 HAIRMOSS. 1192 Polytrichum.

Barren flower on a fruit-ſtalk from the top of the plant.

EMPAL. *Veil* conical; generally woolly; as long as the tip.

CHIVE. *Tip* oblong; mouth fringed; covered with a circular membrane. *Lid* conical.

RECEPT. *Excreſcence* a border under the tip.

SCALYBULB. A cylindrical ſheath; much ſhorter than the fruit-ſtalk.

Fertile flower on a diſtinct plant.

EMPAL. *Common*, coloured; tiled; expanding in form of a little roſe.

POINT. Thread-ſhaped; jointed.

Common 1. Commune HAIRMOSS. Stem ſimple; tip a long ſolid ſquare, or parallelopippedon—*Veil yellow.*

Greater 1. Leaves ſerrated.

Polytrichum quadrangulare vulgare, yuceæ foliis ſerratis. *Dillenius.* 420. t. 54. f. 1.

Polytrichum vulgare et majus capſula quadrangulari. *Ray's Syn.* 90.

Polytrichum aureum majus. *Bauh. pin.* 359. *Park.* 1052.

Muſcus capillaris, ſeu adiantum aureum majus. *Gerard.* 1559.

Great golden Maidenhair.

Leſſer 2. Leaves ſhorter and leſs flexible.

Polytrichum quadrangulare, juniperi foliis brevioribus et rigidioribus. *Dillenius.* 424. t. 54. f. 2.

Polytrichum montanum et minus, capſula quadrangulari. *Ray's Syn.* 90.

Polytrichum aureum medium. *Bauh. pin.* 356.

Downy 3. Leaves ſmaller, and terminating in fine ſoft hairs.

Polytrichum quadrangulare minus, juniperi foliis piloſis. *Dillenius,* 426. t. 54. f. 3.

Woods, heaths, marſhy and barren ground. P. May.

1. In marſhy ſpongy ground. 2. In wet ground on mountains. 3. In very dry ſandy ſituations. *Dillenius.* Hence probably their different appearances.

When the Laplanders ſleep all night in the woods, they make themſelves beds of this moſs; and the Bears collect it for the ſame purpoſe. Squirrels and birds uſe it in making their neſts.

HAIR-

HAIRMOSS. Stem very much branched; fruit-ſtalks at the ends of the branches—*Tips oval.* Mountain 2. Alpinum

Polytrichum alpinum ramoſum, capſulis e ſummitate ellipticis. *Dillenius.* 427. t. 55. f. 4.

Polytrichum urnigerum, *var.* *Hudſon.* 400.

Mountains.

HAIRMOSS. Stem very much branched; fruit-ſtalks from the baſe of the leaves—*but near the ends of the branches.* Tips *when ripe hanging down*; *hardly any* Excreſcence *at the baſe.* Branched 3. Urnigerum

Polytrichum ramoſum, ſetis ex alis urnigeris. *Dillenius.* 427. t. 55. f. 5.

Mountains. June—Aug.

415 MARSHMOSS. 1193 Mnium.

Barren flower on a fruitſtalk.

EMPAL. *Veil* oblong; tapering to a point; oblique; ſmooth.

CHIVE. *Tip* roundiſh: *Mouth* ſurrounded with a fringed ring and covered with a *Lid* ſomewhat conical.

RECEPT. *Excreſcence* none.

Fertile Flower generally on a diſtinct plant.

EMPAL. *Common* compoſed of leaves diſpoſed in form of a ſtar.

POINT. Many, in the centre, collected into a ball.

MARSHMOSS. Stem ſimple; leaves egg-ſhaped—*Veil yellow above, but white below.* Fertile flowers *on ſhort fruit-ſtalks, at the ends of the branches.* Tranſparent 1. Pellucidum

1. Mnium ſerpilli foliis tenuibus et pellucidis. *Dillenius.* 232. t. 31. f. 2. *Common*

Mnium minus non ramoſum, anguſtioribus et pellucidis foliis. *Ray's Syn.* 78.

Smaller, leaves much narrower.

2. Mnium minimum non ramoſum anguſtiſſimis et pellucidis foliis. *Ray's Syn.* 78. *Small. Hud.*

Wet ſhady places, heaths and marſhes. Feb. Apr. July.

Fountain
2.
Fontanum

MARSHMOSS Stem ſimple, with crooked joints—*Tips large; roundiſh ; on long foot-ſtalks.* Fertile flowers *on the ſame plant ; and ſitting at the ends of the branches.*

Bryum paluſtre ſcapis teretibus ſtellatis, capſulis magnis ſubrotundis. *Dillenius.* 340. t. 44. f. 2.

Marſh Bryum. *Hudſon.* 404.

Near ſprings and in bogs. July—Aug.

Wherever this grows, a ſpring of freſh water may be found without much digging.

Common
3.
Paluſtre

MARSHMOSS Stem forked ; leaves awl-ſhaped—*Stem purpliſh. Fertile ſtars yellow.* Tips *purple, growing from the forks of the ſtem.* Veils *yellowiſh*

Mnium majus, ramis longioribus bifurcatis. *Dillenius.* 233. t. 31 f. 3. *Ray's Syn.* 78.

Turf bogs. P. June. July.

Branched
4
Hud.
Ramoſum

MARSHMOSS. Stem *ſomewhat* branched; upright. Fertile flowers on fruit-ſtalks riſing from the baſe of the leaves.—*Numerous.* Hudſon. 403.

Mnium majus, minus ramoſum capitulis pulverulentis crebrioribus. *Ray's Syn.* 78. *Dillenius.* 235. t 31. f. 4.

In marſhes.

OBS. *The barren flowers do not appear to have been diſcovered.*

Upright
5.
Androgynum

MARSHMOSS. Stem branched. Barren and fertile flowers on the ſame plant—*on fruit-ſtalks from the ends of the branches ; upright.* Fruit-ſtalks *of the barren flowers talleſt.* Veils *white.*

Mnium peranguſtis et brevibus foliis. *Ray's Syn.* 78. *Dillenius.* 230. t. 31. f. 1.

Woods and moiſt ſhady places; but ſometimes in dry ſandy places, and heaths. March.

Yellow
6.
Hygrometricum

MARSHMOSS. Stemleſs. Tips nodding; veil four edged; bent back—*Tips inverſely egg-ſhaped ; bright yellow.*

Bryum bulbiforme aureum, calyptra quadrangulari, capſulis pyriformibus nutantibus. *Dillenius.* 407. t. 52. f. 75.

Bryum aureum, capitulis reflexis pyriformibus, calyptra quadrangulari, foliis in bulbi formam congeſtis. *Ray's Syn.* 101.

Muſcus capillaris, ſive adiantum aureum minus. *Gerard.* 1559.

Polytrichon aureum minus. *Park.* 1052.

Yellow Bryum. *Hudſon.* 416.

Woods, heaths, garden walks, walls, old trees, decayed wood, and where coals or cinders have been laid. April.

If the fruit-ſtalk is moiſtened at the baſe with a little water or ſteam, the head makes three or four revolutions : if the head is moiſtened it turns back again.

MARSH-

MARSHMOSS. Stem forked; tips upright; on fruit-stalks rising from the forks of the stem; leaves keel-shaped. Purple 7. Purpureum

Purple Bryum. *Hudson.* 412.

1. Ends of the branches star-like; fruit-stalks purple; leaves spear-shaped; tips scored when dry. *Common*

Bryum tenue stellatum, setis purpureis. *Dillenius.* 336. t. 49. f. 51.

Bryum perangustis foliis et cauliculis, foliis et crebrioribus et circa summitates magis congestis, capitulis erectis e surculis annotinis egredientibus. *Ray's Syn.* 99.

2. Ends of the branches star-like; fruit-stalks rising from near the root; not much longer than the branches; tips edged at the mouth with a long tufted fringe. Leaves spear shaped. *Bearded. Hud.*

Bryum unguiculatum et barbatum, tenuius et stellatum. *Dillenius.* 384. t. 48. f. 48.

3. Leaves awl-shaped; distant. Fruit-stalks not much longer than the branches; tips edged with a long tufted fringe. *Slender. Hud.*

Bryum tenue barbatum, foliis angustioribus et rarioribus. *Dillenius.* 385. t. 48. f. 49.

Bryum perangustis foliis et cauliculis foliis rarioribus cinctis, capitulis erectis, e surculis annotinis egredientibus. *Ray's Syn.* 99.

4. Leaves awl-shaped; distant. Fruit-stalks very long; red; tips long; scored when dry. *Fine leaved. Hud.*

Bryum polytrichoides palustre, setis longioribus rubris sericeis. *Dillenius* 387. t. 49. f. 52.

Bryum perangustis crebrioribus foliis, capitulis erectis longiusculis, pediculis e surculis annotinis nascentibus. *Ray's Syn.* 99.

(1.) Heaths, roads, and walls. (2.) On the ground. (3.) Loose sandy soil, in gravelly pastures. (4.) Marshes. March, April. *Hudson.* (1.) January, February, April, May. *Dillenius.* (3.) Spring and Autumn. *Dillenius.*

MARSHMOSS? Black-headed

MNIUM? 8. Laterale

Stem somewhat branched Lateral branches flowering at the ends. Tips upright. Leaves keel-shaped—*Tips oblong; blackish. ends of the branches star-like. From* Dillenius. Dill.

Bryum palustre brevifolium, capsulis nigricantibus. *Dillenius.* t. 47. f. 39.

Yorkshire, in bogs.

MARSH-

Pale 9. Setaceum MARSHMOSS. Tips upright; lids thread-ſhaped; as long as the tips—*Stems upright*; *very ſhort.* Threads *blackiſh purple*; *zigzag.* Tips *cylindrical*; *lead coloured*; Lid *and* Veil *thread-ſhaped*; *longer and narrower than in any of the other ſpecies.*

Pale Bryum. *Hudſon.* 409.

Common 1. Leaves egg-ſhaped; of a pale green.

Bryum ſtellare nitidum pallidum, capſulis tenuiſſimis. *Dillenius.* 381. t. 48. f. 44.

Bryum trichoides exile pallidum, erectis capitulis e ſurculis annotinis egredientibus. *Ray's Syn.* 96.

Red. Hud. 2. Leaves awl-ſhaped, red; lids and veils rather ſhort.

Bryum ſtellare lacuſtre foliis rubris capillaceis. *Dillenius.* 382. t. 48. f. 45.

Obs. *This variety ſeems to have little affinity to the* Pale Marſhmoſs.

(1.) Ditch banks and walls. Winter and ſpring. March.
(2.) In bogs on mountains. Auguſt.

Starry 10. Cirratum MARSHMOSS. Leaves rolling back as they become dry—*Tips upright.*

Starry Bryum. *Hudſon.* 409.

True 1. Ends of the branches ſtar-like. *Tips* moſtly ſingle. *Lids* needle-ſhaped.

Bryum cirratum et ſtellatum, tenuioribus foliis. *Dillenius.* 379. t. 48. f. 42.

Bryum trichoides exile, erectis capitulis in pediculis longioribus rubris. *Ray's Syn.* 97.

Shortleaved. Hud. 2. Ends of the branches ſtar-like; leaves very ſhort.

Bryum peranguſtis et breviſſimis foliis, extremitatibus ſtellatis. *Ray's Syn.* 98.

Many headed? Hud. 3. Tips numerous; incorporated—*Ends of the branches ſtar-like when dry.*

Bryum cirratum, ſetis et capſulis brevioribus et pluribus. *Dillenius.* 378. t. 48. f. 41.

Bryum trichoides erectis ſublongis capitulis, extremitatibus per ſiccitatem ſtellatis. *Ray's Syn.* 98.

Obs. *Is not this variety, a ſpecies of the* Threadmoss, *and nearly allied to the* Twiſted Threadmoss?

1. Ditch-banks, walls, roofs, and old pales. (3) Mountains. March. *Hud.* (1) April. *Dill.* (3) Aut. *Dill.*

MARSH-

MARSHMOSS. Leaves egg-ſhaped, tapering to a point; transparent. Fruit-ſtalks riſing from near the root. Tips nodding—*Oblong*. Longleaved 11 Annotinum

Bryum annotinum lanceolatum pellucidum, capſulis oblongis pendulis. *Dillenius*. 399. t. 50. f. 68.

Long-leaved Bryum. *Hudſon*. 414.

Woods and moiſt ſhady places. March. In ſummer ſays *Dillenius*.

MARSHMOSS. Tips pendant. Fruit-ſtalks ſomewhat ſerpentine. Shoots undivided. Leaves rough at the edges—*Spear-ſhaped*. Tips *large*; *oblong*, *inclining to oval*. Rough 12 Hornum

Rough Bryum. *Hudſon*. 415.

1. Bryum ſtellare, hornum ſylvarum, capſulis magnis nutantibus. *Dillenius*. 402. t. 51. f. 71. *Swan's Neck.*

Bryum nitidum capitulis majoribus reflexis, calyptra imum virgente, pediculis oblongis e cauliculis nevis egredientibus. *Ray's Syn*. 102.

2. With narrower leaves. *Narrow leaved.* Hud.

Bryum nitidum foliis ſerpilli anguſtioribus, medium. *Ray's Syn*. 103.

Woods and heaths. April—May.

MARSHMOSS. Tips pendant. Leaves egg-ſhaped; keeled; briſtly at the ends. Fruit-ſtalks very long—*riſing from near the root*. Tips *oblong*. Briſtly 13 Capillare

Capillary Bryum. *Hudſon*. 414.

1. Bryum foliis latiuſculis congeſtis, capſulis longis nutantibus. *Dillenius*. 398. t. 50. f. 67. *Common*

Bryum capitulis reflexis, foliolis latiuſculis congeſtis. *Ray's Syn*. 100.

2. Leavebroader; tranſparent. *Broad leaved.* Hud.

Bryum capitulis reflexis foliis congeſtis latioribus et pellucidis. *Ray's Syn*. 101.

Woods and heaths in a gravelly ſoil. March.

MARSHMOSS. Tips pendant. Veils bowed back. Leaves tranſparent—*ſpearſhaped*. Spear-leaved 14

Bryum pendulum hornum molle, foliis et lanceolatis et gramineis. *Dillenius*. 401. t. 50. f. 70. Crudum

Spear-leaved Bryum. *Hudſon*. 415.

Fens in Cambridgeſhire. Sum. *Dillenius*. Mar. *Hudſon*.

MARSHMOSS. Tips pendant: turban-ſhaped. pillar thread-ſhaped. Fertile flowers briſtly.— Golden 15

Bryum trichoides aureum, capitulis pyriformibus nutantibus. *Dillenius* 391. t. 50. f. 60. Pyriforme

Golden Bryum. *Hudſon* 412.

On rocks. April—May.

MARSH-

Hair 16 Polytrichoides — MARSHMOSS. Veil woolly.—*Open; larger than the tip.* Shoot *undivided.*

Dwarf Polytricum. *Hudſon* 400.

Dwarf 1. Tips roundiſh; leaves entire.

Polytricum nanum, capſulis ſubrotundis galeritis, aloes folio non ſerrato. *Dillenius* 428. t. 55. f. 6.

Polytricum minus, capſulis ſubrotundis, calyptra quaſi lacera coronatis. *Ray's Syn.* 91.

Serrated 2. Tips oblong; leaves ſerrated.

Polytricum parvum, aloes folio ſerrato, capſulis oblongis. *Dillenius.* 428. t. 55. f. 7.

Moiſt heaths. From October to March.

Thyme leaved 17 Serpyllifolium — MARSHMOSS. Fruit-ſtalks incorporated. Leaves expanding; tranſparent—

Thyme-leaved Bryum. *Hudſon* 417.

Dotted Punctatum 1. Fruit-ſtalks incorporated. Leaves very entire; inverſely egg-ſhaped; blunt; dotted.

Bryum pendulum ſerpillifolia rotundiore pellucido, capſulis ovatis. *Dillenius* 416. t. 53. f. 81.

Bryum nitidum ſerpilli rotundis et latioribus foliis pellucidis. *Ray's Syn.* 103.

Smooth Glabrum, Hud. 2. Fruit-ſtalks incorporated. Leaves very entire; betwixt ſpear and egg-ſhaped. Tips oblong; pointed. *From Dillenius.*

Bryum pendulum, ſerpillifolio longiore pellucido, capſulis oblongis cuſpidatis. *Dillenius* 486 t. 53. f. 80.

Pointed Cuſpidatum 3. Bryum pendulum foliis variis pellucidis capſulis ovatis. *Dillenius* 413. t. 53. f. 79.

Roſy Proliferum 4. Fruit-ſtalks incorporated. Leaves ſpear-ſhaped; ſharp; diſpoſed in form of a roſe.—*Tips egg-ſhaped.*

Bryum ſtellare roſeum majus capſulis ovatis pendulis. *Dillenius* 411. t. 52. f. 77.

Bryum roſeum majus, foliis oblongis. *Ray's Syn.* 92.

Starry Stellatum Hud. 5. Smaller than the preceding; tips and fruit-ſtalks longer.

Bryum ſtellare roſeum pendulum minus, capſulis et ſetis longiorlbus. *Dillenius* 412. t. 52. f. 78.

Bryum roſeum minus, foliis ſubrotundis. *Ray's Syn.* 92.

Muſcus parvus ſtellaris. *Bauh. pin.* 361. *Gerard.* 1563. *Park.* 1308.

Waved Undulatum 6. Fruit-ſtalks i corporated. Leaves oblong; waved.

Bryum dendroides polycephalon, phyllitidis folio undulato pellucido, capſulis ovatis pendulis. *Dillenius* 410. t. 52. f. 76.

Bryum nitidum foliis oblongis undulatis, capitulis cernuis arbuſculam referens. *Ray's Syn.* 103.

Woods and moiſt heaths. March—April. (4. and 5. Winter *Dill*)

MARSH-

MARSHMOSS. Leaves pointing three different ways; open; betwixt awl and ſpear-ſhaped: keeled—*Stems woolly; about the length of ones finger or longer; growing cloſe together and forming a turf.* Branches *ſeveral, riſing from the end of the ſhoot; three ſquare, open.* Leaves *ſmall; diſtant; tapering to a point; when young of a pale green, but when old brown.* Fruit-ſtalks *ſingle, purple; very long; riſing from the end of the ſhoot.* Tips *either upright, oblique, or pendant; very blunt at the end; yellow.* Veil *awl-ſhaped.* Fertile Flowers *ſtar-like; at the ends of the branches; on the ſame plant with the barren flowers.* Triangular 18 Triquetrum

Triangular Bryum. *Hudſon* 416.

1. Tips oblong, ſomewhat ſwollen. *Swollen*

Bryum annotinum paluſtre, capſulis ventricoſis pendulis. *Dillenius* 404. t. 51. f. 72.

Bryum nitidum foliis ſerpilli anguſtioribus, majus. *Ray's Syn.* 102.

2. Tips oblong, narrower; branches upright, ſlenderer; leaves more thinly ſet. *Long-ſhanked* Hud.

Bryum lanceolatum bimum, ſetis et capſulis longis pendulis. *Dillenius* 405. t. 51. f. 73.

Bryum nitidum foliis ſerpilli pellucidis anguſtis, capitulis tumidis nutantibus, præaltis pediculis ſurculis annotinis egredientibus. *Ray's Syn.* 102.

3. Tips inverſly egg-ſheped. Stems ſhorter. Branches upright. Leaves thinly ſet. *Pear-headed* Hud.

Bryum paluſtre complicatum rubens, capſulis turbinatis pendulis. *Dillenius* 406. t. 51. f. 74.

Bryum nitidum rubens capitulis reflexis, foliis anguſtis pellucidis, cauliculis proliferis. *Ray's Syn.* 102.

Bogs and marſhes. April.

MARSHMOSS. Leaves pointing from two oppoſite lines; very entire; *Fertile flowers on ſhort fruit-ſtalks, at the ends of the branches.* Fern 19 Trichomanis

Mnium trichomanis facie, foliolis integris. *Ray's Syn.* 79. *Dillenius* 236.

Wet ſhady places. February—March.

MARSHMOSS. Leaves pointing from two oppoſite lines; cloven.—*Fertile flowers on ſhort fruit-ſtalks, at the ends of the branches.* Forked 20 Fiſſum

Mnium trichomanis facie, foliolis bifidis. *Ray's Syn.* 79. *Dillenius* 237. t. 31. f. 6.

Shady places. Feb. March, April.

Liver 21 Jungermannia

MARSHMOSS. Leaves pointing from two opposite lines, with little ear-like appendages underneath—*Leaves alternate*; *transparent*; *somewhat egg-shaped*; *smooth*; *embracing the stem.*

Lichenastrum alpinum purpureum foliis auritis et cochleariformibus. *Dillenius* 479. t. 69. f. 1.

Lichenastrum trichomanis facie, prælongum, foliis concavis, unam partem spectantibus. *Ray's Syn.* 112.

Rivulets and marshy places on mountains.

416 THREADMOSS. 1194 Bryum.

Barren Flower on a fruit-stalk, rising from the end of the shoots.

EMPAL. *Veil* oblong; tapering to a point; oblique; mostly smooth.

CHIVE. *Tip* roundish, or somewhat oblong; the mouth edged with a fringed ring, and covered wsth a conical *Lid*.

RECEPT. *Excrescence* none.

SCALY BULB. None; but a *Tubercle* at the base of the fruit-stalk.

OBS. *The* smooth tip'd hairy; *the* pear-headed *and the* beardless *Threadmoss, have no fringe round the mouth of the tip.*

* *Tips sitting.*

Hairy 1 Apocarpum

THREADMOSS. Tips sitting, at the ends of the branches. Veil very small. Stem branched; *Leaves* terminating in hairs; —Tips *inclosed by the* leaves.

Red-tipt Dill. and Lin.

1. Of a dark green; tips red; oblong: mouth fringed.

Spagnum subhirsutum obscure virens, capsulis rubellis. *Dill.* 245. tab. 32. fig. 4.

Smooth-tipt

2. Leaves membranaceous at the ends. Tips greenish, or yellowish; nearly round, but flatted at the top. Mouth not fringed.

Sphagnum nodosum hirsutum incanum. *Dillenius* 246. t. 32. f. 5.

Sphagnum cauliferum et ramosum saxatile hirsutum incanum capitulis virentibus. *Ray's Syn.* 105.

OBS. *In* Dillenius's *figure there is no appearance either of a* Veil, *or* Fringe round the mouth of the Tip. *May not this variety therefore be really a species of* BOGMOSS as Dillenius *has described it?*

On stones and trees. (1) Nov. Dec. Jan. (2) Nov. Dec.

THREAD-

THREADMOSS. Tips *moſtly* ſitting; diſperſed over the branches. Veils ſcored, or covered with hairs pointing upwards—*Nearly allied to the* HAIRMOSS, *but has no excreſcence under the tip*. Stem *branched*. Scored a Striatum

Tree Polytricum. *Hudſon* 401.

1. Tips ſitting, at the ends and ſides of the branches. Leaves awl-ſhaped. *Greater*

Polytrichum bryi ruralis facie, capſulis ſeſſilibus majus. *Dillenius* 430. t. 55. f. 8.

Polytrichum capſulis ſubrotundis pediculis breviſſimis inſidentibus, calyptra ſtriata, arboreum ramoſum, majus. *Ray's Syn.* 91.

2. Tips ſitting, at the ends and ſides of the branches. Leaves egg-ſhaped. Smaller than the preceding. *Leſſer*

Polytricum bryi ruralis facie, capſulis ſeſſilibus minus. *Dillenius* 431. t. 55. f. 9.

Polytrichum capſulis ſubrotundis, pediculis breviſſimis inſidentibus, calyptra ſtriata, arboreum et terreſtre minus ramoſum et breve. *Ray's Syn.* 91.

3. Tips ſitting, moſtly at the ends of the branches. Leaves betwixt egg and ſpear-ſhaped. *Slender*

Polytrichum capſulis ſeſſilibus, foliis brevibus, rectis, carinatis. *Dillenius* 432. t. 55. f. 10.

4. Tips on ſhort fruit-ſtalks, moſtly from the ends of the branches. veils pointed and exceedingly hairy. Leaves awl-ſhaped and curled when dry. *Curled*

Polytricum capillaceum criſpum, calyptris acutis piloſiſſimis. *Dillenius* 433. t. 55. f. 11.

Polytrichum capſulis oblongo rotundis, calyptris piloſiſſimis. *Ray's Syn.* 91.

(1. 3.) Rocks. (2.) Trees, ſtones, walls and earth. (4) Woods. February—March.

** *Tips on fruit-ſtalks upright.*

THREADMOSS. Tips upright; ſpherical—*brown*. Leaves *briſtle-ſhaped*; Fruit-ſtalks *ſolitary*; *purpliſh towards the lower part*. Veil *very ſharp*; *ſmall*; *oblique*; *red*. Round-headed 3 Pomiforme

Bryum capillaceum, capſulis ſphæricis. *Dillenius* 339. tab. 44. fig. 1.

Bryum trichoides vireſcens, erectis majuſcuis capitulis maliformibus. *Ray's Syn.* 97.

On heaths and ſandy banks. March.

THREAD-

Pear-headed 4 Pyriforme

THREADMOSS. Tips upright, inverſely egg-ſhaped; veil awl-ſhaped; ſhoots without ſtems; leaves egg-ſhaped, without hairs.

Bryum ſerpillifolium pellucidum, capſulis pyriformibus *Dillenius.* 345. tab. 44. fig. 6

Bryum parvum erectis pyriformibus majuſculis capitulis, foliolis ſerpylli pellucidis. *Ray's Syn.* 93.

Sandy meadows and ditchbanks. March.

OBS. *The mouth of the tip does not appear to be fringed from* Dillenius's *figure.*

Conic Extinctorium

THREADMOSS. Tip upright; oblong; ſmaller than the veil which is flexible; and equal at the baſe,—*and cut into ſix ſegments which are bent inwards.*

Smaller

1. Bryum calyptra extinctorii forma minus. *Dillenius* 349. tab. 45. fig. 8.

Bryum erectis captiulis, calyptra laxa conica, foliis Serpylli pellucidis anguſtioribus. *Ray's Syn.* 92.

Larger

2. Larger and more branched.

Bryum calyptra extinctorii figura, majus et ramoſum. *Dillenius* 350. tab. 45. fig. 9.

Sandy meadows. Feb. (1. Nov.—Dec. *Dill.*)

Awl-ſhaped 6 Subulatum

THREADMOSS. Tips upright, awl-ſhaped. Shoots without Stems—*Veil brown.*

Bryum capſulis longis ſubulatis. *Dillenius* 350 tab. 45. fig. 10.

Bryum erectis longis et acutis falcatis capitulis, calyptra ſubfuſca, foliis ſerpylli pellucidis. *Ray' Syn.* 92.

Woods and moiſt ditch-banks. March. (Jan.—Feb. *Dill.*)

Grey 7 Rurale

THREADMOSS. Tips nearly upright. Leaves bent back. —*Tips bent like the claw of a bird.* Leaves *terminating in hoary hairs.*

Bryum rurale unguiculatum hirſutum elatius et ramoſius. *Dillenius.* 352. tab. 45. fig. 12.

Bryum majus erectis falcatis capitulis, foliis latiuſculis extantibus, in pilum caneſcentem deſtinentibus. *Ray's Syn.* 94.

Walls, roofs, and trunks of trees. March.

When this takes to grow upon thatched buildings ſo as to cover the thatch, inſtead of laſting but about ten years it will endure for an age.

THREADMOSS. Tips upright. Leaves terminating in hairs, nearly ſtraight, ſhoots ſimple; growing cloſe together, and forming a turf.—*Leaves terminating in hoary hairs.* Wall 8 Murale

Bryum tegulare humile piloſum et incanum. *Dillenius.* 355. tab. 45. fig. 14.

Bryum minus erectis minus falcatis capitulis, foliis latiuſculis congeſtis in pilum caneſcentem definentibus. *Ray's Syn.* 94.

Roofs, tiles, ſtones, and walls. Nov.—March.

THREADMOSS. Tips upright. Leaves not hairy; tranſparent; nearly ſtraight. Shoots moſtly ſimple, growing cloſe together, aad forming a turf. *Green.* From Dillenius. Ground 9 Humile *Dill.*

Bryum humile, pilis carens, viride et pellucidum. *Dillenius* 356. tab. 45. fig. 15.

Bryum murale. *Hudſon.* 406.

On banks of earth, near Streatham in Surry, and Oxford. Winter. *Dill.*

THREADMOSS. Tips nearly upright. Fruit-ſtalks incorporated. Leaves pointing one way; bowed back; ſtem declining.— Broom 10 Scoparium

Bryum reclinatum, foliis falcatis, ſcoparum effigie. *Dillenius.* 357. tab. 46. fig. 16.

Bryum erectis capitulis anguſtifolium, caule reclinato. *Ray's Syn.* 95.

Heaths, woods, and ſhady banks. March—April.

THREADMOSS. Tips nearly upright. Fruit-ſtalks moſtly ſingle. Leaves ſpear-ſhaped; keeled; waved; expanding; ſerrated.—*Tips reddiſh.* Curled 11 Undulatum

Bryum phyllitidis folio rugoſo acuto, capſulis incurvis. *Dill.* 360. tab. 46. fig. 18.

Bryum capitulis oblongis rubentibus, foliis oblongis anguſtis pellucidis rugoſis. *Ray's Syn.* 95.

Woods and ſhady ditchbanks. Dec.—Jan.

THREADMOSS. Tips rather upright. Lid arched. Leaves upright; tiled. Shoots branched.—*Leaves whitiſh.* Lid *exactly awl-ſhaped.* Tips *ſometimes riſing from ſhort lateral branches.* White 12 Glaucum

Bryum albidum et glaucum fragile majus, foliis erectis, ſetis brevibus. *Dillenius.* 362. tab. 46. fig. 20.

Bryum trichoides erectis capitulis, albidum fragile. *Ray's Syn.* 97.

High commons. (Aug.—Sept. *Dil.*) Oct.—Nov. *Hud.*

Tranſparent 13 Pellucidum THREADMOSS. Tips rather upright. Leaves bowed back; ſharp. Stem hairy.—

Bowed 1. Leaves bowed back.—*Tips brown.*

Bryum paluſtre pellucidum, capſulis et foliis brevibus recurvis. *Dillenius.* 364. tab. 46. fig. 23.

Bryum erectis capitulis ſubrotundis fuſcis, foliis minoribus pellucidis rugoſis. *Ray's Syn.* 96.

Bent 2. Leaves bent back.

Bryum erectis capitulis brevibus, foliis reflexis. *Dillenius* 365. tab. 16. fig. 24.

Marſhes and moiſt ſhady places.

Mountain 14 Aciculare THREADMOSS. Tips upright. Lid needle-ſhaped. Leaves upright; ſome of them pointing one ways—*Briſtles yellow*; Veils *very ſharp*; Lid *long and ſlender like a needle.* Tips *moſtly riſing from ſhort lateral branches.*

Bryum montanum hemiheterophyllum, operculis acutis. *Dillenius* 366. tab. 46. fig. 25.

Bryum hypnoides erectum montanum, erectis capitulis acutis. *Ray's Syn.* 94.

On mountains. Auguſt.

Soft 15 Flexuoſum THREADMOSS. Tips upright; leaves briſtle-ſhaped; fruit-ſtalks zigzag.—

Bryum piloſum molle, ſetis intortis. *Dillenius* 373. t. 47. f. 33.

Bryum trichoides capitulis erectis, pediculis intortis tenuibus virentibus. *Ray's Syn.* 97.

On rocks.

Heath 16 Heteromallum THREADMOSS. Tips upright; leaves briſtle-ſhaped; pointing one way—*Very long.* Tips *roundiſh.*

Pencil 1. Bryum heteromallum. *Dillenins* 375. tab. 47. fig. 37.

Bryum trichoides reclinatis cauliculis capitulis erectis acutis. *Ray's Syn.* 96.

Cluſter Hud. 2. Leaves curled inwards; in diſtinct tufts; ſhorter. Tips oblong.

Byrum piloſum interrupte falcatum. *Dillenius* 376. tab. 47. fig. 38.

(1.) Heaths, and ditchbanks. Oct (2) Heaths and mountains. Aug.

THREAD-

THREADMOSS. Tips upright; leaves briſtle-ſhaped, with- Twiſted
out hairs, twiſted back when dry.—*Fruit-ſtalks purpliſh; tips* 17
cylindrical; lid with a creſt of tawny hairs. Tortuoſum

Bryum cirratum, setis et capſulis longioribus. *Dillenius* 3 7. tab. 48. fig. 40.

Bryum trichoides longifolium craſſiuſculis cauliculis, capitulis erectis aduncis acutis. *Ray's Syn.* 98.

On mountains. Auguſt—Oct.

THREADMOSS. Tips upright. Mouth not fringed. Beardleſs
Leaves keeled; open. *Hudſon* 409.—*Riſing from near the ends* 18
of the branches. Imberbe Hud.

Bryum tenue, imberbe et pallidum, foliis crebrioribus. *Dillenius* 382. tab. 48. fig. 46.

Bryum peranguſtis foliis et cauliculis, foliis crebrioribus, et circa extremitates magis congeſtis; capitulis erectis, ad ſummitatem magis egredientibus. *Ray's Syn.* 99.

On ſandy and graſſy places, ditch-banks and walls. March. (Winter *Dillenius*..)

THREADMOSS. Tips upright; oblong; on fruit-ſtalks riſing Bird's-claw
from the baſe of the leaves; leaves upright, pointed, keeled,— 19
Of a yellowiſh green. Stem *upright ſomewhat branched.* Fruit- Unguiculatum
ſtalks *ſhort, riſing from the baſe of the leaves,* (*but at the ends of the* *Hud.*
branches) *Hudſon* 410.—Tips *upright but the veils oblique, which gives it ſome reſemblance to the claw of a bird. Mouth of the tip edged with a tufted fringe.*

Bryum unguiculatum et barbatum, ſurculis in ſummitate craſſioribus. *Dillenius* 383. tab, 48. fig. 47.

Bryum anguſtis viridibus foliis, capitulis erectis brevibus pediculis inſidentibus, calyptra falcata vel avium unguiculos referente. *Ray's Syn.* 96.

On walls and ſandy places. March.

THREADMOSS. Tips upright; roundiſh; ending in a Brown
ſharp point.—*When the lid falls off the* Tip *appears lopped; inverſely* 20
egg-ſhaped, and of a yellowiſh red. Truncatulum

1. Bryum exiguum creberrim capſulis rufis. *Dillenius* 347. *Many-headed*
tab. 45. fig. 7.

Bryum exiguum erectis parvis ſubrotundis creberrimis capitulis rufis, foliolis ſerpilli anguſtis pellucidis. *Ray's Syn.* 94.

2. Tips larger; browniſh. *Headed.* Hud.

Bryum parvum erectis ſubrotundis majuſculis capitulis ſubfuſcis, foliis ſerpilli pellucidis. *Ray's Syn.* 93.

Meadows and paſtures. February.

Haſſelquiſt obſerving this plant growing in great abundance upon the walls of Jeruſalem, conjectures it may be the Hyſſop of Holy Writ, wherein Solomon is ſaid to have known all plants, from the Cedar of Lebanon even unto the Hyſſop, that groweth upon the wall.

Green 21 Viridulum THREADMOSS. Tips upright; egg-ſhaped. Leaves ſpear-ſhaped, tapering to a point, open but ſomewhat tiled.—*Of a bright green.* Tips *yellowiſh*; *fringed.* Leaves *ſomewhat curled when dry.*

Bryum capillaceum breve, pallide et læte virens, capſulis ovatis. *Dillenius.* 380. tab. 48. fig. 43.

Bryum trichoides exile, erectis capitulis in pediculis breviſſimis. *Ray's Syn.* 97.

Sandy ditch-banks, and ſides of hedges, and on moiſt heaths. Dec.—Feb.

Dwarf 22 Paludoſum THREADMOSS. Without any ſtem. Leaves briſtle-ſhaped. Tips very blunt; ſomewhat expanding—*roundiſh*; *brown*; *leaves not curled when dry. Extremely ſmall if not the ſmalleſt of all the Moſſes.*

Bryum trichoides acaulon paluſtre minimum, ſetis et capſulis breviſſimis. *Dillenius* 387. tab. 49. fig. 53.

Marſhes. March.

Heath-leaved 23 Ericæfolium *Dill.* THREADMOSS. Without a ſtem. Leaves awl-ſhaped; blunt; open. Tips oblong; upright.—*covered with a pointed* Lid, *and edged at the mouth with a loug upright tufted fringe.* From Dillenius.

Bryum acaulon, ericœ tenuifoliæ. *Dillenius* tab. 49. fig. 55.

Near Wigmore Herefordſhire. Nov. *Dill.*

Woolly 24 Hypnoides THREADMOSS. Tips upright. Shoot nearly upright. Lateral branches ſhort; bearing flowers.—*Tips covered with a* Lid, *moſtly pointed and edged at the baſe with little teeth.*

Many-headed 1. Tips ſmall; fruit-ſtalks ſhort. Stem not much divided. Lateral branches numerous; leaves ending in a woolly hair.

Bryum hypnoides polycephalon lanuginoſum montanum. *Dillenius* 372. tab. 47. fig. 32.

Bryum hypnoides capitulis plurimis erectis, lanuginoſum. *Ray's Syn.* 97

Small-headed Hud. 2. Tips ſmall; fruit-ſtalks ſhort. Stem pretty much branched: branches long; nearly of an equal length; bearing flowers at the ends. Leaves awl ſhaped.

Bryum hypnoides alpinum ſetis et capſulis exiguis. *Dillenius* 370. tab. 47. fig. 29.

3. Tips

3. Tips large; pointed; lids entire at the base. Stem branched. Leaves ending in a short woolly hair; green. *Green*

Bryum hypnoides hirsutie virescens fasciculare alpinum. *Dillenius* 370. tab. 47. fig. 28.

4. Tips long: mouth edged with a long tufted fringe: on long fruit-stalks rising from the upper part of the shoot. Shoot not much branched. Leaves betwixt awl and egg-shaped. *Heath-like*

Bryum hypnoides ericæ facie, capsulis barbatis alpinum. *Dillenius*. 371. tab. 47. fig. 31.

Bryum hypnoides capitulis plurimis erectis, non lanuginosum. *Ray's Syn*. 478.

5. Tips large; lids blunt at the end. Shoot branched. Leaves awl-shaped. *Blunt-headed* Hud.

Bryum hypnoides, alpinum operculis obtusis. *Dillenius* 371. tab. 47. fig. 30.

6. Tips rather large. Shoots mostly branched; but sometimes simple. Leaves awl-shaped; ending in a wolly ha ir. *Common* Hud.

Bryum hypnoides, hirsute canescens, vulgare. *Dillenius* 368. tab. 47. fig. 27.

Bryum trichoides erectis capitulis, lanuginosum. *Ray's Syn*. 97.

7. Tips large. Veils black; pointed. Shoot somewhat drooping leaves egg-shaped sharp. *Water* Hud.

Bryum hypnoides aquaticum calyptris nigris acutis. *Dillenius* 367. tab. 46. fig. 26.

Bryum hypnoides repens aquaticum, erectis capitulis acutis. *Ray's Syn*. 94.

On mountains. Dec. (2) Aug. (3) Sept.

Obs. *Future observation may determine whether several of these are not distinct species—The fourth and fifth will probably prove so.*

THREADMOSS. Tips upright; fruit-stalks crowding together when dry. Leaves hairy. Shoots all of the same height. *Leaves as if surrounding the stem in whorls.* Whorled 25 Verticilliatum

1. Bryum pilosum verticillatum. *Dillenius*. 374. tab. 47. fig. 35. *Common*

Bryum trichoides brevifolium angustis cauliculis, capitulis erectis parvis et minus aduncis. *Ray's Syn*. 98.

2. Leaves not in whorls; branches longer; of a bright green. *Bogmoss* Hud.

Bryum pilosum, Sphagni subulati facie. *Dillenius* 374. tab. 47. fig. 34.

Mountains and rocks.

Water THREADMOSS. Tips upright, roundiſh; growing from the
26 baſe of the leaves or branches. Leaves awl-ſhaped; diſtant.—
Æſtivum *Stem forked.* Fruit-ſtalks *ſometimes riſing from the diviſions of the ſtem, and ſometimes from the ends of the branches.*

Bryum paluſtre æſtivum confervæ facie. *Dillenius* 375. tab. 47. fig. 36.

Bryum anguſtiſſimis foliis crebrioribus, capitulis erectis brevibus pediculis e ſurculis novis et longis enaſcentibus. *Ray's Syn* 99.

Bryum paluſtre. *Hudſon.* 411.

Marſhes. March.

Obs. *This ſeems to have no ſmall affinity to the* ſlender, *and* fine-leaved *purple* Marshmoss.

Twin THREADMOSS. Tips nearly upright; two growing to-
27. gether. Shoots forked —*Tips at the ends of the branches; ſcored*
Geminatum *when dry.* Lids *ſhort. Blunt. From* Dillenius.
Dill.

Bryum parvum, ſurcuris et ſetis geminatis. *Dillenius* tab. 49. fig. 50.

On hedges in gardens, and on walls. Feb—March.

*** *Tips pendant.*

Silver THREADMOSS. Tips pendant; ſhoots cylindrical, tiled,
28 ſmooth.—*Silvery*; Leaves *minute, numerous, triangular, compact;*
Argenteum *terminating in a ſhort hair*; Tips *blackiſh*; Lid *ſhort.* Fruit-ſtalks *riſing from the baſe of the ſhoots.*

Silky 1. Bryum pendulum julaceum argenteum et ſericeum. *Dillenius* 392. tab. 50. fig. 62.

Bryum capitulis ſabrotundis reflexis, cauliculis teretibus argenteis. *Ray's Syn.* 100.

Green 2. Shoots green; leaves egg-ſhaped; keeled—*Some of the branches with ſtarlike extremities.*—*Q. Whether it may not be a ſpecies of* Marſhmoſs?

Bryum pendulum, ſurculis teretibus viridibus. *Dillenius* 394. tab. 50. fig. 63.

Roofs, walls, (1) ſunny expoſures) (2 garden walls. Dec.

Pincuſhion THREADMOSS. Tips roundiſh, fruit-ſtalks bent back.
29 Leaves hairy—*Fruit-ſtalks green, ſhort, crooked, ſo that the tips*
Pulvinatum *almoſt touch the leaves.*

Bryum orbiculare pulvinatum hirſutie caneſcens, capſulis immerſis. *Dillenius* 395. tab. 50. fig. 65.

Bryum trichoides hirſutie caneſcens, capitulis ſubrotundis reflexis, in perbrevibus pediculis. *Ray's Syn.* 100.

Walls and tiles of houſes. April.

THREADMOSS. Tips pendant. Leaves ſpear-ſhaped, tapering to a briſtly point. Fruit-ſtalks extremely long.---*From the baſe of the ſhoots red about half way up; of a yellowiſh green above.* Matted 30 Cæſpiticium

Bryum pendulum ovatum cæſpiticium et piloſum, ſeta bicolori. *Dillenius*.396. tab. 50. fig. 66.

Bryum trichoides capitulis reflexis, pediculis ima medietate rubris, ſumma luteo-virentibus. *Ray's Syn.* 100.

Old walls, roofs, and gravelly places. March. (Winter and ſpring. *Dill.*)

THREADMOSS. Tips pendant; awl-ſhaped. Fruit-ſtalks riſing from the ends of the ſhoots;—*Shoots moſtly undivided, of a fine green.* Veils *ſhedding.* Lids *blunt.* Briſtly 31 Virens *Dill*

Bryum trichoides læte virens, capitulis cernuis oblongis. *Dillenius* tab. 50. fig. 61.

Woolwich-heath, and Snowdon. Spr. and Sum. *Dill.*

THREADMOSS. Tips pendant; ſomewhat globular. Leaves ſharp alternate—*tranſparent.* Tips *pale red, or fleſh coloured.* Shining 32 Carneum

1. Bryum lanceolatum pellucidum. capſulis rotundis pendulis, carneis. *Dillenius* 400. tab. 5. fig. 69. *Fleſh-coloured*

Bryum nitidnm, foliis ſerpilli pellucidis anguſtioribus reflexis; capitulis ſubrotundis carnei coloris. *Ray's Syn.* 102.

2. Tips green. *Green.* Hud.

Bryum nitidum peranguſtis ſerpillinis pellucidis foliis, reflexis capitulis ſubrotundis virentibus. *Ray's Syn.* 102.

Wet places; drains in meadows. March.

THREADMOSS. Tips pendant oblong: leaves ſpear-ſhaped (*egg-ſhaped tapering to a point.* Hud.) keeled. Shoots branched. fruit-ſtalks riſing from the baſe of the leaves. *Hudſon* 415. Alpine 33 Alpinum *Hud*

Bryum hypnoides pendulum ſericeum, coma inſigni atro-rubente. *Dillenius* 394. t. 50. f. 64.

On the ground, upon rocks.

THREADMOSS. Tips nodding, oblong. Leaves awl-ſhaped. Stem undivided; fruit-ſtalk riſing from the middle of of the ſtem.—*Tips red.* Red 34 Simplex

Bryum rubrum. *Hudſon* 413.

Bryum trichoides, capſulis rubris cernuis. *Dillenius* 390. t. 50. f. 9.

In graſſy places. Wint.

417 FEATHERMOSS. 1195 Hypnum.

Barren Flower on a fruit-ſtalk, riſing from the ſides of the ſhoots, and from the baſe of the leaves.

EMPAL. *Veil* oblong; ſmooth; oblique.

CHIVE. *Tip* rather oblong; *Mouth* fringed; covered with a *Lid*: tapering to a point.

RECEPT *Excreſcence* none.

SCALY BULB. Scaly on the ſides of the ſhoots, ſurrounding the baſe of the *Fruit-ſtalk*.

Fertile Flower on the ſame plant, at the baſe of the leaves.

EMPAL. *Common*, none.

POINT. Little feather-like ſhoots riſing ſingle from the baſe of the leaves.

* *Leaves winged.*

Yew-leaved Taxifolium — FEATHERMOSS. Leaf very ſimple; winged; ſpear-ſhaped. Fruit-ſtalks riſing from the baſe.—*Tips oblong; drooping.*

Hypnum taxiforme minus, baſi capſulifera. *Dillenius*. 263. t. 34. f. 2.

Hypnum repens filicifolium non ramoſum, pediculis brevioribus ad radicem egredientibus. *Ray's Syn*. 88.

Woods, and ſhady banks. Feb. (April. *Dil'*.)

Double-rowed 2 Denticulatum — FEATHERMOSS. Leaf ſimple, winged; with a double row of little leaves on each ſide.—*Fruit ſtalks riſing from the baſe.*

Hypnum denticulatim pinnatum, pinnulis duplicatis recurvis. *Dillenius*. 266. t. 34. f. 5.

Hypnum repens filicifolium non ramoſum, pediculis et capitulis longioribus adaradicem egredientibus, foliolis utrinque duplicatis. *Ray's Syn*. 88.

Woods, and moiſt ſhady places on the ground. April.

Fern 3 Bryoides — FEATHERMOSS. Leaf very ſimple; winged; ſpear-ſhaped. Fruit-ſtalks riſing from the end—*This Moſs is ſmall, but extremely elegant. The tips are crowned with a lid of a lively red, and edged at the mouth with an elegant fringe of the ſame colour.*

Hypnum taxiforme exiguum, verſus ſummitatem capſuliferum. *Dillenius* 262. t. 34. f. 1.

Hypnum repens filicifolium non ramoſum, pediculis brevibus verſus foliorum ſummitatem egredientibus. *Ray's Syn*. 88.

Shady places and ditch-banks. Feb. (Wint. and Spr. *Dill.*)

FEATHER-

FEATHERMOSS. Leaf winged; branched; trailing. Wood
Little leaves ſharp. Fruit-ſtalks riſing from the middle. *Hudſon* 419. 4 Sylvaticum Hud.

Hypnum denticulatum pennatum, pinnulis ſimplicibus rectioribus. *Dillenius* 267 t. 34. f. 6.

Hypnum repens filicifolium ramoſum, pedunculis et capitulis longioribus e foliorum alis egredientibus, foliolis utrinque ſimplicibus. *Ray's Syn.* 88.

Woods; at the roots of trees, and moiſt ſhady places on the ground. April.

FEATHERMOSS. Leaf winged; branched; upright. Upright
Fruit-ſtalks riſing from the middle. 5 Adiantoides

Hypnum taxiforme paluſtre ramoſum majus et erectum. *Dill.* 264. t. 34. f. 3.

Hypnum erectum filicifolium ramoſum, pinnulis acutis. *Ray's Syn.* 87.

In marſhy grounds near water. March.

FEATHERMOSS. Leaf winged; branched: little leaves Flat
tiled; ſharp; folded; compreſſed.—*Gloſſy.* Tips *egg-ſhaped.* 6 Complanatum

1. Hypnum pennatum compreſſum et ſplendens. capſulis ovatis. *Fern*
Dillenius 268. t. 34. f. 7.

Hypnum repens filicifolium ramoſum, ramulis appreſſis et majus complanatis. *Ray's Syn.* 87.

2. Little leaves blunt; reſembling Maidenhair. *Maiden-hair.* Hud.

Hypnum pennatum trichomanoides, ſplendens ramoſum. *Dillenius* 269. t. 34. f. 8.

Hypnum erectum filicifolium ramoſum pinnulis obtuſis. *Ray's Syn.* 87.

On trunks of trees, and (2) ſometimes on the ground. Mar. —April.

* * *Shoots ſtraggling.*

FEATHERMOSS. Shoots branched: branches ſomewhat Shining
winged. Little leaves dotted—*Gloſſy*; *egg-ſhaped.* 7. Lucens

Hypnum pennatum aquaticum lucens, longis latiſque foliis. *Dillenius* 270. t. 34. f. 10.

Hypnum repens filicifolium ramoſum, foliolis majoribus magiſque crebris. *Ray's Syn.* 88.

Wet ſhady places, and in ditches.

FEATHER-

Waved 8 Undulatum

FEATHERMOSS, Shoots branched; branches ſomewhat winged. Little leaves waved and folded,—*ſharp at the ends.* Tips *oblong, point of the veils browniſh yellow.* Fruit-ſtalks *longer than the branches.*

Hypnum pennatum undulatum, Lycopodii inſtar ſparſum. *Dillenius* 271. t. 36. f. 11.

Hypnum repens criſpum, cauliculis compreſſis, Lycopodii in morem per terram ſparſis. *Ray's Syn.* 88.

In woods and ſhady ditches. May.

Curled 9 Criſpum

FEATHERMOSS. Shoots branched; branches ſomewhat winged; little leaves waved; flat,—*blunt at the ends.* Tips *roundiſh.* Fruit-ſtalks *ſhorter than the branches.*

Hypnum pennatum undulatum criſpum, ſetis et capſulis brevibus. *Dillenius* 273. t. 36. f. 12.

Hypnum repens criſpum ramulis compreſſis, filicinorum more diſpoſitis. *Ray's Syn.* 89.

Rocks, chalk-hills, and on the banks of the Thames. Mar.

Triangular 10 Triquetrum

FEATHERMOSS. Branches ſtraggling; bowed backwards. Leaves egg-ſhaped; bent back; expanding—*Of a pale green.*

Hypnum vulgare triangulum, maximum et pallidum. *Dill.* 293. t. 38. f. 28.

Hypnum repens, triangularibus majoribus et pallidioribus foliis. *Ray's Syn.* 80.

Woods and hedges. Feb.

Thready 11 Filiforme Hud.

FEATHERMOSS. Shoots ſtraggling; very much branched; branches thread-ſhaped. Tips oblique—*Lids pointed.*

Hypnum ſericeum ramoſius et tenuius capſulis acuminatis. *Dill.* 327. t. 42. f. 62.

Hypnum repens trichoide terreſtre, priori viridius et minus, capitulis cernuis minus tumidis. *Ray's Syn.* 84.

Green Hypnum. *Hudſon* 421.

On trunks of trees; bones; and rotten-wood. Feb.

FEATHER-

FEATHERMOSS. Branches ſtraggling; partly creeping. Toothed
leaves egg-ſhaped; ſharp; pointed; tiled.—*Veil long, ſhining* 12
Lid *blunt.* Rutabulum
Common

1. Hypnum dentatum vulgatiſſimum, operculis obtuſis. *Dill.* 295. t. 38. f. 29.

Hypnum repens triangularibus anguſtioribus foliis. *Ray's Syn.* 80.

2. Lids pointed and hooked like the beak of a bird. *Beak-headed?*
Hypnum vulgare dentatum, operculis cuſpidatis. *Dillenius* Hud.
297. t. 38. fig. 30.

Hypnum repens, triangularibus minoribus foliis pediculis et capitulis brevioribus et tumidioribus, minus. *Ray's Syn.* 80.

3. Short; upright. Leaves ſhort; narrow; ſet cloſe together. *Marſh?*
Ray. Hud.

Hypnum paluſtre erectum breve, foliis brevibus anguſtis tenuibus contertis. *Ray's Syn.* 83.

Obs. *Are not the two latter varieties diſtinct ſpecies?*

Woods, (1. 2) and hedges; about the roots and trunks of trees. Heaths. Dec. (3.) marſhes.

FEATHERMOSS. Branches ſtraggling; ſhoot trailing. Butchers-
Leaves egg-ſhaped; blunt; tiled—*keeled.* Tips *roundiſh; oblique.* Broom
Lids *pointed; ſomewhat hooked.* Branches *moſtly upright; ſome-* 13
what cylindrical. From Dillenius. Ruſcifolium
Hypnum foliis ruſciformibus, capſulis ſubrotundis. *Dillenius* *Dill.*
t. 38. f. 31.

In rivulets and waters in Yorkſhire. *Dill.*

FEATHERMOSS. Shoots ſtraggling; trailing. Leaves Yellow
awl-ſhaped. Tips oblique. *Hudſon* 421. 14
Hypnum ſericeum, ſurculis longioribus et rectioribus, capſulis Luteſcens
incurvis. *Dillenius* 325. t. 42. f. 60. Hud.

Chalk hills. Jan.—Feb.

FEATHERMOSS? Water

HYPNUM? 15
Fluitans
Dill.

Branches ſtraggling, either floating or aſcending. Leaves awl-ſhaped; open.—No flowers appear to have been diſcovered, but Dillenius has arranged it as a ſpecies of the *Feathermoſs. From Dillenius.*

Hypnum erectum, aut fluitans, foliis oblongis peranguſtis acutis. *Dill.* t. 38. f. 33.

In the bogs in Ireland. In a ditch going to Marſton near Oxford. Hackney, and other places near London. *Dill.*

* * * *Shoots*

* * * *Shoots winged.*

Bog 16 Filicinum *Curled* FEATHERMOSS. Shoots winged; little branches distant; leaves tiled, bent inwards; sharp—*pointing one way.*

1. Hypnum repens filicinum crispum. *Ray's Syn.* 85. *Dill.* 282. t. 36. f. 19.

Fine-leaved Hud. 2. Leaves narrower, mostly straight: in some of the branches pointing one way.

Hypnum repens filicinum, trichoides palustre. *Ray's Syn.* 83. *Dillenius* 286. t. 35. f. 21.

Brown. Hud. 3. Tips bowed inwards.

Hypnum repens trichoides palustre vernum fuscum capitulis oblongis incurvis. *Ray's Syn.* 85.

Wet marshy places. March. (1)-Sum. (2) early in the Spr. *Dill.*

Proliferous 17 Proliferum FEATHERMOSS. Shoots proliferous; nearly flat and winged; fruit-stalks incorporated—*This species is of a very singular structure, one shoot growing out from the center of another.* Veil *yellow shining.* Lid *with a kind of long bill.* Leaves *not shining, sometimes of a yellowish, and sometimes of a deep green.*

Hypnum filicinum, tamarisci foliis minoribus non splendentibus. *Dillenius* 276. t. 35. f. 14.

Hypnum repens filicinum minus, luteo-virens. *Ray's Syn.* 86.

Woods and heaths. Feb. *Hud.* (Aut.—March and April. *Dill.*)

This Moss covers the surface of the earth in the thickest woods through which the sun never shines, and where no other plant can grow.

Wall 18 Parietinum FEATHERMOSS. Shoots nearly flat and winged, undivided for a considerable length; fruit-stalks incorporated.—*Leaves shining. The old shoots do not branch out into new shoots as in the preceding species.*

Hypnum filicinum tamarisci foliis majoribus splendentibus. *Dillenius* 274. t. 35. f. 13.

Hypnum repens filicinum, veluti spicatum. *Ray's Syn.* 86.

Muscus filicinus major. *Gerard.* 1561. *Park.* 1309. *Bauh. pin.* 360.

Woods and shady places. March.

Both this and the preceding species are used to fill up the chinks in the walls of wooden houses.

FEATHE.

FEATHERMOSS. Shoots ſomewhat winged; drooping. Trailing
Little branches remote; leaves egg-ſhaped; tips on crooked 19
fruit-ſtalks.— Prælongum

Hypnum filicinum, triangularibus parvis foliis, prælongum. *Ray's Syn.* 80. *Dillenius.* 278. t. 35. f. 15.

Trunks of trees: rotten wood; and on the ground. Feb.—March.

FEATHERMOSS. Shoots winged. Little branches near Creſted
together; bent back at the ends—*Shining*; *yellowiſh.* 20

Hypnum filicinum, criſtam caſtrenſem repræſentans. *Dill.* Criſta Caſtrenſis
284. t. 36. f. 20.

Muſcus filicifolius luteus, folio craſſo et undulato. *Ray's Syn.* 86. ſub num. 32.

On walls and chalk cliffs.

FEATHERMOSS. Shoots winged; compreſſed. Leaves Shrubby
ſharp; bowed inwards; tips nearly upright, egg-ſhaped.—*Leaves* 21
ſhining. Hudſon 423. Compreſſum *Hud.*

Hypnum filicinum ſericeum, molle et pallidum mucronibus aduncis *Dillenius* 286. t. 36. f. 22.

On ſticks and roots in woods. April.

FEATHERMOSS. Shoots winged; ſomewhat cylindrical; Fir
remote; unequal.—*Leaves egg-ſhaped yellowiſh.* Fruit-ſtalks *from* 22
the middle of the rib of the ſhoot: ſingle; purple; ſtraight; as long as Abietinum
the ſhoot. Veil *upright*; *awl-ſhaped*; *pale.* Tips *yellowiſh red, more bowed in than any of the reſt: edge of the mouth entire, with a ſhort open fringe within.*

Hypnum luteſcens, alis ſubulatis tenacibus. *Dillenius* 280. t. 35. f. 17.

Hypnum repens filicinum trichoides montanum, ramulis teretibus luteſcentibus, non diviſis. *Ray's Syn.* 86.

On hills.

FEATHERMOSS. Shoots winged; creeping. Branches Winged
crowded together. Leaves tiled; awl-ſhaped. Tips upright.— 23
On ſhort fruit-ſtalks. Plumoſum

Hypnum repens filicinum plumoſum. *Ray's Syn.* 86. *Dillenius* 280. t. 35. f. 16.

Trunks and roots of trees, brick-walls and tiles.

* * * * *Leaves bent back.*

Cypreſs 24 Cupreſſiforme — FEATHERMOSS. Shoots ſomewhat winged. Leaves pointing one way; bowed back; awl-ſhaped towards the points.—

Curled — 1. Hypnum criſpum cupreſſiforme, foliis aduncis. *Dillenius* 287. t. 37. f. 23.
Hypnum repens criſpum cupreſſiforme. *Ray's Syn.* 89.

Ground. Hud. — 2. Hypnum repens trichoides terreſtre, foliolis uno verſu diſpoſitis. *Ray's Syn.* 85.
In woods, at the roots of trees. Feb.

Hooked 25 Aduncum — FEATHERMOSS. Shoots nearly upright; ſomewhat branched. Leaves pointing one way; awl-ſhaped; bowed back; branches hooked at the ends.

Hypnum paluſtre erectum ſummita tibus aduncis. *Ray's Syn.* 82 *Dillenius* 292. t. 37. f. 26.

Marſhy places. May.

Scorpion 26 Scorpioides — FEATHERMOSS. Branches ſtraggling; trailing; bowed back; leaves pointing one way, tapering to a point.—*Branches brown; hooked; and yellow at the ends.*

Hypnum ſcorpioides paluſtre magnum, Lycopodii inſtar ſparſum. *Dillenius* 290. t. 37. f. 25.

Marſhes.

Mountain 27 Viticuloſum — THREADMOSS. Shoots creeping. Branches ſtraggling; cylindrical. Leaves expanding; tapering to a point.—*Tips and* Branches *upright.*

Hypnum ſubhirſutum, viticulis gracilibus erectis, capſulis teretibus. *Dillenius* 307. t. 39. f. 43.

Hypnum repens trichoides arboreum majus, capſulis et ſurculis erectis, minus ramoſis. *Ray's Syn.* 85.

Stumps of trees; heaths, and chalk hills. March.

Common 28 Squarroſum — FEATHERMOSS. Branches ſtraggling. Leaves ſpear-ſhaped, keeled, and almoſt folded; pointing five different ways, bent back—*Branches ſcurfy all over.*

Larger — 1. Hypnum repens, triangularibus reflexis foliis, majus. *Ray's Syn.* 82. *Dillenius* 303. t. 39. f. 38.

Smaller — 2. Hypnum repens triangularibus reflexis foliis, minus. *Ray's Syn.* 82. *Dillenius* 303. t. 39. f. 39.
Moiſt Meadows.

FEATHER-

FEATHERMOSS. Shoots creeping, branches crowded; upright. Leaves egg-ſhaped; pointing one way. Tips nearly upright.— Marſh 29 Paluſtre

Hypnum heterophyllum aquaticum, polycephallum, repens. *Dillenius* 293. t. 37. f 27.

Hypnum repens filicifolium ramoſum ramulis ſurrectis et minus complanatis. *Ray's Syn.* 87.

Banks of rivers. Jan.—Feb.

FEATHERMOSS. Shoots taking root. Branches ſtraggling; upright. Leaves pointing one way. Tips roundiſh— Ground 30 Loreum

Hypnum loreum montanum, capſulis ſubrotundis. *Dillenius* 305. t 39. f. 40.

Hypnum repens, ſurculis magis erectis, foliis reflexis longioribus cinctis, operculo capituli magno. *Ray's Syn.* 82.

Muſcus terreſtris vulgaris. *Gerard* . 1370. alter. *Park.* 1306.

On hills.

* * * * * *Shoots ſhrub-like; or the branches collected into bundles.*

FEATHERMOSS. Shoots upright. Branches in bundles, terminating; the ſhoot rather ſimple. Tips upright.—*Leaves yellowiſh; ſhining.* Tree 31 Dendroides

Hypnum dendroides ſericeum, ſetis et capſulis longioribus erectis. *Dillenius* 313. t. 40 f. 48.

Hypnum erectum arbuſculum referens, ramulis ſubrotundis, confertim naſcentibus. *Ray's Syn.* 81.

Moiſt woods. March.

FEATHERMOSS. Shoot upright. Branches in bundles; terminating the ſhoot; ſub-divided; tips rather nodding—*In this ſpecies the ſtem-leaves ſtand open, but in the preceding they lie cloſe to the ſtem.* Leaves *dark-green.* Foxtail 32 Alopecurum

Hypnum dendroides obſcurius, ſetis et capſulis brevioribus nutantibus. *Dillenius* 315. t. 41. f. 49.

Hypnum paluſtre erectum, arbuſculam referens, ramulis ſubrotundis. *Ray's Syn.* 81.

Woods, and moiſt ſhady places at the roots of trees. April.

* * * * * * *Shoots nearly cylindrical.*

FEATHERMOSS. Shoots ſtraggling; cylindrical. Leaves egg-ſhaped; ſharp; open. Tips pendant. Pendulous 33 Curtipendulum

Hypnum dentatum curtipendulum, viticulis rigidis. *Dillenius* 333. t. 43. f. 69.

Hypnum arboreum repens, capſulis reflexis, brevibus pediculis inſidentibus. *Ray's Syn.* 89.

Roots of trees and ſtony places. February.

FEATHER-

Meadow 34 Purum — FEATHERMOSS. Shoots winged and ſcattered; awl-ſhaped. Leaves egg-ſhaped; blunt; approaching.—*Tips oblique; on long* Fruit-ſtalks.

Common — 1. Hypnum cupreſſiforme vulgare, foliis obtuſis. *Dillenius* 309. t. 40. f. 45.

Hypnum terreſtre erectum, ramulis teretibus, foliis inter rotunda et acuta medio modo ſe habentibus. *Ray's Syn.* 81.

Finer. Hud. — 2. Branches ſlender and flatter; points of the leaves open.

Hypnum cupreſſiforme tenuius et compreſſius. *Dillenius* 312. t. 40. f. 47.

Hypnum longum erectum, foliis anguſtis caulibus appreſſis. *Ray's Syn.* 83.

Woods and paſtures. February.

Geniculated 35 Illecerbum — FEATHERMOSS. Shoots and branches ſtragling; cylindrical; nearly upright; blunt.—*Leaves egg-ſhaped; tiled; whitiſh.*

Hypnum cupreſſiforme rotundius, vel illecebræ æmulum. *Dillenius* 311. t. 40. f. 46.

Hypnum terreſtre erectum, ramulis teretibus, foliis ſubrotundis albo-virentibus cinctis. *Ray's Syn.* 81.

Heaths. March, April.

White 36 Albicans *Dill.* — FEATHERMOSS. Shoots and branches upright; cylindrical. Leaves awl-ſhaped; tips roundiſh; oblique—*Leaves whitiſh.* From *Dillenius.*

Hypnum ſericeum gracile albicans capſulis ſubrotundis. *Dillenius* t. 42. f. 63.

On a looſe ſandy ſoil, on heaths, and where little graſs grows. *Dillenius.*

Floating 37 Riparium — FEATHERMOSS. Shoots cylindrical; branched. Leaves ſharp; open; diſtant—*Some of the* Shoots *are very long; others not more than an inch; very ſlender; red.* Tips *red; hooked; very ſhort.*

Hypnum aquaticum, flagellis et teretibus et pennatis. *Dillenius* 308. t. 40. f. 44.

Hypnum ramoſum fluitans pennatum. *Ray's Syn.* 81.

In rivers and on the banks of rivers. December—February.

Pointed 38 Cuſpidatum — FEATHERMOSS. Shoots ſcattered; the ends tapering to a point, formed by the edges of the leaves rolled inwards.—*Leaves egg-ſhaped; open; ſhining; of a pale yellowiſh green inclined to white.* Fruit-ſtalks *long.* Tips *oblong; oblique.* Lids *blunt.*

Hypnum paluſtre extremitatibus cuſpidatis et pungentibus. *Dillenius* 300. t. 39. f. 34.

Hypnum repens paluſtre, foliis triangularibus per caules expanſis, extremitatibus conſolatis et acuminatis. *Ray's Syn.* 82.

Marſhy places. March.—April. (ſpring and ſummer. *Dill.*)

* * * * * * * *Shoots*

* * * * * * * *Shoots crowded together.*

FEATHERMOSS. Shoot creeping. Branches crowded; Silky
upright. Leaves awl-ſhaped. Tips upright; oblong. *Lids* 39
pointed. *Leaves* yellowiſh green, ſometimes deep green. Sericeum

Hypnum vulgare ſericeum recurvum, capſulis erectis cuſpidatis. *Dillenius* 323. tab. 42. fig. 59.

Hypnum repens trichoides terreſtre, luteo-virens, vulgare majus, capitulis erectis. *Ray's Syn.* 84.

On walls, trunks of trees and paſtures. Dec.—Jan.

For a microſcopic obſervation of its flowers, ſee Martyn's Philoſ. Tranſ. vol. x. p. 758.

FEATHERMOSS. Shoot creeping. Branches crowded; Velvet
upright. Leaves awl-ſhaped. Tips ſomewhat nodding. 40
Velutinum

1. Hypnum velutinum, capſulis ovatis cernuis. *Dillenius*, 326. *Common*
tab. 42. fig. 61.

Hypnum repens trichoides terreſtre viridius minus, capitulis tumidioribus cernuis. *Ray's Syn.* 84.

2. Smaller and ſhorter, and of a paler green. *Small. Hud.*

Hypnum repens trichoides terreſtre minimum et breve, capitulis tumidioribus cernuis. *Ray's Syn.* 84.

Shady places and hedges. Feb.

FEATHERMOSS. Shoots creeping. Branches thread- Creeping
ſhaped. Leaves hardly perceptible.—*Tips upright, oblong; on* 41
long fruit-ſtalks. Serpens

Hypnum trichoides ſerpens, ſetis et capſulis longis erectis. *Dillenius* 329. tab. 42. fig. 64.

Hypnum repens trichoides terreſtre minimum, capitulis majuſculis oblongis erectis. *Ray's Syn.* 85.

On the ground; in hedges; at the roots of trees, on old wood, ſtones, and bones. April.

FEATHERMOSS. Shoots upright; branched; bent in- Squirrel-tail
wards.—*Fruitſtalks when ripe twiſted like a rope.* Fertile flowers 42
on diſtinct ſhoots. Leaves *ſhining.* Sciuroides

Hypnum arboreum ſciuroides. *Dillenius* 319. tab. 41. fig. 54.

Hypnum trichoides erectum, ramulis recurvis, obſcuri coloris. *Ray's Syn.* 83.

Trunks of Trees. Feb.—April.

Mousetail 43 Myosuroides — FEATHERMOSS. Shoots very much branched; branches awl-shaped; somewhat cylindrical, but tapering each way. —*Branches slender; pointed at the ends; tips oblique; leaves egg-shaped; sharp pointed.*

Slender — 1. Hypnum myosuroides tenuius, capsulis nutantibus. *Dillenius.* 317. tab 41. fig. 51.

Hypnum polyanthon, triangularibus angustis foliis. *Ray's Syn.* 83.

Thick. Hud. — 2. Branches short and thick. Tips oblique. Leaves egg-shaped; blunt.

Hypnum myosuroides brevius et crassius, capsulis cernuis. *Dillenius* 318. tab. 41. fig. 52.

Greater. Hud. — 3. Branches long; thick, pointed at the ends; tips upright; lids tapering, sharp-pointed; leaves egg-shaped, blunt.

Hypnum myosuroides crassius, capsulis erectis. *Dillenius* 316. tab. 41. fig. 50.

Hypnum repens triangularibus angustis foliis, ramulis subrotundis. *Ray's Syn.* 83.

Hooked — 4. Branches slender; shining; somewhat hooked at the points. Leaves awl-shaped, hooked, mostly pointing one way. Tips upright.

Hypnum myosuroides sericeum tenuius, capsulis erectis. *Dillenius* 318. tab. 41. fig. 53.

(1. 3.) In woods, at the roots of trees, and on stones. (2. 4.) On walls. (1.) Jan. Feb. Mar. (2.) Aut. Dec. Jan.

Bird'sfoot 44 Ornithopoides *Hud.* — FEATHERMOSS. Shoots creeping. Branches in bundles; cylindrical; nearly upright. Tips upright; egg-shaped, *Hudson.* 430.

Hypnum gracile ornithopoides. *Dillenius* 320. tab. 41. fig. 55.

On beech-trees and rocks. April. (Feb. *Dil.*)

Club 45 Clavellatum — FEATHERMOSS. Creeping. Branches upright, crowded close together; tips bowed inwards. Lids bent inwards.—*Fruit-stalks short.*

Hypnum clavellatum parvum repens, setis brevibus. *Dillenius* 551. tab. 85. fig. 17.

Trunks of trees and rotten wood. Dec.

******** *Obscure Species.*

FEATHERMOSS. Shoot upright; somewhat branched. Leaves betwixt awl and bristle-shaped; points bowed back;—*Shoot yellow above, but blackish below.* Flowers *not discovered. From* Dillenius. Woolly 46 Trichoides *Dill.*

Hypnum palustre erectum, coma lutea, basi nigricante. *Dillenius* tab. 39. fig. 36.

In bogs, in the northern parts of Yorkshire, and in moist places on Mount Glyder. *Dillenius.*

Obs. *It seems to have a greater affinity to the* Threadmoss

HYPNUM? Starjelly Tremelloides *Dill.*

FEATHERMOSS? Shoots branched; somewhat winged. Leaves sharp; pointing from two opposite lines; of a jelly-like substance.—*No flowers hitherto discovered. Future observation may perhaps determine whether it be a* Feathermoss, Livermoss, *or* Jellywort.

Tremella palustris gelatinosa tenerrima, Musci filicifolii facie. *Dillenius,* tab. 10. fig. 11.

In the ditches of a field near Chichester in Suffolk, towards the last gate. June.

418 WATERMOSS. 1190 Fontinalis.

Barren Flower nearly sitting.

Empal. *Veil* smooth; conical; sitting.

Chive. *Tip* oblong; fringed at the mouth; covered with a *Lid*; tapering to a point.

Recept. *Excrescence* none.

Scalybulb. Swelling out; tiled; inclosing the tip and the vei'

Fertile Flower.

WATERMOSS. Leaves keeled; almost doubled; pointing three ways; sharp. Tips on the sides of the branches.—*Stalks purple.* Greater 1 Antipyretica

Fontinalis triangularis major, complicata e foliorum alis capsulifera. *Dillenius* 254. tab. 33. fig. 1.

Fontinalis major, foliis triangularibus complicatis capitulis in foliorum alis sessilibus. *Ray's Syn.* 79.

Brooks, rivulers, slow streams and ponds. P. May.

The Scandinavians line the inside of their chimnies with this to defend them against the fire, for contrary to the nature of all other Moss, this is hardly capable of burning.

Lesser 2 Minor WATERMOSS. Leaves egg-shaped; concave; pointing three ways; sharp; often in pairs. Tips at the ends of the branches.—*Leaves keeled and almost doubled; on the thicker branches in pairs.*

Fontinalis triangularis minor carinata, e cymis capsuliferis. *Dillenius* 257. tab. 33. fig. 2.

Fontinalis minor, foliis triangularibus minus complicatis, capitulis in summis ramulis sessilibus. *Ray's Syn.* 79.

On the brinks and sides of rivers. P. Oct.—Nov.

Scaly 3 Squamosa WATERMOSS. Leaves tiled; betwixt awl and spear-shaped. Tips on the sides of the branches.—*Leaves shining; of a blackish green.*

Fontinalis squamosa tenuis sericea atrovirens. *Dillenius* 250. tab. 33. fig. 3.

In rivulets in mountainous countries. P. Aug.—Sept.

Feathered 4 Pennata WATERMOSS. Leaves pointing two ways; expanding. Tips on the sides of the branches---*tawny yellow.*

Common 1. Leaves waved; scaly-bulb scaly.

Sphagnum pennatum undulatum, vagina squamosa. *Dillenius* 250. t. 32. f. 9.

Hairy. Hud. 2. Leaves waved, scaly-bulb hairy.

Sphagnum pennatum undulatum, vagina pilosa. *Dillenius* 249. t. 32. f. 8.

Flat. Hud. 3. Leaves flat.

Sphagnum pennatum planum. *Dillenius* 249. t. 32. f. 7.

Trunks of trees. P. Aug.—Sept.

OBS. *Are not these rather distinct species?*

419 STARTIP. 1196 Jungermannia.

Barren Flower on a long straight fruit-stalk rising from the empalement.

EMPAL. *Scaly Bulb* tubular.

BLOSS. None.

CHIVE. *Tip* egg-shaped; opening into four valves; expanding; equal; permanent.

Fertile Flower mostly on the same plant; sitting.

EMPAL. None, at least visible to us.

BLOSS. None.

SEEDS. Roundish, either single, or set close together.

OBS. *Many species of this genus are beautiful microscopic objects.*

* *Leaves*

* *Leaves winged; pointing one way.*

STARTIP. Leaves simply winged. Little leaves egg-shaped; somewhat fringed---*Stems purple, flowering at the ends.* Spleenwort Asplenioides

1. Little leaves distinct from each other. *Thinset*
Lichenastrum Asplenii facie, pinnis laxioribus. *Dillenius* 482. t. 69. f. 5.

2. Little leaves set close to each other. *Thickset.*
Lichenastrum asplenii facie, pinnis confertioribus. *Dillenius* 483. t. 69. f. 6.
Lichenastrum trichomanis facie, capitulis e foliorum summitate enascentibus majus. *Ray's Syn* 112.
Woods and moist shady places. Feb.

STARTIP. Leaves simply winged; little leaves flat, naked, strap-shaped---*Flowering at the base and middle of the branches.* Straggling 2 Viticulosa
Lichenastrum trichomanis facie, e basi et medio florens. *Dillenius* 484. t. 69. f. 7.
Lichenastrum capitulis nudis, trichomanis facie, foliolis densius congestis majus. *Ray's Syn.* 111.
Shady woods, hedge banks, and wet places. April.

STARTIP?
JUNGERMANNIA?
Leaf simply winged; little leaves roundish; distinct—*Leaf mostly simple; sometimes branched; sweet-scented.* Flowers *not discovered.* From Dillenius. Sweet scented 3 Odorata *Dill.*
Lichenastrum trichomanoides aquaticum odoratum fontis S. Winifredæ. *Ray's Syn.* 112. *Dillenius* t. 69. f. 8.
St. Winifred's well Flintshire; Dartmere in Cornwall, and in Devonshire. *Dill.*

STARTIP. Leaf simply winged, flowering at the base; little leaves sharp. *Hudson* 491 ---*Awl-shaped; flowers numerous, set close together.* Manyflowered 4 Multiflora *Hud.*
Lichenastrum multiflorum exile, foliis angustissimis. *Dillenius* 481. t. 69. f. 4
Woods. March,

STARTIP. Leaves simply winged; little leaves very entire; tiled; convex—*Flowers numerous, rising from the middle of the stem and branches.* Imbricated 5 Polyanthos
Lichenastrum trichomanis facie, polyanthemum breve et repens. *Dillenius* 486. t. 70. f. 9.
In wet places, and in rivulets. April.

Spear-leaved 6 Lanceolata — STARTIP. Leaves ſimply winged; ſpear-ſhaped; flowering at the ends. Little leaves very entire—*of a beautiful green.*

Lichenaſtrum trichomanis facie minus, ab extremitate florens. *Dillenius* 486. t. 70. f. 10.

Lichenaſtrum capitulis nudis, trichomanis facie, foliolis denſius congeſtis, minus. *Ray's Syn.* 112.

Moiſt ſhady places. Oct. Nov. Dec. Feb.

Cloven 7 Bidentata — STARTIP. Leaves ſimply winged; flowering at the ends. Little leaves marked at the end with two teeth.—

Large. 1. Larger; little leaves ſharper and hollow.

Lichenaſtrum pinnulis acutioribus et concavis, bifidis, majus. *Dillenius* 487. t. 70. f. 11.

Lichenaſtrum trichomanis facie, foliolis bifidis, majus. *Ray's Syn.* 113.

Small. Dill? 2. Smaller; little leaves blunter. Flowers not diſcovered.

Lichenaſtrum pinnulis obtuſioribus bifidis minus. *Dillenius* t. 70. f. 12.

Woods, moiſt heaths, and ſhady banks. Oct. Nov.

Forked 8 Bicuſpidata — STARTIP. Leaves ſimply winged; flowering in the middle; little leaves marked at the end with two teeth.—

Lichenaſtrum pinnulis acutiſſime bifidis, minimum. *Dillenius* 488. t. 70. f. 13.

Lichenaſtrum trichomanis facie, foliolis bifidis, minimum. *Ray's Syn.* 113.

Moiſt ſhady places. March.

Three-toothed 9 Tridentata *Dill.* — STARTIP?

JUNGERMANNIA?

Leaves ſimply winged. Little leaves marked at the end with three teeth—*The lower ones ſomewhat ſerrated.* Flowers *not diſcovered.* Leaf *branched.* From Dillenius.

Lichenaſtrum ramoſius, foliis trifidis. *Dillenius* t. 70. f. 15.

Snowden. Summer. *Dill.*

Toothed 10 Quinquedentata — STARTIP. Leaves winged; branched; flowering at the ends. Little leaves marked at the end with five teeth—*Shoot ſometimes ſimple, taking root as it creeps along. The* Tips *appear in autumn, and bloſſom in the ſpring.*

Lichenaſtrum multifidum majus, ab extremitate florens. *Dillenius* 494. t. 11. f. 23.

Lichenaſtrum trichomanis facie, foliolis multifidis, capitulis e ſummis ramulis naſcentibus. *Ray's Syn.* 113.

Toothed Lichenaſtrum.

Woods. April.

JUNGER-

STARTIP ? Serrated
JUNGERMANNIA ? 11 Serrata *Dill.*

Leaves simply winged; little leaves serrated—*alternate. Serratures sharp along the inner margin.* Leaf *simple.* Flowers *not discovered.* From Dillenius.

Lichenastrum pinnulis alternis, quasi spinosis. *Dillenius* t. 70. f. 16.

Caernarvon. *Dill.*

* * *Leaves winged; little leaves covered on the upper side, with small ear-like scales.*

STARTIP. Leaves doubly winged above; flowering at the ends; little leaves roundish; very entire; waved.— Eared 12 Undulata

Lichenastrum pinnulis auriculatis majoribus et non crenatis. *Dillenius* 490. t. 71. f. 17.

Shady places. March.

STARTIP. Leaves doubly winged above; flowering at the ends. Little leaves fringed—*inversely egg-shaped; somewhat tiled.* Wood 13 Nemorea

Lichenastrum auriculatum, pinnis minoribus crenatis. *Dillenius* 490. t. 71. f. 18.

Wood Lichenastrum.

Moist woods. March.

STARTIP ? Bird's-foot
JUNGERMANNIA ? 14 Ornithopoides *Dill.*

Leaves doubly winged above, little leaves betwixt egg and spear-shaped; fringed; distinct—*Shoot somewhat branched.* Flowers *not discovered.* From Dillenius.

Lichenastrum auriculatum, Ornithopodii minoris pinnulis ciliatis. *Dillenius* t. 71. f. 21.

Snowden. *Dill.*

STARTIP. Leaves doubly winged above; flowering towards the base. Little leaves somewhat scolloped; tiled; round. Curled 15 Resupinata

Lichenastrum auriculatum, pinnulis rotundis crispum. *Dillenius* 491. t. 71. f. 19.

Lichenastrum trichomanis facie, capitulis e foliorum summitate enascentibus, medium. *Ray's Syn.* 112.

Clefts of rocks. April.

The under side of this resembles the under side of a staircase.

White 16 Albicans STARTIP. Leaves doubly winged above; flowering at the ends. Little leaves ſtrap-ſhaped; bowed back.—

Lichenaſtrum auriculatum, pinnulis anguſtis, planis, recurvis. *Dillenius* 492. t. 71. f. 20.

Lichenaſtrum trichomanis facie, capitulis e foliorum ſummitate enaſcentibus minus. *Ray's Syn.* 112.

Moiſt woods, and ſhady places. March.

Trifid 17 Trilobata STARTIP. Leaves doubly winged underneath. Little leaves almoſt ſquare; with three imperfect lobes—*Rib of the leaf jointed.*

Lichenaſtrum pinnulis obtuſe trifidis nervo geniculato. *Dillenius* 493. t. 71. f. 22.

Wales.

Creeping 18 Reptans STARTIP. Leaves doubly winged underneath; taking root at the ends. Little leaves with four teeth—*flowers at the baſe.* Tips *black*; *ſhining.*

Lichenaſtrum multifidum exiguum ad baſin florens, per ſiccitatem imbricatum. *Dillenius* 494. t. 71. f. 24.

Lichenaſtrum trichomanis facie, foliolis multifidis, capitulis ex imis cauliculis naſcentibus. *Ray's Syn.* 113.

Wet ſhady places. Ditch-banks. April.

*** *Leaves tiled.*

Flat 19 Complanata STARTIP. Shoots creeping; flowering at the ends. Leaves doubly tiled with little ſcales underneath. Branches of an equal breadth throughout—*Leaves of a yellowiſh green*; *flat.*

Lichenaſtrum imbricatum majus, ſquamis compreſſis et planis. *Dillenius* 496. t. 72. f. 26.

Lichenaſtrum imbricatum majus. *Ray's Syn.* 111.

In ſhady places, at the roots of trees, damp ſtones and walls. March—April.

Scaly 20 Dilatata STARTIP. Shoots creeping; flowering at the ends. Leaves doubly tiled, with little ſcales underneath. Branches broader towards the ends—*Leaves convex, ſmaller, and* Shoots *narrower than in the preceding ſpecies.*

Lichenaſtrum imbricatum minus, ſquamis convexo-concavis. *Dillenius* 497. t. 72. f. 27.

Lichenaſtrum imbricatum minus. *Ray's Syn.* 111.

Trunks of trees, and ſtones. March. April.

Small 21 Minima *Dill.* STARTIP. Shoots creeping, flowering at the baſe. Leaves doubly tiled; with little ſcales underneath—*flat*; *aſhcoloured.*

Lichenaſtrum imbricatum minimum ſquamis planis cinereis. *Dillenius* t. 72. f. 28.

Bagley wood near Oxford. *Dill.*

STAR-

STARTIP. Leaves tiled in a double row. The upper leaves Tamariſk
circular; convex; blunt; four times as large as the other.— 22
Dark purple. Tamariſci

Lichenaſtrum imbricatum tamariſci Narbonenſis facie. *Dillenius* 499. t. 72. f. 31.

Rocks and trunks of trees. March.

STARTIP. Shoots trailing; tiled underneath. Leaves Wall
heart-ſhaped; ſharp. The little ſcales diſpoſed along the rib of 23
the ſhoot; wedge-ſhaped—*Leaves dark green.* Platyphylla

Lichenaſtrum Arboris Vitæ facie, foliis minus rotundis. *Dillenius* 501. t. 72. f. 32.

Walls and trunks of trees. March. April.

STARTIP? JUNGERMANNIA? Arbor Vitæ
Shoots trailing; tiled underneath. Leaves heart-ſhaped; 24
the little ſcales diſpoſed along the rib of the ſhoot; awl-ſhaped— *Dill.*
Leaves rather rounder than thoſe of the preceding ſpecies. Flowers Arboris Vitæ
not diſcovered. From Dillenius.

Lichenaſtrum Arboris Vitæ facie, foliis rotundioribus. *Dillenius* t. 72. f. 33.

On trees. *Dill.*

STARTIP. Shoots creeping; leaves doubly tiled; with Fern
little ſcales and fringed appendages underneath—*The upper leaves* 25
are often cloven. The under ſurface appears hairy from the quantity Ciliaris
of fringe upon the little leaves and appendages.

Lichenaſtrum filicinum pulchrum villoſum. *Dillenius* 503. t. 73. f. 35.

Lichenaſtrum filicinum criſpum. *Ray's Syn.* 111.

Wet moſſy places near rivulets, on heaths. April.

STARTIP. Shoots nearly upright; tiled; and pointing two Upright
ways. Leaves deeply divided.----*The* bloſſoming ſhoots *are up-* 26
right, and furniſhed with narrow leaves pointing every way. Thoſe Varia
which do not bloſſom, or the fertile ſhoots, *creep along the ground, edged with roundiſh leaves pointing only two ways.*

Lichenaſtrum foliis variis. *Ray's Syn.* 113. *Dillenius* 505. t. 73. f. 36.

Woods, heaths, and moiſt ſhady places. March.

* * * * *Shoots tiled on every ſide, little leaves ſcattered.*

STARTIP. Shoots cylindrical. Leaves tiled on every ſide; Round
flowers on fruit-ſtalks.— 27

Lichenaſtrum alpinum Bryi juliacei argentei facie. *Dillenius* Julacea
506. t. 73. f. 38.

Wet rocks. Aug.—Sept.

STAR-

Rock 28 Rupestris

STARTIP. Shoots cylindrical. Leaves awl-shaped, pointing one way.—*This plant is extremely short; blackish; nearly upright and seldom branched.* Flowers *on short fruit stalks.*

Lichenastrum alpinum nigricans, foliis capillaceis reflexis. *Dillenius* 507. t. 73. f. 40.

Moist rocks. April—May.

Hairy 29 Tricophylla

STARTIP. Shoots cylindrical. Leaves hair-like; equal.—*Flowers at the ends of the branches.*

Lichenastrum trichoides minimum, in extremitate florens. *Dillenius* 505. t. 73. f. 37.

Boggy Heaths.

Mountain 30 Alpina

STARTIP. Shoots cylindrical. Leaves egg shaped; expanding; cups tiled, *dark reddish colour.* Valves *of the tips rolled back when out of blossom.*

Lichenastrum alpinum attro-rubens teres, calycibus squamosis. *Dillenius* 506. t. 73. f. 39.

Marshy places on mountains. In the hollow parts of rocks Sept.—Oct.

**** *Without stems. Leaves simple.*

Broad-leaved 31 Epiphylla

STARTIP. Without a stem. A little leaf growing upon the large one.—*The little leaves which arise from the middle of the large ones, are hooded, and contain in their cavities a little globular substance which is the* Pointal; *and round this about seven small* Chives. *At length the pointal is furnished with a fruit-stalk and then the chives fall off.* Tips *globular.*

Lichenastrum capitulis rotundis, e foliorum medio enascentibus. *Ray's Syn.* 110. *Dillenius* 508. t. 74. f. 41.

Lichen petræus, cauliculo calceato. *Baub. pin.* 362.

Lichen alter minor, folio calceato. *Park.* 1214.

Wet shady places near brooks, and rivers. March.

Jagged 32 Pinguis

STARTIP. Without a stem. Leaf oblong; indented; greasy to the touch.—*Fruit stalks rising from the hollows of the leaf* Sheath *long; cylindrical.* Tips *oblong. The plant which bears the* Barren Flowers *is much smaller and more jagged, than that which bears the* Fertile *ones, which grow together, and form a close upright tuft.*

Lichenastrum capitulis oblongis, juxta foliorum divisuras enascentibus, *Ray's Syn.* 110. *Dillenius* 509. t. 74. f. 42.

Marshes. April.

TARTIP.

STARTIP. Without a ſtem, Leaves with doubly winged clefts. Valves of the tips fringed at the ends with hair,—*Strap-ſhaped. Tips riſing from near the baſe of the leaves.* Dwarf 33 Multifida

Lichenaſtrum ambroſie diviſura. *Ray's Syn.*. 111 *Dillenius* 511. t. 74. f. 43.

Woods, and moiſt ſhady places. March—April.

STARTIP. Without a ſtem. Leaves with doubly winged clefts. Valves of the tips entire; ſpear-ſhaped—Tips *riſing from the hollows of the leaf.* Leaf *not ſo finely divided as in the prceding ſpecies.* From Dillenius. Germander 34 Chamedryfolia *Dill.*

Lichenaſtrum chamedryos multifidæ diviſura. *Dillenius* t. 74. f. 44.

At the head of Elm-cragg-well under Bill-bank. *Dill.*

STARTIP. Without a ſtem. Leaves ſtrap-ſhaped; branched; forked at the ends and bluntiſh.—*Scaly bulb globular; edged with teeth; hairy.* Globe 35 Furcata

Licehnaſtrum tenuifolium furcatum, thecis globulis piloſis, *Dill.* 512. t. 74. f. 45.

Ulva ſaxatilis furcata, latiuſculis et tenerioribus ſegmentis. *Ray's Syn.* 63.

Wet rocky places, on the ground, and trees in woods. April.

STARTIP. Without a ſtem. Leaves ſlightly divided into winged clefts. Lobes tiled. Scaly bulb plaited—*The little head before it unfolds is black and nearly globular.* Valves *of the tip wedge-ſhaped.* Shining 36 Puſilla

Lichenaſtrum exiguum, capitulis nigris lucidis, e cotylis parvis naſcentibus. *Dill.* 513. t. 74. f. 46.

On heaths. March—April.

420 LIVERWORT. 1198 Marchantia.

Barren Flower upon a long, ſtraight fruit-ſtalk; riſing out of a little cup.

EMPAL. *Cup, common*; target-ſhaped; with four, five, or ten clefts; very large; target-ſhaped; containing under it as many florets as there are ſegments. *Segments* equal; bent downwards at the ſides.

BLOSS. One petal; turban-ſhaped; ſtraight; ſhorter than the empalement.

CHIVE. *Thread* one; longer than the bloſſom; ſimple. *Tip* ſomewhat egg-ſhaped; opening at the top into as many ſegments as there are in the common empalement. *Duſt* fixed to a little hairy lock.

Fertile Flower upon the ſame plant; ſitting.

EMPAL. *Cup* conſiſting of a membranaceous margin; upright; open; entire; permanent; of one leaf.

BLOSS. None, unleſs the cup.

SEEDS. Many; roundiſh; compreſſed; naked; at the bottom of the cup.

OBS. *In ſome ſpecies the barren and fertile flowers are upon diſtinct plants.*

Common
1
Polymorpha

LIVERWORT. Common cup with ten clefts—*Leaves broad; irregular; of a duſky green. A yellowiſh ſubſtance reſembling a lock of wool proceeds from the tips, appearing to move within them whilſt the duſt is falling out.*

Larger 1. Lichen fontanum major ſtellatus æque ac umbellatus et cyathophorus. *Dillenius* 523. t. 76. f. 6.

Lichen petræus latifolius, ſive hepatica fontana. *Bauh. pin.* 362. *Ray's Syn.* 115.

Smaller 2. Smaller.

Lichen domeſticus minor ſtellatus æque ac umbellatus et cyathophorus. *Dillenius* 527. t. 77. f. 7.

Lichen petræus ſtellatus. *Bauh. pin.* 362. *Ray's Syn,* 115.

1. Shady moiſt rocks, and near ſprings and wells. 2. Shady garden-walks and in the interſtices of walls and ſtones towards the north. Auguſt.

LIVER-

LIVERWORT. Common cup with four diviſions; ſegments tubular—*Florets with four diviſions.* Fertile Flower *creſcent-ſhaped.* Leaves *of a fine green.* Croſs-headed 2 Cruciata
Lichen feminifer lunulatus; florifer pileatus, tandem cruciatus. *Dillenius.* 521. t. 75. f. 5.
Lichen pileatus parvus, capitulo crucis inſtar, ſe expandente. *Ray's Syn.* 115.
Shady places. July—Auguſt.

LIVERWORT. Common cup with five clefts; hemiſpherical. Scaly bulb none.—*Florets nearly globular; edged with teeth.* Marſh 3 Hemiſphærica
Lichen pileatus parvus, foliis crenatis. *Ray's Syn.* 114. *Dillenius* 519. t. 75. f. 2.
In wet marſhy places. April—May.

LIVERWORT. Common cup ſomewhat egg-ſhaped; divided into five cells—*Leaves in large cluſters, indented; blunt; green; with ſeveral white tubercles.* Wart 4 Conica
Lichen vulgaris major pileatus et verrucoſus. *Dillenius* 516. t. 75. f. 1.
Lichen petræus pileatus. *Park.* 1314. *Ray's Syn.* 114.
Lichen petræus cauliculo pileolum ſuſtinente. *Bauh. pin.* 362.
On the banks of rivulets in ſhady places. April.

421 VETCHCAP. 1197 Targionia.

EMPAL. Somewhat globular; with two valves.
CHIVE. *Tip* bell ſhaped; ſitting at the bottom of the empalement.

VETCHCAP. As there is only one ſpecies known Linnæus gives no deſcription of it.—*Leaf betwixt heart and ſpear-ſhaped. Green at firſt, afterwards dark purple; blackiſh underneath.* Flower *at the end on the under ſide.* Cup *black.* Dotted Hypophylla
Lichen petræus minimus, fructu orobi. *Bauh. pin.* 362. *Dill.* 532. t. 78. f. 9.
Lichenaſtrum capitulo oroboide. *Ray's Syn.* 110.
Lichen petræus minimus acaulos. *Park.* 1315.
In moſſy places.

422 HORNFLOWER. 1201 Anthoceros.

Barren Flower ſitting.

EMPAL. One leaf; ſomewhat cylindrical; lopped; entire.

BLOSS. None.

CHIVE. *Thread* none. *Tip* ſingle; awl-ſhaped; exceedingly long; with two valves. *Duſt* fixed to a looſe hair-like receptacle.

Fertile Flower ſitting, generally upon the ſame, but ſometimes upon a diſtinct plant.

EMPAL. One leaf; with ſix diviſions; expanding.

SEEDS. About three; naked; roundiſh; at the bottom of the empalement.

OBS. *Diviſions of the empalement ſometimes only four, and then there are but two ſeeds.*

Spotted Punctatus HORNFLOWER. Leaves undivided, indented, dotted.—*Tips an inch long.* Tip and Fruit-ſtalk *from two to three inches.*

Anthoceris fol is minoribus magis laciniatis. *Dill.* 476. t. 68. f. 1.

Lichenaſtrum gramineo pediculo et capitulo oblongo biſurco *Ray's Syn.* 109.

Moiſt ſhady places and on heaths. April.

423 LEATHERCUP. 1199 Blaſia.

Barren Flower.

EMPAL. One leaf; egg-ſhaped at the baſe; ſomewhat cylindrical in the middle; expanded towards the end, and lopped.

CHIVES. A number of little grains lying looſe in the cup.

Fertile Flower.

EMPAL. Hardly any.

S. VESS. Roundiſh; ſingle; buried in the leaves.

SEEDS. Several; roundiſh.

OBS. *It is a doubt which of theſe is to be called the* Barren *and which the* Fertile *flower.*

LEATHER-

LEATHERCUP. As there is only one ſpecies known Linnæus givs no deſcription of it.—*Leaves in a circle, from one to two inches in Diameter; deep purple at the baſe; green at the edges; jagged.* Seeds *ſo ſmall that their form is ſcarce to be diſtinguiſhed by the naked eye.* Dwarf Puſilla

Mnium Lichenis facie. *Dill.* 237. t. 31. f. .

On the ſides of ditches and rivers in a ſandy ſoil.

424 GRAINWORT. 1200 Riccia.

EMPAL. None; except a hollow bladder within the ſubſtance of the leaf.

BLOSS. None.

CHIVE. *Tip* cylindrical; fitting on the ſeedbud; opening at the end.

POINT. *Seedbud* turban-ſhaped. *Shaft* thread-ſhaped; perforating the tip.

S.VESS. Globular; with one cell; crowned with the ſhrivelled tip.

SEEDS. Many; hemiſpherical; on little foot-ſtalks.

GRAINWORT. Leaves ſmooth; deeply divided; ſharp at the ends—*Spreading on the ground.* Small 1

Lichen omnium minimus, foliolis ſuper terram expanſis. *Ray's Syn.* 115. *Dillenius* 534. t. 78. f. 11. Minima

In grounds that are overflowed and on wet heaths. Nov.

GRAINWORT. Leaves ſmooth; chanelled, with two lobes—*blunt at the end.* Marſh 2

Lichen minimus, foliis venoſis bifariam vel trifariam ſe dividendo progredientibus. *Dillenius* 533. t. 78. f. 10. Glauca

Wet commons. March.

GRAINWORT. Leaves forked; between thread and ſtrap-ſhaped.— Float 3

Lichenaſtrum aquaticum fluitans tenuifolium furcatum. *Dillenius* 514. t. 74. f. 47. Fluitans

Ulva paluſtris furcata, anguſtioribus et firmioribus ſegmentis. *Ray's Syn.* 63.

In ſtagnant waters.

GRAINWORT. Leaves inverſely heart-ſhaped; fringed.— Fringed

Lichen parvus vermus cordiformis, ima parſe fimbriatus, Lentis paluſtris modo aquæ innatans. *Ray's Syn.* 116. *Dill.* 536. t. 78. f. 18. 4 Natans

In ſtagnant waters.

GRAIN-

Ear-like 5 Auriformis Dill. GRAINWORT. Leaves inverſely heart-ſhaped; indented; tiled.—*Broader than thoſe in the preceding ſpecies; not fringed; of a blackiſh green; ſpreading horizontally tuto a circular tuſt.* From Dillenius.

Lichenoides gelatinoſum atro-virens, auriculatum et granoſum. *Dillenius* t. 19. f. 24.

Garden gravel walks. Spring. *Dill.*

Rock 6 Sinuata Hud. GRAINWORT. Leaves branched; ſegments indented; broader and ſcolloped at the ends. *Hudſon* 441.

Lichenoides gelatinoſum, foliis latioribus tuniformibus. *Dill.* 142. t. 19. f. 29.

On rocks.

425 CUPTHONG. 2102 Lichen.

Barren Flowers numerous; reſiding in a *receptacle* which is moſtly circular; very large; ſhining; gummy; either flat. convex, or concave.

Fertile Flowers and *Seeds* like meal, ſcattered over the ſame, or a diſtinct plant.

This Genus is ſubdivided into the

A. Tubercled; or thoſe which conſiſt of a ground or thin cruſt ſtudded with convex receptacles, or tubercles.

B. Saucer-like; a cruſt ſtudded with concave receptacles reſembles little bowls or ſaucers.

C. Tiled; leaves ſpreading flat, and cloſely adhering to the ſubſtance on which they grow.

D. Leafy. Leaves looſe from the ſubſtance on which they grow.

F. Leatherlike. Of a ſubſtance reſembling leather.

F. *Sooty*; adhering to the ſubſtance on which they grow only in one point; and the ſurface as if ſprinkled with ſoot.

G. Glass-shaped. Receptacles expanding into the form of a drinking glaſs.

H. Shrubby. Shooting into branches reſembling a ſhrub.

I. Threadlike. Shooting into long thread-like branches.

Obs. *Mr. Hudſon calls this Genus* Liverwort, *but as the Marchantia of Linnæus is commonly known by the name of Liverwort, it was thought better to continue the old name to that and find a new one for this.*

A. Tubercled.

CUPTHONG. A whitiſh ground, with black branching lines reſembling written characters— Lettered 1 Scriptus

Lichenoides cruſta tenuiſſima, peregrinis velut litteris inſcripta. *Ray's Syn.* 71. *Dillenius* 125. t. 18. f. 1.

On the barks of trees.

CUPTHONG. A yellowiſh ground, with black lines reſembling a map.— Map 2 Geographicus

Lichenoides nigro-flavum tabulæ geographicæ inſtar pictum. *Dillenius* 126. t. 18. f. 5.

On rocks.

CUPTHONG. A whitiſh ground, diverſified with ſimple black lines and dots, ſet thick together.— Wrinkled 3 Rugoſus

Lichenoides punctatum et rugoſum nigrum. *Dillenins* 125. t. 18. f. 2.

Barks of trees.

CUPTHONG. A greeniſh ground inclining to aſh colour, with black tubercles—*Large; nearly globular; not ſhining; of a deep black without, but red within if broken.* Black knobbed 4 Sanguinarius

Lichenoides leproſum, cruſta cinereo-vireſcente, tuberculis nigerrimis. *Ray's Syn.* 71. *Dillenius* 126. t. 18. f. 3.

On rocks and trunks of trees.

CUPTHONG. Ground brown with black turbercles— Wall 5 Fuſco-ater

On old walls and rocks.

CUPTHONG. Ground a clear white, with black tubercles— Dyers 6 Calcarius

Lichenoides tartareum tinctorium candidum, tuberculis atris. *Dillenius* 128. t. 18. f. 8.

On lime-ſtone rocks.

This ſpecies is ſo peculiar to limeſtone rocks, that wherever that ſtone occurs among others, it may be diſtinguiſhed at the firſt view by this plant growing upon it—When dried, powdered and ſteeped in urine it is uſed to dye ſcarlet, by the Welch and the inhabitants of the Orkneys. The colour is ſaid to be very fine.

CUPTHONG. A greeniſh ground, with orange coloured tubercles. *Hudſon* 443. Orange 7 Flavo rubeſcens Hud.

Trunks of trees.

CUPTHONG. A whitiſh ground, with tubercles the colour of ruſty iron. *Hudſon* 444. Ruſty 8 Ferrugineus *Hud.*

Lichenoides leproſum, tuberculis fuſcis et ferrugineis. *Dill.* 126. t. 18. f. 4. t. 55. f. 8.

On the bark of trees, and ſtones.

White 9 Fagineus CUPTHONG. A white ground, with white mealy tubercles—

On the trunk and branches of Elm and Hornbeam (*Hudſon*) On the trunks of the Beech. (*Linnæus.*)

Hornbeam 10 Crapineus CUPTHONG. An aſh coloured ground, with whitiſh wrinkled tubercles—

On Hornbeam, Aſh and Hazel.

Powdered 11 Byſſoides CUPTHONG. Ground a fine powdery cruſt; tubercles nearly globular; on fruit-ſtalks.—*Cruſt of a greeniſh aſh colour.* Tubercles *reſembling little Muſhrooms, of a bluiſh red at the top.* Fruit-ſtalks *whitiſh.*

Coralloides fungiforme ex ungula equina, livide rubeſcens. *Dill.* 78. t. 14. f. 5.

Fungi parvi globoſi ex ungue equino putreſcente. *Ray's Syn.* 13. t. 1. f. 3.

In gravelly ſoil, and on horſe-dung.

Heath 12 Ericetorum CUPTHONG. Ground a clear white, with fleſh coloured tubercles,—*On fruit-ſtalks, reſembling little Muſhrooms, convex at the top.* Fruit-ſtalks *fleſh-coloured.*

Lichenoides fungiforme, cruſta leproſa candida, capitulis et pediculis incarnatis. *Ray's Syn.* 70.

Coralloides fungiforme carneum baſi leproſa. *Dill.* 76. t. 14. f. 1.

On wet heaths. October—April.

Brown 13 Rufus *Hud* CUPTHONG. Ground brown, with browniſh flat tubercles on fruit-ſtalks. *Hudſon* 443.

Coralloides fungiforme, ſaxatile, pallide fuſcum. *Dill.* 78. t. 14 f. 4.

On old walls, and rocks. December—February.

Warty 14 Pertuſus CUPTHONG. A white or aſh-coloured ground; covered with a number of angular ſmooth warts, ſet chequer wiſe, and pierced with one or two cylindrical holes.—*The ground itſelf is thin, but the cruſt formed by the warts pretty thick.* Warts *greeniſh.*

Lichenoides verrucoſum et rugoſum, cinerum, glabrum. *Dill.* 128. t. 18. f. 9.

Lichen verrucoſus. *Hudſon.* 445.

Walls, ſtones, and barks of trees.

B. SAU-

B. Saucerlike.

CUPTHONG. A deep yellow cruſt, with pale yellow ſau- Yellow
cers.— 15

1. Lichen candelarius. *Hudſon* 444. Candelarius
Walls and bark of trees. *Common*

It gives a yellow colour. In Sweden they bruiſe it and mix it with ſuet to make yellow candles for feſtivals.

2. A greeniſh cruſt inclining to yellow, with deep yellow ſaucers. *Hudſon*, 445. *Greeniſh.* Hud.
Lichen flaveſcens. *Hudſon*, 445.
Lichenoides cruſtoſum, orbiculis et ſcutellis flavis. *Dillenius*, 136. tab. 18. fig. 18.
On walls.

CUPTHONG. A greeniſh cruſt with a tinge of white; with yellowiſh ſaucers white at the edge.— Stonerag 16

Lichenoides cruſtaceum et leproſum acetabulis majoribus luteis, limbis argenteis. *Ray's Syn.* 71. *Dillenius*, 132. tab. 18. fig. 13. Tartareus

Welſh Liverwort.

On rocks.

It is common in Derbyſhire, and incruſts moſt of the ſtones at Urſwic Mere. It is gathered for the dyers, by peaſants who ſell it for a penny a pound. They can collect twenty or thirty pound a day. It gives a purple colour.

CUPTHONG. A whitiſh cruſt, with wan-coloured ſaucers. Pale

1. Saucers ſmall, ſometimes browniſh, or of a lead-colour, with an aſh-coloured margin. *Common* 17 Palleſcens
Lichenoides cruſtoſum orbiculare incanum. *Dillenius*, 135. tab. 18. fig 17.
Lichenoides cinereum mere cruſtaceum eleganter expanſum. *Ray's Syn.* 71.

2. Saucers greeniſh, inclining to aſh-colour. *Greeniſh.* Hud.
Lichenoides cruſtaceum et leproſum, ſcutellis cinereo vireſcentibus. *Ray's Syn.* 71.
Walls and trunks of trees.

CUPTHONG. A whitiſh cruſt with black ſaucers; the new formed ones hollow and diſtended.— Black-ſpangled. 18

1. A whitiſh cruſt with black wrinkled ſaucers. Ater. *Hud.*
Lichenoides cruſtaceum et leproſum, ſcutellis nigricantibus majoribus et minoribus. *Ray's Syn.* 71. *Dillenius* 133. tab. 18. fig. 15. *Wrinkled*
Lichen Subfuſcus. *Syſt. Natur.*
On trees, walls and ſtones.

Brown-ſpangled 19 Subfuſcus *Sp. Pl.* and *Hud.*

CUPTHONG. A whitiſh cruſt, with browniſh ſaucers and an aſh-coloured margin ſomewhat notched.—

Lichenoides cruſtaceum et leproſum, ſcutellis ſubfuſcis. *Ray's Syn.* 71. *Dillenius* 134. tab. 18. fig. 16.

On the bark of trees, walls and rocks.

Red-ſpangled 20 Rufeſcens *Dill.*

CUPTHONG. A livid coloured cruſt, with ſaucers of a reddiſh brown, and a fine thin margin. *From Dillenius.*

Lichenoides tartareum lividum, ſcutellis rufis, margine exili. *Dillenius*, tab. 18. fig. 14.

On Pentir rocks in Wales. *Dill.*

Meal-ſpangled 21 Albeſcens. *Hud.*

CUPTHONG. An aſh-coloured cruſt, with white flat mealy ſaucers. *Hudſon*, 445.—*Cruſt very thin, margin edged with a greeniſh blue ſtripe.*

Lichenoides candidum et farinaceum, ſcutellis fere planis. *Dillenius*, 131. tab. 18. fig. 11.

Lichenoides cruſta tenuiſſima, ſulcis cochleæ-formibus inſignita. *Ray's Syn.* 71.

Walls, and barks of trees.

Crab's-eye 22 Parellus. *Lin.* and *Dill.*

CUPTHONG. A white cruſt, with hollow blunt pale targets.—

Lichenoides leproſum tinctorum, ſcutellis lapidum cancri figura *Dillenius*, 130. tab. 18. fig. 10.

Walls, rocks, and ſtones; at all times. *Dill.* On Malvern hills.

C. Tiled.

Spreading 23 Centrifugus

CUPTHONG. Tiled: leaves with many imperfect clefts; ſmooth; whitiſh; pointing from a center. Saucers reddiſh brown.—*Leaves greeniſh, ſaucers cheſnut-coloured.* Dill. *It is ſometimes too, of a dark cheſnut with chocolate coloured ſaucers, but upon putting it into water the cheſnut will change in a few minutes to a browniſh green, or true olive, and if ſuffered to remain ſometime longer, the chocolate ſaucers will turn to a bright bay, and the whole appear exactly as* Dillenius *has deſcribed it.—When dry, it reſumes its former colour.*

Lichenoides imbricatum viridans, ſcutellis badiis. *Dillenius*, 180. tab. 24. fig. 75.

On large ſtones.

Flat-ſpangled 24 Cartilagineus *Dill.*

CUPTHONG. Tiled: leaves with many imperfect clefts; pointing from a center. Saucers flat; tawny,—*pretty large.* Leaves *of a griſtly ſubſtance.* From Dillenius.

Lichenoides cartilagineum, ſcutellis fulvis planis. *Dillenius*, tab. 24. fig. 74.

On ſtones, and rocks in Angleſey, and on Glyder-hill. *Dill.*

CUPTHONG.

CUPTHONG. Tiled: leaves bluntly notched; pointing Notched
from a center. Saucers notched at the edge.—*Leaves ſoft and* 25
thin. Has ſomewhat of the appearance of Agaric. From Dillenius. Crenulatus.
Lichenoides tenue et molle, Agarici facie. *Dillenius*, tab. 24. *Dill.*
fig. 73.
On trees near Pentir and Bangor. *Dill.*

CUPTHONG. Tiled: leaves indented; rough: pitted. Stone
Saucers cheſnut coloured.—*Leaves bluiſh grey; black and hairy* 26
underneath. Saucers *ſometimes of a blackiſh hue.* Saxatilis

1. Lichenoides vulgatiſſimum cinereo-glaucum, lacunoſum et *Blue.*
cirroſum. *Dillenius*, 188. tab. 24. fig. 83.
Lichenoides cruſta folioſa, ſuperne cinereo-glauco, inferne
nigra et cirroſa, ſcuttellis nigricantibus. *Ray's Syn.* 72.

2. Leaves ſmooth, and purpliſh, with larger ſaucers. *Purple.*
Lichenoides arboreum purpuraſcens tenue et læve, ſcutellis Hud.
majoribus. *Ray's Syn.* 73.
On ſtones, and trunks of trees.
It is uſed by the inhabitants of the North to dye purple.

CUPTHONG. Tiled: leaves with many clefts; ſmooth; Purple
blunt; hoary; ſprinkled with ſcattered riſing dots; —*purple.* 27
Lichen petræus purpureus Derbienſis. *Park.* 1315. Omphalodes
Lichenoides ſaxatile tinctorium, foliis piloſis purpureis. *Ray's*
Syn. 74. *Dillenius*, 185. t. 24. fig. 80.

Cork, or Arcell. Kenkerig *Welſh.*
On rocks.
It dyes wool of a dull but durable crimſon or purple colour.
It has been uſed as a ſtyptic.

CUPTHONG. Tiled: leaves lobed; ſhining; of a livid hue; Olive
—*Targets orbicular; membranaceous; raiſed above the ſurface of* 28
the leaf, and their margins unconnected with it. Olivaceus

1. Targets warty; larger. *Warty*
Lichenoides olivaceum, ſcutellis amplioribus verrucoſis. *Dillenius*, 184. tab. 24. f. 78.

2. Targets ſmooth. *Smooth.* Hud.
Lichenoides olivaceum, ſcutellis lævibus. *Dillenius*, t. 24. f.
77.
Lichenoides cruſta folioſa ſcutellata, pullum. *Ray's Syn.* 72.

3. Lobes broader; of a browniſh green. *Broad.* Hud.
Lichenoides arborum et ſaxatile, cruſta folioſa tenui, fuſco-virente, in ſegmenta latiora plerumque diviſa. *Ray's Syn.* 74.
On rocks, and trees.

Green 29 Virens *Dill.* CUPTHONG. Tiled: leaves lobed; indented; of a bright green. Saucers reddiſh yellow; downy underneath.—*Leaves large; broad.* Saucers *numerous.* From Dillenius.

Lichenoides læte virens, ſcutellis fulvis. *Dillenius*, tab. 25. fig 98.

On the aſh in Ireland, on ſtones at Comb Floyd near Biſhop's Caſtle, and on oaks between Carno and Mayne Loin in Merionethſhire. *Dillenius.*

Jagged 30 Laciniatus *Hud.* CUPTHONG. Tiled: leaves lobed; ſmooth; indented; *of a pale bluiſh colour.* Saucers reddiſh yellow; downy underneath—*Leaves ſtill larger than the preceding*; lobes *narrower.* From *Hudſon.*

Lichenoides ſubglaucum cumatile, foliis tenacibus, eleganter laciniatis. *Dill.* 197. t. 26. f. 99.

On the ſide of the road between Carnoe and Mayne Loin in Merionethſhire. Aug. *Dill.*

Obs. *Hudſon has characteriſed it* "Leafy, creeping," &c. *and has accordingly arranged it amongſt the* Leafy Cupmoſſes.—*Future obſervation may perhaps determine whether this and the preceding may not be only varieties of one common ſpecies.*

Ruſt-ſpangled 31 Fuliginoſus *Dill.* CUPTHONG. Tiled: leaves indented; ſprinkled all over with a black mealineſs like ſoot. Saucers ruſt-coloured; downy underneath. *From Dillenius.*

Lichenoides fuliginoſum, ſcutellis rubiginoſis. *Dill.* t. 26. f. 100.

Obs. *This and the two preceding ſpecies I have ventured to arrange amongſt the* Tiled *cupmoſſes. Dillenius characteriſes them all as* cloſely adhering to the ſubſtance on which they grow. *Perhaps too the* caperatus *of Linnæus might more properly find a place here, as he very juſtly remarks it can hardly be called* leafy. *Accurate examinations however of the plants themſelves can alone determine it.*

Curled 32 Criſpus CUPTHONG. Tiled: leaves lobed; lopped; ſcolloped; blackiſh green Saucers of the ſame colour—*Of a jelly-like ſubſtance.* Saucers *nearly as broad as the leaves.*

Lichenoides gelatinoſum atro virens, criſpum et rugoſum. *Dill.* 139. t. 19. f. 23.

Shady ſtony places, and at the bottom of walls.

Jelly 33 Gelatinoſus *Dill.* CUPTHONG. Tiled: leaves cloven: ſegments lopped; notched. Saucers ſmaller than the leaves.—*Of a jelly-like ſubſtance.*

Lichenoides tenue criſpum, foliis parvis depreſſis. *Dill.* t. 19. f. 33.

Wigmore Cliffs Herefordſhire. *Dill.*

CUP-

CUPTHONG. Tiled: edged with fringe like teeth. Targets larger than the leaves.—*Of a jelly-like ſubſtance. Leaves narrow, of a bluiſh lead colour.* Saucers *brown.* Teeth *ſometimes blunt.* Creſted 34 Criſtatus

1. Lichenoides gelatinoſum, foliis imbricatis et criſtatis. *Dill.* 140. t. 19. f. 26. Common

Lichenoides gelatinoſum tenerius laciniatum, ex plumbeo colore cæruleſcens, ſcutellis fuſcis. *Ray's Syn.* 72.

2. Leaves hairy, thicker, of a dark green. *Hairy.* Hud.

Lichenoides foliis piloſis eraſioribus, obſcure virentibus ſcutellis fuſcis. *Ray's Syn.* 74.

On rocks and walls.

CUPTHONG. Tiled: leaves curled; deep yellow. Targets the ſame colour—*Leaves ſometimes of a yellowiſh green, with targets of a pale yellow. And ſometimes of a bright orange.* Gold-ſpangled 35 Parietinus

Lichenoides vulgare ſinuoſum, foliis et ſcutellis luteis. *Dill.* 180. t. 24. f. 76.

Lichenoides cruſta folioſa ſcutellata, flaveſcens. *Ray's Syn.* 72.

Common Liverwort.

On trees, wood, ſtones, walls, and tiles.

CUPTHONG. Tiled: ſegments blunt; ſomewhat inflated—*The ends of the leaves appear ſwelled, and when cut tranſverſely are hollow as if blown up. Aſh coloured on the upper; black and ſmooth on the under ſurface* Inflated 36 Phyſodes

Lichenoides ceratophyllon obtuſius et minus ramoſum. *Ray's Syn.* 76. *Dill.* 154. t. 20. f. 49.

On heaths.

CUPTHONG. Tiled: leaves oblong; aſh-coloured; divided into narrow ſegments. Saucers blackiſh brown. Starry 37 Stellatis

Lichenoides cinereum, ſegmentis argute ſtellatis, ſcutellis nigris. *Dill.* 176. t. 24. f. 70.

Lichenoides arboreum, cruſta folioſa albo-cinerea, tenuiter et eleganter diſſecta, ſcutellis nigris. *Ray's Syn.* 74.

On aſh and blackthorn.

CUPTHONG. Tiled: leaves oblong; green; divided into narrow ſtraddling ſegments. Saucers blackiſh brown. *From Dillenius.* Diſtorted 38 Diſtortus *Dill.*

Lichenoides viride, ſegmentis anguſtis diſtortis, ſcutellis pullis. *Dill.* t. 24. f. 72.

About Oxford. *Dill.*

Saw-toothed 39 Runcinatus *Dill.* CUPTHONG. Tiled: leaves oblong; narrow; betwixt indented and toothed; brown; segments blunt. Saucers brownish black; notched at the edge. *From Dillenius.*

Lichenoides angustifolium fuscum, scutellis pullis. *Dill.* t. 24. f. 69.

Bangor rocks. *Dill.*

Circular 40 Orbicularus *Dill.* CUPTHONG. Tiled: leaves oblong: betwixt indented and toothed; sea green; segments pretty broad. Saucers black; somewhat notched at the edge.—*Expanding into a circular tuft.* From Dillenius.

Lichenoides glaucum orbiculare, segmentis latiusculis, scutellis nigris. *Dill.* t. 24. f. 71.

On Oak, Beech, Elm, &c. *Dill.*

D. Leafy.

Fringed 41 Ciliaris CUPTHONG. Leafy, somewhat upright; segments strap-shaped fringed; saucers on fruit-stalks; scolloped—*Black.*

Larger 1. Of a brownish ash-colour, with sometimes a tinge of sea-green.

Lichenoides hispidum majus et rigidius, scutellis nigris. *Dill.* 150 t. 20. f. 45.

Lichenoides arboreum foliosum cinereum, scutellis nigris, foliorum extremitatibus hispidis et pilosis. *Ray's Syn.* 73.

Trunks of trees.

Smaller Dill. 2. Smaller, more flexible, ash-coloured; saucers more numerous, pretty entire at the edge. From *Dillenius.*

Lichenoides hispidum minus et tenerius, scutellis nigris. *Dillenius* t. 20. f. 46.

Trees and stones.

Horny 42 Corneus *Dill.* CUPTHONG. Leafy (or rather horny) somewhat upright; segments oblong, betwixt toothed and fringed; diversified with roundish warts. From *Dillenius.*

Lichenoides corneum, marginibus eleganter fimbriatis. *Dill.* t. 29. f. 116.

On Snowdon, and on Berwyn, Derbyshire. *Dillenius.*

Obs. *It appears to have a very great affinity with the* Fringed Cupmoss.

CUP-

CUPTHONG. Leafy, ascending, jagged: borders raised, fringed—*Targets round, entire, on the surface of the leaves; whitish, brown, or purple. Sometimes the targets are at the ends of the leaves.* Eryngoleaved 43 Islandicus

1. Lichenoides rigidum, eryngiifolia referens. *Ray's Syn.* 77 *Dillenius* 209. t. 28. f. 111. *Common*

2. Leaves finer, and more curled. *Finer*
Lichenoides eryngii folia referens, tenuioribus et crispioribus foliis. *Dillenius* 212. t. 28. f. 112.

3. Smaller branches cylindrical, hollow within, not fringed—resembling black lace—*and yet notwithstanding this so remarkable a difference, they are only varieties of the same species. A variety of specimens has shewn how they insensibly run into one another.* *Blacklace*
Coralloides tenuissimum nigricans, mundi muilberis instar textum. *Dillenius* 113. t. 17. f. 32.
Heaths and mountains.
The Icelanders boil it in broth, or dry it and make it into bread. They likewise make gruel of it to mix with milk; but the first decoction is always thrown away, for it is apt to purge.

CUPTHONG. Leafy, ascending, jagged, curled, smooth, pitted, white; border raised;—*generally of a snowy whiteness; never with any tinge of yellow.* Tubercles *brown.* Snow 44 Nivalis
Lichenoides lacunosum candidum glabrum, Endiviæ crispæ facie. *Dillenius* 162. t. 21. f. 56.
Gravelly heaths.

CUPTHONG. Leafy; jagged; blunt; smooth. Pitted above, downy underneath.—*Targets at the edges.* Lungwort 45 Pulmonarius
Lichen arborum. *Gerard.* 1566.
Lichenoides pulmoneum reticulatum vulgare, marginibus peltiferis. *Dillenius* 212. t. 29. f. 113.
Lichenoides peltatum arboreum maximum. *Ray's Syn.* 76.
Muscus pulmonarius, sive Lichen arborum. *Park.* 1311.
Tree Liverwort. Tree Lungwort.
On trunk of trees, particularly Oak and Ash, on rocks, and sometimes on heaps of stones in shady places.
It is reckoned very efficacious in consumptive cases; this opinion merits a further investigation—Boiled with woollen cloth it gives it a brown orange colour.

Mealwarted 46 Scrobiculatus *Dill.*

CUPTHONG. Leafy ; jagged ; border indented ; curled ; mealy ; ſmooth above, diverſified with targets, and mealy warts ; downy underneath.—*Of a ſea green, inclining to a ſulphur colour above, ſomewhat pitted ; light brown underneath* ; Down *of a dirty brown. The leaf appears ſometimes pierced with a number of holes, as if eaten through, which ſeem to have been the places where former warts have grown. It has a near affinity to the pceceding ſpecies. From* Dillenius.

Lichenoides pulmoneum villoſum, ſuperficie ſcrobiculata et peltata.. *Dill.* t. 29. f. 114.

Found by Mr. Ran on *Stones* at Cockbuſh on the Sea ſide in Suſſex,—at Dolgelle in Merionethſhire. *Dill.* and on *Rocks* on Malvern-hills, juſt above Great Malvern (with the preceding ſpecies.) St.

Spongy 47 Spongioſus *Dill.*

CUPTHONG. Leafy ; jagged ; ſegments ſtrap-ſhaped ; blunt. Targets above ; ſpongy underneath.—*Segments very much reſembling the horns of a Stag.* From Dillenius.

Lichenoides Damæ cornua referens, ſubtus ſpongioſum. *Dill.* t. 29. f. 115.

On trees. *Dill.*

Branny 48 Furfuraceus

CUPTHONG. Leafy ; drooping ; as if ſprinkled with bran : ſegments ſharp ; black and pitted underneath ;—*White, or aſh coloured above.*

Lichenoides cornutum amarum, deſuper cinereum, inferne nigrum. *Dill.* 157. t. 21. f. 52.

Barks of trees, and pales.

Tranſparent 49 Decumbens *Hud.*

CUPTHONG. Leafy ; drooping ; jagged : ſegments cloven at the end into three parts. Saucers browniſh. *Hudſon* 449.—*Of a jelly-like ſubſtance.*

Lichenoides gelatinoſum, fuſcum, Jacobææ maritimæ diviſura. *Dill.* 140 t. 19. f. 25.

On rocks.

Blackiſh 50 Nigreſcens *Hud.*

CUPTHONG. Leafy ; wrinkled ; lobed, of a blackiſh green. Saucers tawney red, ſet thick together. *Hudſon* 450.—*Small.* Leaf *between jelly-like and membranaceous* ; *thin. Dillenius compares it to a bat's wing.*

Lichenoides gelatinoſum membranaceum, tenue nigricans. *Dill.* 138. t. 19. f. 20.

Lichenoides ſaxatile membranaceum gelatinoſum tenue nigreſcens. *Ray's Syn.* 72.

Trunks of trees, rocks, and walls.

CUP-

CUPTHONG. Leafy; rather flat; lobed; scolloped tar- Bladder
gets globular; inflated. 51
Lichenoides tinctorium glabrum vesiculosum. *Dill.* 188. t. Ampullaceus
24. f. 82.
Lichenoides saxatile tinctorium, foliis latioribus non pilosis, vesiculas proferens. *Ray's Syn.* 74.
In hilly pastures.

CUPTHONG. Leafy; upright; compressed; branched; Mealy
with mealy warts on the edge—*dispersing a quantity of dust.* 52
Farinaceus

1. Leaves of a whitish ash-colour, with narrow segments. *Whitish*
Lichenoides segmentis angustoribus, ad margines verrucosis et pulverulentis. *Dill.* 172. t. 23. f. 63.
Lichenoides arboreum ramosum, cinereo-candicans, segmentis angustioribus. *Ray's Syn.* 76.

2. Leaves of a greenish ash-colour, with narrow segments. *Greenish.* Hud.
Lichenoides arboreum ramosum, cinereo-virescens segmentis angustioribus, circa margines crispis. *Ray's Syn.* 76.

3. Leaves with broader segments; rough: with larger warts, *Broad-leaved*
containing a great quantity of dust. Hud.
Lichenoides non tubulosum Platydasyphyllon tuberculis pulverulentis donatum. *Ray's Syn.* 66.
Trunks of trees; whitest on the sloe.

CUPTHONG. Leafy; upright; strap-shaped; branched; Channelled
pitted; convex; with sharp points at the ends,—*Saucers nume-* 53
rous; terminating the branches. Leaves *of a greenish ash-colour.* Calicaris
Lichenoides coralliforme rostratum et canaliculatum. *Dill.* 170. t. 23. f. 62.
Lichenoides arboreum ramosum, segmentis, angustioribus cinereo-virescentibus. *Ray's Syn.* 76.
Rocks, and barks of trees.

CUPTHONG. Leafy; straight; oblong; spear-shaped; some- Ash
what jagged; pitted; smooth; saucers on short fruit-stalks,—*mostly* 54
of a pale straw colour, but sometimes of the same colour with the leaf. Fraxineus.
Leaf *of a greenish ash-colour, of a rigid consistence.*
Lichenoides longifolium rugosum rigidum. *Dill.* 165. t. 22. f. 59.
Lichenoides arboreum ramosum scutellatum majus et rigidius, colore virescente. *Ray's Syn.* 75.
Trunks of trees, on Poplar, Apple, &c. but chiefly on Oak and Ash,

CUP-

Jersey 55 Cæsarienfis *Dill.*

CUPTHONG. Leafy; nearly ftraight; jagged; fomewhat pitted. Segments ftrap-fhaped; blunt.—*Often cloven at the ends. Saucers fcattered over the furface, and along the edges of the leaf.* From Dillenius.

Lichenoides fuciforme tinctorium, corniculis longioribus et obtufioribus. *Dill.* t. 22. f. 60.

Jerfey rocks. *Dill.*

Blackthorn 56 Prunaftri

CUPTHONG. Leafy; rather upright; pitted; white and downy underneath;—*Branched like the horns of a Deer.*

Common

1. Lichenoides cornutum, bronchiale molle fubtus incanum. *Dill.* 160. t. 21. f. 55.

Lichenoides arboreum ramofum majus et mollius, colore candicante. *Ray's Syn.* 75.

White

2. White; fegments narrower.

Lichenoides corniculatum candidum molle, fegmentis anguftis. *Dill.* 159. t. 21. f. 54.

(1) On moft trees, but chiefly on the Blackthorn. On the Willow foft, Whiteft on the Blackthorn. (2) Lippock Heath Hampfhire. Aug. *Dill.*

It has a remarkable property of imbibing and retaining odours and is therefore the bafis of many perfumed powders.

Ragged 57 Lacerus *Dill.*

CUPTHONG. Leafy: rather upright; pitted; fomewhat jagged.—*Margin divided into a number of narrow fegments.* From Dillenius.

Lichenoides lacunofum lacerum, latius et anguftius. *Dill.* t. 21. f. 57.

About Beddington, and Carfhalton Surry; Slinford Suffex; and Oxford Woods. *Dill.* And near Worcefter. St.

Juniper 58 Juniperinus

CUPTHONG. Leafy; jag ed; curled; deep yellow; with livid coloured targets.—*Refembles the* Gold-fpangled CUPTHONG *but is of a yellowifh green with brown* targets, *and the* Leaves *loofe and fomewhat upright.*

On boughs and trunks of trees.

The Rufties in Sweden efteem it a fpecific for the Jaundice.—They ufe it to dye their woollens yellow.

CUPTHONG. Pale green; wrinkled, and waved at the Roſe-leaved
edge,—*Of a yellowiſh hue inclining to ſea green; black and ſmooth* 59
underneath. Can hardly be called leafy, *adhering cloſe to the ſub-* Caperatus
ſtance on which it grows. Saucers *ſeldom to be found.*

Lichenoides caperatum, roſaceæ expanſum, e ſulphureo virens. *Dill.* 193. t. 25. f. 97.

Lichenoides cruſta folioſa, ex cinereo et luteo vireſcente, inferne nigra et lævi. *Ray's Syn.* 73.

On ſtones, and trees.

It dyes woolens of an orange colour.

CUPTHONG. Leafy; depreſſed, lobed; ſmooth; border Shining
curled; mealy.—*Leaves thinner than paper; of a mixture of* 60
white, aſh-colour, and ſea green; black underneath. Glaucus

Lichenoides endiviæ foliis criſpis ſplendentibus, ſubtus nigricantibus. *Dill.* 192. t. 25 f. 96.

Hilly paſtures and heaths.

CUPTHONG. Leafy; depreſſed; of a jellylike ſubſtance; Browniſh
divided into lobes thicker than the reſt of the leaf. Of a green
browniſh green.—Tubercles *diſperſed near the margin.* From 61
Dillenius. Fuſco-virens *Dill.*

Lichenoides gelatinoſum lobis craſſioribus fuſco-viridibus. *Dill.* t. 19. f. 22.

Rocks near Marſton in the neighbourhood of Oxford. Winter and Spring. *Dill.*

CUPTHONG. Leafy; of a jellylike ſubſtance. Tubercles Cluſter
turban-ſhaped; in cluſters; larger than the leaves; root leaves 62
very ſmall, like clear jelly.—Tubercles *on ſhort fruit-ſtalks; edged* Faſcicularis
with a blunt border.

Lichenoides gelatinoſum palmatum, tuberculis conglomeratis. *Dill.* 141, t. 19. f. 27.

Woods and garden walks.

E. Leather-like.

CUPTHONG. Leathery, creeping, lobed. Targets on the Livid
edge, facing downwards—*Reddiſh brown.* Leaves *brown.* 63
Lichenoides fuſcum, peltis poſticis ferrugineis. *Dill.* 206. t. Reſupinatus
28 f. 105.

Lichenoides Saxatile fuſcum, peltis in averſa foliorum ſuperficie locatis. *Ray's Syn.* 77.

On trees, pebbles, and on theground.

CUP-

Seagreen 64 Aphtosus CUPTHONG. Leathery; creeping; flat; with blunt lobes; sprinkled with watery excrescences; targets horizontal; on the edge—*black.*

When dry the colour is sea-green, but when fresh a full green.

Ths country people make an infusion of it in milk and give it to children that have the Thrush. In large doses it operates by purging and vomiting, and destroys worms.

Liverwort 65 Caninus CUPTHONG. Leathery; creeping; flat; with blunt lobes; woolly and veiny underneath. Targets on the edge, ascending.—*Leaves covered with a kind of ash-coloured mealiness.*

Lichenoides digitatum cinereum, Lactucæ foliis sinuosis. *Dill*, 200. t. 27. f. 102.

Lichenoides peltatum terrestre cinereum majus, foliis divisis. *Ray's Syn.* 76.

Ash coloured ground Liverwort.

Heaths, woods and hedges.

This is the basis of the famous powder recommended by Doctor Mead to prevent madness, aud recorded in our dispensatories under the name of Pulvis Antilyssus.

Lettuce-leaved 66 Lactucæfolius *Dill.* CUPTHONG. Leathery; creeping; with blunt lobes; border curled; wooly underneath.; targets on the edge, ascending.—*Leaf of a reddish colour above, not veiny underneath.* From Dillenius.

Lichenoides digitatum rufescens, foliis Lactucæ crispis. *Dill.* t. 27. f. 103.

Heaths, hedges, and woods. *Dill.*

Smallshield 67 Sylvaticus CUPTHONG. Leathery; creeping; jagged; pitted. Targets on the edge; ascending. *Small.*

Lichenoides polyschides villosum et scabrum, peltis parvis *Dill.* 199. t. 27. f. 101.

In woods.

Many-fingered 68 Polydactylos *Dill.* CUPTHONG. Leathery; creeping; jagged; veiny and wooly underneath; ends of the segments divided into many shallow clefts, supporting as many ascending targets—*Leaf ash-coloured.* From *Dillenius.*

Lichenoides cinereum polydactilon. *Dillenius* t. 27. f. 107.

About Perfedgoed-House, near Bangor. *Dill.*

Double fingered 69 Didactylos *Dill.* CUPTHONG. Leathery; membranaceous; transparent; jagged; segments cloven, each segment supporting two targets—*Veiny and wooly underneath.* From *Dillenius*

Lichenoides membranaceum pellucidum, peltis digitatis geminatis. *Dill.* t. 27. f. 108.

Rumbies-Mere near Kelwick, Yorkshire. *Dill.*

CUP-

CUPTHONG. Leathery; creeping; flat; not veiny un- Flat ſhield
derneath; targets on the edge; horizontal—*Brown.* 70
Lichenoides ſubfuſcum, peltis horizontalibus planis. *Dill.* Horizontalis.
205. t. 28. f. 104.
In woods. Winter and Spring. *Dill.*

CUPTHONG. Leathery, creeping; lobed; ſmooth; black Pearl
underneath. Saucers on fruit-ſtalks entire.—*Wooly underneath;* 71
of a ſea green, or aſh colour above. Perlatus
Lichenoides glaucum perlatum, ſubtus nigrum et cirroſum.
Dill. 147. t. 20. f. 39.
Trunks of trees.

CUPTHONG. Leathery; creeping; circular; targets Pouched
ſunk below the ſurface of the leaf, forming a kind of bag 72
beneath. Saccatus
Lichenoides Lichenis facie, peltis acetabulis immerſis. *Dill.*
221. t. 30. f. 121.
Lichen immerſus. *Hudſon* 453.
On rocks.

F. SOOTY,—*adhering to the ſubſtance on which they grow only in one point.*

CUPTHONG. Hunched; dotted; deep yellow underneath Cloudy
—*Smooth; aſh coloured on the upper ſurface.* 73
Lichenoides coriaceum nebuloſum cinereum punctatum ſubtus Miniatus
fulvum. *Dill.* 223. t. 30. f. 127.
On rocks and mountains.

CUPTHONG. Wrinkled; blackiſh brown; ſprinkled with Black-warted
black warts; wooly underneath.—*Of a rigid conſiſtence.* From 74
Dillenius. Verrucoſus
Lichenoides rugoſum durum pullum peltis albis verrucoſis. *Dill.*
Dill. t. 30. f. 118.
Snowdon and Berwyn Derbyſhire. *Dill.*

CUPTHONG. Very hairy underneath—*Leaf very broad;* Fleecy
ſprinkled with warts. 75
Lichenoides coriaceum, látiſſimo folio umbilicato et verrucoſo. Velleus
Dill. 545. t. 82. f. 5.
On rocks.
The Canadian Indians eat it when preſſed by hunger.

Singed 76 Pustulatus — CUPTHONG. Pitted underneath: sprinkled over with a black meal—*Ash-coloured above; black underneath.*

Lichenoides pustulatum cinereum et veluti ambustum. *Dill.* 226. t. 30. f. 131.

On rocks facing the South.—Malvern Hills.

This may be converted into an exceedingly black paint; and a beautiful red colour may be prepared from it.

Smutted 77 Deustus — CUPTHONG. Of an even surface on both sides.—*Ash-coloured above with black flatted warts; so brittle that unless in a moist state it can hardly be separated entire from the rocks on which it grows.*

Lichenoides coriaceum cinereum, peltis atris compressis. *Dill.* 218. t. 30. f. 117.

Lichenoides saxatile, foliis minus divisis, cinereo-fuscis. *Ray's Syn.* 73.

St. Vincent's Rock near Bristol.

Smooth 78 Polyphyllus — CUPTHONG. Composed of several leaves; even on both sides; greenish black and scolloped—

Lichenoides tenue pullum, foliis utrinque glabris. *Dill.* 225. t. 30. f. 129.

On high mountains.

Lead-coloured 79 Luridus *Dill.* — CUPTHONG. Composed of several leaves; of an even surface on both sides; lead-coloured; downy; indented underneath.—*Very much resembles the preceding species.* From Dill.

Lichenoides imbricatum luridum. *Dill.* t. 30 f. 128.

On stones in a rivulet near a mill by Perfedgoed-house near Bangor, also in a rivulet in the county of Radnor. *Dill.*

Fibrous 80 Polyrrhizos — CUPTHONG. Composed of several leaves, of an even surface on both sides; fibres numerous.—*Leaves dirty brown, or greenish ash-colour; black and hairy underneath.*

Lichenoides pullum superne et glabrum, inferne nigrum et cirrosum. *Dill.* 226. t. 30. f. 130.

On high mountains.

G. Glass-shaped.

Mashroom 81 Pezizæformis *Dill.* — CUPTHONG. Cup simple; imperfect; convex at the top; with a hollow in the middle; brown.—*Has a near affinity to the* Powdered, Heath, and Brown Cupthongs *but the tubercles do not rise from a thin ground or crust, but from* Leaves *like those of the* Glass-shaped Cupthongs, *to which it seems naturally to belong.*

Coralloides fungiforme fuscum, basi foliacea. *Dill.* t. 14. f. 2.

Hampstead heath. *Dill.*

CUPTHONG. Cup ſimple; very entire; fruit-ſtalk cylindrical; tubercles ſcarlet.—*Cups greeniſh grey, ſometimes ſpringing out of one another.* Red 82 Cocciferus

Coralloides ſcyphiforme, tuberculis coccineis. *Dill.* 82. t. 14. f. 7.

Lichenoides tubuloſum pyxidatum, tuberculis amene coccineis. *Ray's Syn.* 69.

On heaths and in ſhady places. November—February.

The ſcarlet tubercles on the affuſion of Ley, ſtrike a durable purple.

CUPTHONG. Cup ſimple; ſhorter than the leaves. Tubercles ſcarlet.—*Cups fringed with leaves; not always ſhorter than the leaves.* Radiated 83 Cornucopioides

Coralloides ſcyphiforme, marginibus radiatis et foliatis. *Dill.* 85. t. 14. f. 9.

Lichenoides pyxidatum, marginibus eleganter foliatis. *Ray's Syn.* 69.

On heaths. Feb.

CUPTHONG. Cup ſimple; ſhorter than the leaves; brown within; on very ſhort fruit-ſtalks;—*margin entire.* Short-footed 84 Humilis *Dill.*

Coralloides ſcyphis humilibus, intus fuſcis. *Dillenius* t. 14. f. 11.

Charlton, Woolwich. January, February. *Dill.*

CUPTHONG. Cup ſimple; a little ſcolloped; tubercles brown. Common 85 Pyxidatus

1. Coralloides ſcyphiforme, tuberculis fuſcis. *Dillenius* 79. t. 14. f. 6. *Simple*

   Lichenoides tubuloſum pyxidatum cinereum. *Ray's Syn.* 68.

   Muſcus pyxoides terreſtris. *Bauh. pin.* 361.

   Heaths, woods, walls, and ditch-banks.

2. Cups ſpringing out of one another, ſometimes from the edge, and ſometimes from the center. *Proliferous*

   Lichenoides tubuloſum pyxidatum proliferum. *Ray's Syn.* 69.

3. Cups ſpringing out of one another, with browniſh black tubercles. *Black-tipt. Hud.*

   Lichenoides tubuloſum pyxidatum, tuberculis fuſco nigris, proliferum. *Ray's Syn.* 69.

   Heaths, woods, walls and ditch-banks.

Indented 86 Fimbriatus

CUPTHONG. Cup ſimple; edged with little teeth; fruit-ſtalk cylindrical.—

Simple

1. Coralloides ſcyphiforme gracile, marginibus ſerratis. *Dill.* 84. t. 14. f. 8.

Lichenoides tubuloſum pyxidatum, marginibus ſerratis. *Ray's Syn.* 69.

Proliferous Hud.

2. Cups ſpringing out of one another.

Lichenoides pyxidatum proliferum, marginibus ſerratis. *Ray's Syn.* 69.

Woods and heaths.

Trumpet 87 Filiformis *Hud.*

CUPTHONG. Cup ſimple: very entire; long and ſlender. Tubercles brown. *Hudſon* 456. leaves jagged.

Lichenoides tubuloſum pixidatum exiguum, ſuſco-virens. *Ray's Syn.* 70.

Coralloides ſcyphis gracilibus tubiformibus, pedicularis folio. *Dillenius* 85. t. 14. f. 10.

Walls and heaths.

Proteus 88 Polymorphos *Dill.*

CUPTHONG. Branched; leafy—*Leaves nearly upright; jagged; curled.* From Dillenius. *This plant aſſumes very different forms, ſometimes conſiſting of a ſimple leaf, with a number of little cups riſing from its ſurface. At other times it appears like a ſhrub, the ends of the branches running into cups, and edged with tubercles. Again, reſembling each, and yet ſtill different, it has a ſtem which branches into leaves. This proves them to be only varieties of one common ſpecies. From this circumſtance I have ventured to call it the* Lichen Polymorphos *or the* Proteus Cupthong, *though Hudſon has already named one of its varieties, ſince Linnæus it is probable, would not allow of it, as it is the name of one of the ſubdiviſions,—the* Leafy [*Foliaceus.*]

Branched. Dill.

1. Very much branched; leaves at the diviſions of the ſtem; branches terminating in cups; cups toothed, edged with tubercles.

Coralloides ſcyphiforme, foliis alcicorniformibus cartilaginoſis. *Dill.* t. 14. f. 12. B.

Snowdon. *Dill.*

Branching Dill.

2. Stem branched; branches running into leaves.—*Leaves upright, with winged cleſts.*

Dillenius ib. D.

Trowbridge. *Dillenius.*

Leafy Foliaceus Hud.

3. Leaves nearly upright, jagged, curled; cups very ſhort, conical, on the ſurface of the leaf. *Hudſon* 457.

Dillenius ib. A.

Lichenoides cartilaginoſum, tubulis et pyxidulis exiguis. *Ray's Syn.* 70.

Blackheath. *Dill.* Mountainous heaths. *Hud.*

CUR-

CUPTHONG. Branched; thread-ſhaped; cups edged with little teeth—*Some of the branches tapering to a point, others terminated by a cup, tipt with bright brown tubercles.* Tall 89 Gracilis

Coralloides ſcyphiforme ſerratum elatius, caulibus gracilibus glabris. *Dillenius* 88. *t.* 14. f. 13.
Lichenoides pyxidatim cinereum elatius, ramulis pyxidatum deſinentibus. *Ray's Syn.* 69.
Mountainous heaths, and ſtoney places.

CUPTHONG. Branched; branches cylindrical; cups pretty entire, edged with round brown tubercles—*Tubercles numerous. Cups ſometimes ſimple.* From Dillenius. Branching 90 Ramuloſus *Dill.*
Coralloides parum ramoſum, tuberculis fuſcis. *Dillenius* t. 15 f. 20.
Woolwich-heath. Feb. *Dill.*

CUPTHONG. Very much branched; branches cylindrical; cups entire, edged with round tubercles.—*Tubercles numerous, ſcarlet.* Fingered 91 Digitatus
Coralloides ramuloſum, tuberculis coccineis. *Dillenius* 96. t. 15. f. 19.
Lichenoides coralliforme, apicibus coccineis. *Ray's Syn.* 68.
Heaths and woods, and decayed roots of trees. Feb.

CUPTHONG. Very much branched; inflated; cups toothed; tubercles brown. *Hudſon* 457. *Stem* nearly cylindrical, expanding into a cup, which branches out into a number of ſub-diviſions, which terminate in their turn in other cups, divided into teeth, and tipt with brown tubercles. Cruſted 92 Ventricoſus Hud.
Coralloides cornucopioides incanum, ſcyphis criſtatis. *Dill.* 94. t. 15. f. 17.
Lichenoides tubuloſum cinereum, valde cruſtaceum, ramulis brevioribus et crebrioribus ex acetabulis enaſcentibus. *Ray's Syn.* 68.
In woods.

CUPTHONG. Nearly ſimple; ſomewhat inflated; cups entire.—*Aſh-coloured. Some of the branches tapering to a point, others terminating in imperfect cups. Stems moſtly of an equal thickneſs throughout.* Plain 93 Cornutus
Coralloides vix ramoſum, ſcyphis obſcuris. *Dillenius* 90. t. 15. f. 14.
Lichenoides tubuloſum cinereum, non ramoſum. *Ray's Syn.* 68.
On heaths. Nov.—Feb.

Serrated 94 Deformis CUPTHONG. Neatly simple; somewhat inflated; cups toothed;—*slenderest at the base; somewhat hoary.*

Coralloides crassius subincanum, calycibus dentatis. *Dillenius* 95. t. 15. f. 18.

Lichenoides tubulosum minus ramosum cauliculis crassioribus difformibus. *Ray's Syn.* 68.

On heaths.

Horntipt 95 Cornigerus *Dill.* CUPTHONG. Nearly simple; somewhat inflated; cups with many clefts;—*Segments mostly awl shaped pretty long; upright; some tapering to a point, others tipt with tubercles.* Stems *sometimes branched; cylindrical; tapering towards the base.* From Dillenius.

Coralloides scyphiforme cornutum. *Dillenius* t. 15. f. 16.

In woods, in Enfield-Chace. *Dill.*

H. SHRUBBY.

Reindeer 96 Rangiferinus CUPTHONG. Shrubby; perforated; very much branched; little branches nodding.—*There is an opening or perforation between every division of the stem and branches. Soft when young, but afterwards of a stoney hardness.*

White 1. With whitish tubercles.

Coralloides montanum fruticuli specie, ubique candicans. *Dillenius* 107. t. 16. f. 29.

Lichenoides tubulosum ramosissimum, fruticulis specie, candicans. *Ray's Syn.* 66.

Brown-tipt 2. With brown tubercles.

Coralloides fruticulis specie candicans, corniculis rufescentibus. *Dillenius* 110. t. 16. f. 30.

Lichenoides tubulosum ramosissimum, fruticulis specie, candicans, corniculis rufescentibus. *Ray's Syn.* 67.

Woods, heaths, and mountains.

The Laplanders could not exist without this plant. It is the food of the Rein-deer, and the Rein-deer supplies every necessary of life for the contented people of that inhospitable climate.

CUPTHONG. Shrubby; perforated; little branches very Short
ſhort; ſharp— 97
Uncialis

1. Small: not much more than an *Inch* long, whence its Latin *Smaller*
name.
Coralloides perforatum minus, molle et tenue. *Dillenius* 99. t. 16. f. 22.
Lichenoides tubuloſum, cauliculis mollioribus et craſſioribus, minus. *Ray's Syn.* 67.

2. Conſiderably larger; more than twice as large; ſtem and *Larger*
branches thicker; in every other reſpect almoſt exactly *Dill.*
ſimilar.
Coralloides perforatum majus, molle et craſſum. *Dill.* t. 16. f. 21.
Heaths and hills. (2.) Leath-hill, Surry, and Peter's-field-heath, Hampſhire. *Dill.*

CUPTHONG. Shrubby; tubular; very much branched; Prickly
thorny. Little branches with finger-like diviſions, tipt with 98
tubercles. *Hudſon* 459.—Of an aſh-coloured brown. Spinoſus.
Hud.

Coralloides ſparſum, caulibus tortuoſis et ſpinoſis. *Dillenius* 101. t. 16. f. 25.
Lichenoides tubuloſum ramoſum rigidius, majus et craſſius, cinereo-fuſcum. *Ray's Syn.* 16.
Muſcus coralloides ſaxatilis cervi cornua referens. *Bauh. pin.* 361.
Heaths.

CUPTHONG. Shrubby; tubular; branched; branches up- Forked
right; forked. *Hudſon* 458.—*Aſh-coloured; ſometimes covered* 99
with cruſtaceous leaves. Furcatus
Hud.

Coralloides e paniculis brevioribus et crebrioribus. *Dillenius* 104. t. 16. f. 27.
Lichenoides tubuloſum cinereum, ramoſius et cruſtaceum. *Ray's Syn.* 67.
Heaths and high hills.

CUPTHONG. Shrubby; ſomewhat forked: branches un Horned
divided; awl-ſhaped—*Tubercles ſmall, brown, globular, at the* 100
ends of the branches. Stem *ſometimes fringed with a few ſcattered* Subulatus
cruſtaceous leaves.
Coralloides corniculus longioribus et rarioribus. *Dillenius* 102. t. 16. f. 26.
Lichenoides tubuloſum cinereum, minus cruſtaceum, minuſque ramoſum. *Ray's Syn.* 67.
Muſcus corniculatus. *Gerard.* 1560. *Park.* 1308.
Heaths.

Madrepore 101 Madreporæformis *Dill.*

CUPTHONG. Shrubby; little branches very ſhort; blunt —*Stems brittle, tubular, ſcarcely divided; growing cloſe together, and forming a tuft reſembling* Organ Coral. From *Dill.*

Coralloides minimum fragile, madreporæ inſtar naſcens. *Dill.* t. 16. f. 28.

Bagſhot near Farnham. Spring and Winter. *Dill.*

Round-headed 102 Globiferus

CUPTHONG. Shrubby; ſolid; ſmooth. Tubercles globular; hollow; at the ends of the branches—*black within, aſh-coloured without.* Little branches *nearly cylindrical; ſharp and forked at the ends. Sometimes very much branched.*

Coralloides cupreſſiforme, capitulis globoſis. *Dill.* 117. t. 17. f. 35.

Lichenoides non tubuloſum, ramulis ſcutellis nigris terminatis. *Ray's Syn.* 66.

Lichen globoſus. *Hudſon* 460.

Rocks and mountains.

Oarweed 103 Fucoides *Dill.*

CUPTHONG. Shrubby; ſolid; ſmooth. Tubercles flattiſh; at the ends of the branches.—*Black.* From *Dillenius.*

Coralloides corniculatum, fuci tenuioris facie. *Dill.* t. 17. f. 37.

On Snowden. *Dill.*

Thorny 104 Aculeatus *Dill.*

CUPTHONG. Shrubby; ſolid; ſtem and branches prickly. —*Brown.*

Coralloides fruticuli ſpecie fuſcum, ſpinoſum. *Dillenius* t. 17. f. 31.

Stieporſtone-hill Shropſhire, heaths near London, and Cambridgeſhire hills. *Dill.*

Woody 105 Paſchalis

CUPTHONG. Shrubby; ſolid; covered with little cruſtaceous leaves—*Which make a beautiful appearance, eſpecially if viewed through a glaſs that magnifies a little.*

Coralloides criſpum et botryforme alpinum. *Dill.* 114. t. 17. f. 33.

Lichenoides non tubuloſum cinereum ramoſum, totum cruſtaceum. *Ray's Syn.* 66.

High mountains.

Brittle 106 Fragilis

CUPTHONG. Shrubby; ſolid; little branches cylindrical, blunt—*So brittle as not to be gathered but when it is moiſt.* Branches *greyiſh brown.* Tubercles *black.*

Coralloides alpinum, corallinæ minoris facie. *Dill.* 116. t. 17. f. 34.

Lichenoides non tubuloſum ramoſiſſimum, fruticuli ſpecie cinereo fuſcum. *Ray's Syn.* 65.

Mountains and high moorlands.

CUP-

CUPTHONG. Shrubby; ſolid; ſomewhat branched. Tu- Podded
bercles hollow; on the ſides of the branches. *Hudſon.* 460.— 107
No appearance of leaves. Siliquoſus

Coralloides faſciculare verrucoſum et veluti ſiliquoſum. *Dill.* 119. t. 17. f. 38.

On rocks.

I. Threadshaped.

CUPTHONG. Thread-ſhaped; pendant: branches matted Hairy
together. Saucers radiated.— 108

Uſnea vulgaris, loris longis implexis. *Dill.* 56. t. 11. f. 1. Plicatus

Muſcus arboreus, Uſnea officinarum. *Bauh. pin.* 361. *Ray's Syn.* 64.

Muſcus quernus. *Gerard.* 1558.

Muſcus arboreus, vulgaris et quercinus. *Park.* 1311.

Tree Moſs.

On trees in thick woods.

CUPTHONG. Thread-ſhaped; pendant; ſomewhat jointed. Bearded
Branches expanding—*Aſh-coloured.* 109

Uſnea barbata, loris tenuibus fibroſis. *Dill.* 63. t. 12. f. 6. Barbatus

Muſcus capillaceus longiſſimus. *Bauh. pin.* 361.

On trees in woods.

CUPTHONG. Thread-ſhaped; pendant; compreſſed at Fennel
the diviſions of the branches.—*Blackiſh grey.* 110

Uſnea jubata nigricans. *Dill.* 64. t. 12. f. 7. Jubatus

Muſcus corallinus ſaxatalis fæniculaceum. *Ray's Syn.* 65.

Woods and rocks.

It is uſed to cure ulcerations of the ſkin; and it is eaten by Reindeer.

CUPTHONG. Thread-ſhaped; very much branched; Woolly
drooping; matted, opake,—*Like black Wool.* 111

Uſnea lanæ nigræ inſtar, ſaxis adhærens. *Dill.* 66. t. 13. f. 8. Lanatus

Muſcus coralloides lanæ nigræ inſtar ſaxis adhærens. *Ray's Syn.* 65.

On rocks.

CUPTHONG. Thread-ſhaped; branched; drooping. Wiry
Branches ſtraddling; ſerpentine and ſomewhat matted— 112

Uſnea rigida, horſum vorſum extenſa. *Dill.* 6. t. 13. f. 10. Chalybeiformis

Muſcus caule rigido inſtar fili chalybei. *Ray's Syn.* 65.

Trunks of trees, ſtones, and decayed wood.

Hair spangled 113 Scutelliserus *Dill.* CUPTHONG. Thread-shaped; branched; drooping; rough with hair; with black saucers on the sides, and at the ends of the branches. From Dillenius.

Lichenoides subhirsutum teres, scutellis parvis nigris. *Dill.* t. 21. f. 51.

On the Sloe-tree near the sea. *Dill.*

Rough Hirtus CUPTHONG. Thread-shaped; very much branched; upright. Tubercles mealy; scattered.—

Usnea vulgatissima tenuior et brevior, sine orbiculis. *Dill.* 67 t. 13. f. 12.

Muscus ramosus. *Gerard.* 1372. *Ray's Syn.* 65.

Woods and hedges.

Gold 114 Vulpinus CUPTHONG. Thread-shaped; very much branched; upright; branches nearly of the same length; angular; angles unequal.—*Of a Citron colour.*

Usnea capillacea citrina, fruticuli specie. *Dill.* 73. t. 13. f. 16.

Muscus aureus tenuissimus. *Ray's Syn.* 65.

In woods.

In Norway they mix this plant with powdered glass and strew it upon dead carcases to poison Wolves.—It dyes woollens yellow.

Jointed 115 Articulata CUPTHONG. Thread-shaped; jointed; little branches very slender, dotted—

Usnea capillacea et nodosa. *Dill.* 60. t. 11. f. 4.

Muscus arboreus nodosus. *Bauh. pin.* 361. *Ray's Syn.* 65.

Muscus arboreus nodosus, sive geniculatus. *Park.* 1311.

In woods, on Beech and Hazel.

Flowering 116 Floridus CUPTHONG. Thread-shaped; branched; upright; saucers radiated—

Usnea vulgatissima tenuior et brevior, cum orbiculis. *Dill.* 69. t. 13. f. 13.

Muscus arboreus cum orbiculis. *Bauh. pin.* 361. *Ray's Syn.* 65.

Muscus ramosus floridus. *Gerard.* 1560.

Woods, thickets, and old hedges.

426 POWDERWORT. 1208 Byſſus.

Fibres ſimple ; uniform ; like ſoft wool, or duſt.

* *Thready.*

POWDERWORT. Threads downy ; ſwimming upon water Paper
—and forming a kind of thin cruſt upon the ſurface. Colour green 1
or white. It riſes in the middle of ſummer, and mixes with the water Flos Aquæ
ſo as to render it turbid and unfit to drink ; at night it falls down towards the bottom.
Byſſus latiſſima, papyri inſtar ſupra aquam expanſa. *Ray's Syn.* 57. *Dill.* 1.
Stagnant waters.

POWDERWORT. Woolly ; Violet coloured ; growing Violet-coloured
upon wood—
Byſſus lanuginoſa violacea, lignis adnaſcens. *Dill.* 4. t. 1. 2
f. 6. Phoſphorea
Byſſus pulverulenta violacea lignis adnaſcens. *Ray's Syn.* 56.
On rotten wood.

POWDERWORT. Hair-like ; green. Threads branched— Velvet
Byſſus tenerrima viridis velutum referens. *Ray's Syn.* 56. *Dill.* 3
7. t. 1. f. 14. Velutina
On the barks of trees, on the ground in ſhady places and gravel walks. Winter and Spring.

POWDERWORT. Hair-like ; powdery ; flowers ſcattered. Saffron-co-
Threads ſimple and branched.—*Grows in little globular tufts.* loured
When dry it turns of an aſh colour. 4
Byſſus petræa, crocea glomerulis lanuginoſis. *Dill.* 8.t.1.f.16. Aurea
Byſſus aureus Derbienſis humifuſus. *Ray's Syn.* 56.
On rocks.

POWDERWORT. Threads branched ; of a yellowiſh Brown
brown. *Hudſon* 487.—*Retains its colour when dry. Of a more* 5
rigid conſiſtence than the preceding. Fulva
Byſſus arborea crocea fibroſa. *Ray's Syn.* 57. *Dill.* 9. t. 1. *Hud.*
f. 17.
On rotten moiſt wood.

POWDERWORT. Threads branched, ſtiff, black ; grow- Black
ing on ſtone. *Hudſon* 487. 6
Byſſus petræa nigerrima fibroſa. *Ray's Syn.* 57. *Dill.* t. 1. Nigra
f. 18. *Hud.*
On rocks.

POWDER-

Bearded 7 Barbata *Hud.*

POWDERWORT?

BYSSUS?

Threads upright; ends branched, *Hudſon* 488.—*Of a tawny colour.*

Byſſus arborea barbata, fulvi coloris. *Ray's Syn.* 57. *Dill.* 9. t. 1. f. 19.

On rotten wood.

Grey 8 Candida *Hud.*

POWDERWORT. Threads very much branched; branches whitiſh, in bundles. *Hudſon* 488.—*Branching in form of a tree or ſhrub. Of a ſubſtance reſembling* Mould; *of a livid white.*

Byſſus tenerrime villoſa et elegantiſſime ramuloſa. *Ray's Syn.* 476. t. 23. *Dill.* 7. t. 1. f. 15.

On rotten wood and leaves.

Briſtle 9 Cryptarum

POWDERWORT. Hair-like; perennial; aſh-coloured; *adhering tenaciouſly to rocks or walls.*

Byſſus albida brevis ſetacea. *Dill.* 10. t. 1. f. 20.

In vaults.

* * *Powdery.*

Black powde 10 Antiquitatis

POWDERWORT. Powdery; black—

On very old walls.

Stone 11 Saxatilis

POWDERWORT. Powdery; aſh-coloured; covering the ſurface of rocks.—

On rocks.

It will grow upon the bareſt rocks and ſtones.

Yellow 12 Candelaris

POWDERWORT. Powdery: yellow. Growing upon wood.—*At firſt ſight one might ſuppoſe it to be the* Gold ſpangled Cupthong *in its younger ſtate, but it has been obſerved to continue the ſame for many years. Like that too in ſhady places, it is ſometimes of a greeniſh colour.*

Byſſus pulverulenta flava lignis adnaſcens. *Ray's Syn.* 56. *Dill.* 3. t. 1. f. 4.

Barks of trees, and wood.

Green 13 Botryoides

POWDERWORT. Powdery; green—

Byſſus botryoides ſaturate virens. *Ray's Syn.* 56. *Dill.* 3. t. 1 f. 5.

On barks of trees, walls, pales, and on the ground in damp ſhady ſituations.

POWDER-

POWDERWORT. Powdery; white; like ſcattered meal.— Mealy
Byſſus pulverulenta incana, farinæ inſtar ſtrata. *Ray's Syn.* 14
56. *Dillenius*, 3. tab. 1. f. 3. Incana
On a gravelly ſoil, on the ſides of roads and ditches, on Moſs, and in damp ſhady ſituations, in Autumn and Winter.

POWDERWORT. Betwixt powdery and cruſtaceous; very White
white. 15
Byſſus candidiſſima, calcis inſtar Muſcos veſtiens. *Dillenius*, Lactea
t. 1. f. 2.
On moſs, and barks of trees.

427 STARJELLY. 1204 Tremella.

SUBSTANCE. Uniform; tranſparent; membranaceous; jelly-like or leafy.
FLOWERS. Scarce perceptible, in a jelly-like ſubſtance. *Syſt. Nat.*

OBS. *It differs from the Cupthong in having neither tubercles nor ſaucers.*

STARJELLY. Plaited and waved.—*Greeniſh or yellowiſh.* Starfall
Tremella terreſtris ſinuoſa pinguis et fugax. *Dillenius*, 52. 1
t. 10. f. 14. Noſtoc
Ulva terreſtris pinguis et fugax. *Ray's Syn.* 64.
Tar-ſlough.
In paſtures after rainy ſeaſons, in gravelly ſoil, on the tops of hills, and on gravel walks. Spring and Autumn.

OBS. *Micheli deſcribes the ſeeds as lying in the form of little ſtrings of beads coiled up within the folds of the plant, and only to be diſcovered in the microſcope.—It is ſuppoſed by the country people to be the remains of a meteor or falling ſtar.*

STARJELLY?
TREMELLA?—A white fungous matter, of a uniform ſub- Snowy
ſtance, like large flakes of ſnow, or fleeces of the whiteſt wool, 2
—*hanging down from the beams of wine-vaults. It is of a moiſt* Nivea.
watery nature, and upon diſſolving and drying it by heat, it runs Ray
into a tough membranaceous matter of a fungous ſmell, which at length turns to a ſubſtance like touchwood, and crackles in the fire. From Ray.
Fungus niveus aqueus, lignis cellarum vinariarum. *Ray's Syn.* 26.
Common in moſt wine-vaults in London.

OBS. *This ſubſtance ſeems to partake of a middle nature between the* Funguſſes *and the Starjelly.*

STAR-

Witches-butter
3
Nigricans
Dill.

STARJELLY. Pitted; blackiſh; of a firmer conſiſtence, and longer duration than the preceding. *From Dillenius*

Tremella arborea nigricans, minus pinguis et fugax. *Dillenius* t. 10. f. 15.

On the bark of trees. *Dill.*

Tender
4
Tenerrima
Dill.

STARJELLY. Curled and wrinkled; extremely tender; green. *From Dillenius.*

Tremella terreſtris tenera, criſpa. *Dillenius*, t. 10. f. 12.

On the ground, on walls and hedges in the ſhade. Jan.—Feb.

Bluehorn
5
Ceranoides
Dill.

STARJELLY. Flattiſh; wrinkled; jagged at the edge. Segments indented,—*reſembling ſtag's horns.* From Dillenius.

Tremella paluſtris gelatinoſa damæ cornuum facie. *Dillenius*, t. 10. f. 10.

Conferva gelatinoſa damæ cornua repræſentans. *Ray's Syn.* 60.

In the ditches of a field near Chicheſter, Suſſex, without the Eaſt gate. *Dill.*

Jagged
6
Laciniata
Dill.

STARJELLY. Jagged; pitted; ſegments long; narrow; waved—*ſomewhat reſembling in form the* Black Thorn Cup-thong. *From Dillenius.*

Tremella terreſtris cornuta. *Dillenius*, t. 10. f. 13.

On the ground, Endfield Chace, and near Southgate. April. *Dill.*

Tranſparent
7
Lichenoides

STARJELLY. Upright; flat; curled and a little jagged at the edge,—*browniſh, and ſo brittle as hardly to bear ſeparation from the ſubſtance on which it grows.*

Endive-leaved

1. Lichenoides pellucidum, Endiviæ foliis tenuibus criſpis. *Dillenius*, 143. t. 19. f. 31.

Lichenoides ſaxatile tenue ruſeſcens. *Ray's Syn.* 77.

Small
Lin. and *Dill.*

2. Leaves very ſmall; upright; growing cloſe together in a tuft; wedge-ſhaped: ſometimes cloven; lopped at the ends and notched. *From Dillenius.*

Lichenoides tenue criſpum, foliis exiguis ſurrectis. *Dillenius*, t. 19. f. 34.

Prickly Lin. Dill.

3. branched; ſhrub-like; branches ſharp; upright; growing in a tuft. *From Dillenius.*

Lichenoides tenuiſſimum criſpum et veluti aculeatum. *Dillenius*, t. 19. f. 35.

Woods, heaths, and ſhady places. (1.) On moſs. (1.) All the year. (3.) Winter.

STAR-

STARJELLY. Upright; flat; jagged; ſegments cloven at the end, blunt;—*growing in a kind of tuft. Brown.* From Dillenius — Horned 8 Corniculata *Dill.*

Lichenoides pellucidum fuſcum corniculatum. *Dillenius*, t. 19. f. 30.

Heaths. Winter and Spring. *Dill.*

STARJELLY. Flat; ſtrap-ſhaped; forked; ſegments broader at the ends; lopped; edged with a few imperfect teeth. *From Dillenius.* — Forked 9 Dichotoma *Dill.*

Lichenoides gelatinoſum foliis anguſtioribus tuniformibus. *Dillenius*, t. 19 f. 28.

Malham Cove Rivulet, Yorkſhire, and Snowden torrents. Summer. *Dill.*

STARJELLY. Solid; wrinkled, tubercled; *ſomewhat globular; of a blackiſh yellow.* — Warty 10 Verrucoſa

Tremella fluviatilis gelatinoſa et uterculoſa. *Dillenius*, 54. t. 10. f. 16.

On ſtones in brooks and rivers.

STARJELLY. Roundiſh; full of hollows, aſſuming a variety of forms, of a jelly-like ſubſtance? — Curled? 11 Difformis?

Lichenoides maritimum gelatinoſum craſſum, inteſtinorum gyros referens. *Dillenius*, t. 19. f. 19.

Lichen maritimum gelatinoſum, inteſtinorum gyros referens. *Ray's Syn.* 478.

1. Thick and ſhort; of a firmiſh jelly-like ſubſtance and of a greeniſh brown colour.

In the ſea, and ſalt marſhes about Delkey, Chicheſter, Suſſex. *Dill.*

Obs. *Uncertain whether this plant of Dillenius's be the* Tremella difformis *of Lin. But it ſeems to correſpond pretty exactly to his deſcription of it.*

STARJELLY. Hemiſpherical; ſcattered.— — Sea 12 Hemiſpherica

On ſea river-weeds, and oar-weeds.

STARJELLY. Nearly globular; ſitting; ſingle; ſmooth,—*moſtly of a pale purple, or fleſh-colour; but ſometimes of a beautiful carmine.* — Purple 13 Purpurea

Lichenoides tuberculoſum amoene purpureum. *Ray's Syn.* 71. *Dillenius*, 127. t. 18. f. 6.

On the decayed bark and branches of trees.

STAR-

Lentil STARJELLY?
14 Nigra TREMELLA?
Dill.

Globular or oblong; ſomewhat flatted; ſitting; ſingle; black. —*Larger than the preceding, and of a firmer conſiſtence.* From *Dillenius.*

Lichenoides tuberculoſum compreſſum nigrum, lignis putridis adnaſcens. *Ray's Syn.* 71. *Dill.* t. 18. f. 7.

On the bark of decayed wood. Winter. *Dill.*

OBS. *Some may be inclined, with Dillenius, to refer this and the preceding ſpecies, to the* CUPTHONG, *or with Linnæus to the* JELLYWORT, *but they are neither tranſparent, nor of a jelly-like conſiſtence. Nor do they form a ground, or cruſt, as the* Cupthong, *but riſe diſtinct from each other. Do they not rather belong the* FUNGUSSES? *and are they not in reality ſpecies of the* CLUBTOP?

Juniper? 15 Juniperina? *Ray* STARJELLY. Sitting; membranaceous; ear-like; of a yellowiſh brown—

Fungus gelatinus dentatus, ſabinæ adaſcens, fulvi coloris. *Ray's Syn.* 16.

On Savine.

Flat 16 Flana *Ray* STARJELLY. Round and flat; of a jelly-like ſubſtance—

Fungus rotundus planus ligno putrido adnaſcens. *Ray's Syn.* 17.

On rotten wood.

428 LAVER. 1206 Ulva.

FLOWERS in a ſemi tranſparent bladder-like membrane; with no appearance of a regular leaf.

Turkey-Feather 1 Pavonia LAVER. Flat; kidnep-ſhaped; ſitting; ſcored *a-croſs and a-croſs*—

Fucus maritimus gallopavonis pennas referens. *Ray's Syn.* 43.

Alga maritima gallopavonis plumas referens. *Baub. pin.* 364.

Fucus pavonicus. *Hudſon* 472. *Sp. Pl.* 1630.

On rocks and ſtones in the ſea.

Navel 2 Umbilicalis LAVER. Flat; round; ſitting; target-ſhaped; of a leathery ſubſtance—*ſomewhat hollow. Border indented; fixed only by a point in the middle, to the ſubſtance on which it grows; of a dark ſooty colour, ſhining.*

Ulva marina umbilicata. *Ray's Syn.* 62.

Fucus umbilicus marinus dictus. *Baub. pin.* 364.

Tremella marina umbilicata. *Dillenius* 46. t. 8. f. 3.

On rocks and ſtones in the ſea.

LAVER. Tubular; ſimple.—*Floats moſtly near the ſurface of the water, though ſometimes adhering to ſtones.* Gut 3
Ulva marina tubuloſa, inteſtinorum figuram referens. *Ray's Syn.* 62. Inteſtinalis.
Fucus cavus. *Bauh. pin.* 364.
Tremella marina tubuloſa inteſtinorum figuram. *Dillenius* 47. t. 9. f. 7.
Moſtly in ditches near the ſea, but ſometimes in freſh water ditches.

2. Branched. *Branched*
Fucus herbaceous cavus fluitans ramoſus, calami anſerina fere craſſitudine. *Ray's Syn.* 2d. edit. 340. *Ray.*
Moſtly in ditches near the ſea, but ſometimes in freſh water ditches.

LAVER. Tubular; branched; compreſſed—*Irregular, zig-zag; cells of different ſizes, but communicating with one another; branches ſcattered, but little divided; ſometimes nearly cylindrical.* Flat 4 Compreſſa
Ulva marina tenuiſſima et compreſſa. *Ray's Syn.* 63.
Tremella marina tenuiſſima et compreſſa. *Dill.* 48. t. 9. f. 8.
On rocks and ſtones in the ſea.

LAVER. Strap-ſhaped; branched; nearly upright; forming a turf;—*Segments moſtly cloven, rather blunt.* From *Dill.* Cluſter 5
Tremella marina cæſpitoſa, ſegmentis tenuibus. *Dill.* t. 10. f. 9. Cæſpitoſa *Dill.*
Side rocks Angleſy. *Dill.*

LAVER. Thread-ſhaped: jointed; joints alternately, compreſſed; *tubular.* Riverweed 6
Conferva tubuloſa. *Hudſon* 483. Confervoides
Conferva marina fiſtuloſa. *Dill.* 34. t. 6. f. 39.
1. Conferva marina geniculata albicans, diaphragmatis diſtincta. *Ray's Syn.* 60. *Whitiſh.* Hud
2. Conferva marina nodoſa, ex albo rubeſcens mollis, ſed minus lubrica. *Ray's Syn.* 61. *Reddiſh* Hud.
On the ſea-ſhore.

LAVER. Tubular; very much branched; jointed; joints cylindrical; branches oppoſite—*Purple. Hudſon* 476. Jointed 7
Corallina lenta purpurea compreſſa. *Ray's Syn.* 34. Articulata *Hud.*
On the ſea-ſhore.

LAVER.

Horned LAVER?
8 ULVA?
Ceranoides *Dill.* Nearly cylindrical; very much branched; branches irregular; tapering and forked at the ends.—*Of a ſlippery conſiſtence.* From *Dill.*

Conferva marina lubrica et mucoſa. *Dill.* t. 4. f 24.
Bagnor and Cockbuſh-ſhore, Suſſex. June. *Dill.*

Branched LAVER. Leaf branched; flat; waved; purple. *Hudſon* 476.
9 On the ſea-ſhore.
Ramoſa *Hud.*

Graſs leaved LAVER. Leaf forked; green. *Hudſon* 476.
10 Fucus membranaceus dichotomus gramineus. *Ray's Syn.* 45.
Dichotoma *Hud* On the ſea-ſhore.

Curled LAVER. Oblong: flat waved: membranaceous; green, but
11 ſometimes purple;—*very long and very broad.*
Latiſſima Fucus longiſſimo, latiſſimo, tenuique folio. *Ray's Syn.* 40.
Alga longiſſimo, latiſſimo tenuique folio. *Bauh. pin.* 364.
On the ſea-ſhore.

Oyſter LAVER. Hand-ſhaped; proliferous membranaceous; ſeg-
12 ments narrower towards the baſe.—*Leaves crowded together;*
Lactuca *pale; ſegments waved; inverſely egg-ſhaped; blunt; tranſparent; each growing into leaves.*

Sea 1. Ulva marina lactucæ ſimilis. *Ray's Syn.* 62.
Tremella marina vulgaris, lactucæ ſimilis. *Dill.* 42. t. 8. f. 1.
Fucus marinus, lactuca marina dictus. *Park.* 1293.
Muſcus marinus lactucæ ſimilis. *Bauh. pin.* 364.
Lichen Marinus. *Gerard.* 1377.

Freſh water Hud. 2. Smaller and more tender.
Ulva paluſtris lactucæ marinæ ſimilis ſed multo minor et tenerior. *Ray's Syn.* 63.
Tremella paluſtris, vulgari marinæ ſimilis, ſed minor et tenerior. *Dill.* 43. t. 8. f. 2.
Green Laver. Oyſter Green.
(1.) In the ſea. (2) in freſh water ditches.

This is eſteemed by the inhabitants of the ſea-coaſt as a wholeſome and pleaſant food, being gently opening and antiſcorbutic. It is frequently ſent to London in earthen pots.

Leek LAVER. Spear-ſhaped; flat.
13 Tremella marina, porri folio. *Dill.* 46. t. 9. f. 5.
Lanceolata On rocks in the ſea.

LAVER

LAVER. Leaf oblong; bliſtered—*Folded; margin curled; indented.* Ribbon 14 Linza
Ulva marina faſciata. *Ray's Syn.* 62.
Tremella marina faſciata. *Dill.* 46. t. 9. f. 6.
Muſcus lactucæ marinæ ſimilis. *Bauh. pin.* 364.
On the ſea-ſhore.

LAVER. Globular; growing in cluſters—*About the ſize of hemp-ſeed; reſembling little muſhrooms.* Bladder 15 Granulata *Sp. Pl.*
Tremella paluſtris, veſiculis ſphæricis fungiformibus. *Dill.* 55. t. 10. f. 17.
Lichenoides fungiforme, capitulis vel veſiculis ſphæricis, aqueo humore repletis. *Ray's Syn.* 70.
On the ſides of ditches, and in ditches dried up. Autumn.

OBS. *The* U. Granulata *of the Syſt. Nat. and of the Mant. Pl.* I. *p.* 136. *appears to be a diſtinct ſpecies. It is found in Sweden in great quantities on the ſides of ſprings and marſhes, floating in the water, of a green colour, of the ſize of hemp-ſeed, and full of a viſcid pulp.*

429 OARWEED. 1205 Fucus.

Barren Flowers?

Little Bladders; ſmooth, hollow; interſperſed within with ſoft hairs interwoven together.

Fertile Flowers?

Little bladders; ſmooth; full of a jelly-like pulp; ſprinkled with grains buried in the ſubſtance of the bladders; ſomewhat prominent at the points. SEEDS ſolitary.

OBS. *See Plate* 1. *fig.* G.

OARWEED. Stem ſomewhat cylindrical; compreſſed; leaves oblong; very entire. *Hudſon* 468. Oval-leaved 1 Ovatus *Hud.*
On the Yorkſhire coaſt.

OARWEED. Leaf flat; forked; partly toothed and partly ſerrated. Flowers tubercled; at the ends of the branches.— Serrated. 2 Serratus *Common*

1. Fucus, ſeu Alga latifolia major dentata. *Ray's Syn.* 42.
Sea Wrack.

2. Fucus marina humilis, latifoliæ ſerratæ ſimilis. *Ray's Syn.* 42. *Dwarf* Hud.
On ſtones and rocks in the ſea.

Oak-leaved 3 Vesiculosus OARWEED. Leaf flat; forked; very entire; the bladders at the divisions of the leaf in pairs; those at the ends of of the branches tubercled—

Common 1. Fucus, seu Alga marina latifolia vulgatissima. *Ray's Syn.* 40.

Fucus maritimus vel Quercus maritima vesiculas habens. *Bauh. pin.* 363.

Quercus marina herbacea et varietas. *Park.* 1293.

Quercus marina. *Gerard* 1567.

Swollen 2. With the ends of the branches swollen.

Fucus maritimus seu Quercus maritima foliorum extremis tumidis. *Bauh pin.* 365.

Sea Oak.

On rocks and stones in the sea.

Shrubby 4 Divaricatus OARWEED. Leaf flat; forked; very entire. Branches straddling; the bladders at the divisions of the leaf in pairs— *Thinner than the preceding. The bladders at the divisions of the leaf in pairs, the others solitary.*

Sea shore.

Narrow leaved 5 Linearis *Hud.* OARWEED. Leaf flat; forked; strap-shaped; sharp. Bladders egg-shaped; scattered. *Hudson* 467.

Fucus longo angusto crassoque folio. *Bauh. pin.* 364. *Ray's Syn.* 43.

Fucus marinus secundus. *Park.* 1293.

Quercus marina 2. *Gerard.* 1115.

Sea Thongs.

Sea-shore.

For an elegant engraving of it see the Gent. Mag. for 1756. Page 64 This plant at its first appearance so much resembles a Fungus, that some authors have mistaken it for one. Mr. Ray seems to have described it as a distinct species under the name of "Fucus Fungis affinis." *Syn.* p. 43. n. 15. from the center of the little Fungus-like substance 3 or 4 shoots arise, and extending by degrees into branches, constitute the perfect plant. The little Fungus still continues and forms a kind of fence or cup at the base of the stem.

Furrowed 6 Excisus OARWEED. Leaf strap-shaped; forked; channelled on one side. Branches straddling; dotted—*Segments hollowed on one side, and swelling out on the other: yellowish green.*

Fucus pumilus dichotomus, segmentis ex una parte gibbosis, ex altera excavatis. *Ray's Syn.* 43.

Sea shore.

OARWEED. Leaf flat; forked; very entire; dotted; spear-shaped; flowers tubercled; cloven; at the ends of the branches.—*Hollow dots dispersed upon both surfaces.* Bucks-horn 7 Ceranoides

1. Edge of the leaf entire. *True*

2. Edge of the leaf ragged, as if fringed. *Torn.* Sp. Pl.
Fucus lacerus. Sp. Plant. 1627.
Fucus membranaceus ceranoides varie dissectus. *Ray's Syn.* 44.

3. Stems somewhat cylindrical; ends of the branches broader, flatter and ragged. *Small.* *Hud.*
Fucus cauliculis teretibus, summitatibus membranaceis dilatatis et laceratis. *Ray's Syn.* 44.

4. When dry of a greenish cast. *Green.* Hud.
Fucus membranaceus ceranoides ramosus, per siccitatem obsolete virescens. *Ray's Syn.* 44.

5. Whitish; ends of the branches star-like. *White.* Hud.
Fucus ceranoides albidus, ramulorum apicibus stellatis. *Ray's Syn.* 44.
Sea-shore.

OARWEED. Leaf flat; forked; very entire, dotted; strap-shaped and channelled towards the base. Flowers in pairs; tubercled.—*On fruit-stalks, at the ends of the branches, oblong and thickish. Whilst growing in the Sea it is twisted spirally.* Twisted 8 Spiralis

1. Fucus spiralis maritimus major. *Ray's Syn.* 41. *Larger*
2. Fucus spiralis maritimus minor. *Ray's Syn.* 42. *Smaller.* Hud.
Sea-shore.

OARWEED. Leaf flat; forked: thread-shaped. *Hudson* 472. Flat 9 Filiformis *Hud.*
On the Lancashire Coast.

OARWEED. Leaf compressed; forked. Little leaves pointing two ways; very entire. Bladders in the substance of the leaf, solitary; dilated,—*and egg-shaped, those in the middle of the branches, broader.* Little Leaves *spear-shaped, blunt, disposed in two rows on the angles of the great one.* Knotted 10 Nodosus
Fucus maritimus nodosus. *Bauh. pin.* 365. *Ray's Syn.* 48.
On rocks and stones in the sea.

Podded 11 Siliquoſus

OARWEED. Leaf compreſſed; branched. Little leaves pointing two ways; alternate; very entire. Flowers on fruit-ſtalks; oblong; ſharp-pointed. *Leaf* thread-ſhaped; zigzag; compreſſed; marked with little teeth at the angles on each ſide. Flowers ſpindle-ſhaped; alternate; furniſhed with a bill, and larger than the little leaves, which are ſpear-ſhaped.

Fucus anguſtifolius veſiculis longis ſiliquarum æmulis. *Ray's Syn.* 48.

Fucus maritimus alter, tuberculis pauciſſimis. *Baub. pin.* 368.

Quercus marina quarta. *Gerard.* 1569.

On the ſea-ſhore.

Jointed 12 Abrotanifolius

OARWEED. Leaf thread-ſhaped; compreſſed; doubly winged. The ends bladder-like; dilated; terminated with flowers tubercled on one ſide.

On rocks in the ſea, on the Suſſex ſhore.

Fennel-leaved 13 Fæniculaceus

OARWEED. Leaf thread-ſhaped; very much branched: bladders egg-ſhaped; terminated with little leaves divided into many blunt ſegments bearing fruit at the ends.—*About the length of the* Hairy Cupthong. Little Branches *very numerous; hair-like; ending in a bladder or bladders; the laſt of which terminates in ſeveral little leaves with tubercles at the ends.*

Common 1. Fucus radicibus arborum fibroſis ſimilis. *Ray's Syn.* 49.

Fucus foliculaceus: fæniculi folio. *Bauh. pin.* 365.

Bearded Barbatus 2. With oblong bladders. Little leaves at the ends ſtrap-ſhaped.

Fucus folio tenuiſſimo diviſo ſiliquatus. *Ray's Syn.* 48.

Fucus folliculaceus foliis abrotani. *Bauh pin.* 365.

Fucus maritimus foliis tumidis barbatis. *Bauh. pin.* 365.

Muſcus marinus abrotanoides. *Park.* 1290.

Fucus barbatus. *Hudſon* 469. *Sp. Plant.* 1629.

(1.) On the Yorkſhire and Lancaſhire ſhores. (2.) On the ſhores of Suſſex and Cornwall.

Warty 14 Verrucoſus *Hud.*

OARWEED. Leaf thread-ſhaped; branched. Branches undivided, with round little bladders on the ſides. *Hudſon* 470.

Fucus marinus purpuraſcens parvus, caule et ramulis ſeu foliolis teretibus. *Ray's Syn.* 50.

Sea-ſhore.

Prickly 15 Aculeatus

OARWEED. Leaf thread-ſhaped, compreſſed; very much branched; edged with awl-ſhaped, alternate, upright teeth.—*Reſembles a Horſe's tail.*

Fucus anguſtifolius, foliis dentatis. *Ray's Syn.* 48.

Fucus uſneoides. *Hudſon* 469.

On the ſhores of Cornwall, and in the Britiſh Channel.

OARWEED. Leaf cylindrical; branched; little branches Black
tiled; ſegments pointing one way; bowed inwards. *Hudſon* 16
470. *Little branches beſet with briſtles pointing upwards. Ray.* Incurvus
Pinus maritima, ſeu Fucus teres, cujus ramuli ſetis ſurſum *Hud.*
tendentibus ſunt obliti. *Ray's Syn.* 50.

Sea Pine.

On the Suſſex ſhore.

Obs. *Is not this the* F. Lycopodioides *Syſt. Nat. p.* 717. *n.* 35. (55)

OARWEED. Leaf cylindrical, very much branched; branch- Thorny
es ſcattered; beſet alternately with ſoft thorns. 17
On the Yorkſhire Coaſt. Muſcoides

OARWEED. Leaf thread-ſhaped; cylindrical; forked; Hairy
covered over with a very fine hairineſs,—*But ſo fine as hardly to* 18
be diſtinguiſhable to the naked eye. It has ſomewhat of the the ap- Hirſutus
pearance of Sponge, and bears no ſmall affinity to the Tamariſk Oarweed. *The* Stems *are about a foot long, of the thickneſs of coarſe thread; very much branched; the branches blunt, cylindrical, and covered over with the hairineſs above deſcribed, which is extremely ſhort and thick, and very much reſembles the ſkin of a Fawn.*

In the ſea.

OARWEED. Stem cylindrical; very much branched. Tamariſk
Leaves awl-ſhaped; alternate. Bladders round; tiled; ſitting in 19
the baſe of the leaves. *Hudſon* 469. Ericoides
Thread-ſhaped; very much branched, rough with hair. *Sp. Pl.*
Linnæus Sp. Plant 1631. n. 33.

Fucus foliis Ericæ, ſeu Tamariſci. *Ray's Syn.* 49.

Fucus tamariſco ſimilis maritima. *Bauh. pin.* 365.

Fucus tamariſci folius. *Hudſon* 469.

On the ſhores of Yorkſhire and Cornwall.

Obs. *The* F. Ericoides *is not to be found in the* Syſt. Nat.

OARWEED. Leaf thread-ſhaped; rather cylindrical; White
ſomewhat forked; joints a little ſwollen: branches diſtant, 20
ſharp. *Hudſon* 470.—*White.* Albus
Fucus teres albus tenuiſſime diviſus. *Ray's Syn.* 50.

On the Suſſex ſhore.

OARWEED. Leaf cylindrical; very much branched; lit- Rough
tle branches alternate, in bundles; very ſhort; tubercled; rough. 21
Hudſon 474. Confervoides
On the Yorkſhire coaſt. *Hud.*

Thread 22 Filum OARWEED. Leaf thread-ſhaped; ſomewhat brittle; opake—*Not floating upon the water but a little below the ſurface.*

Fucus chordam referens teres prælongus. *Ray's Syn.* 40.

Alga nigro capillaceo folio. *Bauh. pin.* 364.

On rocks in the ſea.

Matted 23 Plicatus Hud. OARWEED. Leaf hair-like; uniform; very much branched; matted; nearly ſemitranſparent. *Hudſon.* 470.—*Of a gold colour; the ends of the branches forked.*

Golden 1. Fucus trichoides noſtras aurei coloris, ramulorum apicibus furcatis. *Ray's Syn.* 45.

Coral 2. Upright.

Fucus coralloides erectus. *Ray's Syn.* 51.

Sea-ſhore.

Upright 2 Scorpioides Hud. OARWEED. Leaf cylindrical; branched: branches alternate, very much branched; bent inwards at the points. *Hudſon.* 471.

Fucoides erectum fruticuli ſpecie, ſummitatibus inflexis. *Ray's Syn.* 38. t. 2. f. 6.

Sea-ſhore,—on the Suſſex coaſt.

Round 25 Rotundus Hud. OARWEED. Leaf cylindrical; ſomewhat forked; uniform; branches blunt; covered with wart-like ſubſtance. *Hud.* 471.

Fucus kali geniculato ſimilis, non tamen geniculatus. *Ray's Syn.* 43.

On rocks in the ſea.

Forked 26 Faſtigiatus OARWEED. Leaf thread-ſhaped; forked; very much branched. Branches nearly of the ſame length; blunt.—*Bladders at the ends of the branches, cloven.*

Fucus ſeu alga exigua dichotomus foliorum ſegmentis longiuſculis, craſſis et ſubrotundis. *Ray's Syn.* 45.

Sea ſhore.

Sharp-pointed 27 Furcellatus OARWEED. Leaf thread-ſhaped; forked; very much branched; branches tapering to a point,—*equal. Nearly allied to the preceding, but longer, and the branches thicker.*

Fucus parvus ſegmentis prælongis teretibus acutis. *Ray's Syn.* 45.

Fucus forcellata lumbricalis ſpecies. *Bauh, pin.* 366.

Fucus lumbricalis. *Hudſon* 471.

Sea-ſhore,—and in the ſea. *Ray's*

OARWEED. Leaves ſomewhat cylindrical; branched; of a jelly-like ſubſtance. *Hudſon* 471. Jelly 28 Gelatinoſus *Hud.*

Fucus ſpongioſus nodoſus. *Gerard.* 1570. *Ray's Syn.* 49.
Spongia ramoſa altera anglica. *Park.* 1304.
Sea-ſhore.

OARWEED. Leaf hand-ſhaped; flat— Handed 29 Palmatus

Fucus membranaceus ceranoides. *Ray's Syn.* 46.
Dulleſh. *Iriſh*, Dills. *Scotch*, Dulls. Dulſe in *Northumberland*. *Ray.*
Sea-ſhore.

After being ſoaked in freſh water it is eaten either boiled, or dried, and in the latter ſtate has ſomething of a violet flavour.

OARWEED. Leaf hand-ſhaped; little leaves ſword-ſhaped; ſtalk cylindrical—*very ſhort.* Fingered 30 Digitatus

1. Fucus arboreus polyſchides edulis. *Bauh. pin.* 364. *Ray's Syn.* 46. *Common*

Fucus phaſganoides et polyſchides. *Gerard.* 1570.
Fucus maximus polyſchides. *Park.* 1292.
2. Fucus membranaceous polyphyllus major. *Ray's Syn.* 46. *Many-leaved*
3. Fucus ſcoticus latiſſimus edulis dulcis. *Ray's Syn.* 46. *Broad-leaved* Hud.
Fucus alatus, ſeu phaſganoides. *Bauh. pin.* 362.
1. Sea-girdle.
On rocks in the ſea.

OARWEED. Leaf almoſt ſimple; ſword-ſhaped; ſtalk cylindrical; very ſhort—*It is often four feet long and two broad; nearly oval or oblong, of a leather-like ſubſtance, waved at the edge, and narrowing towards the baſe.* Sweet 31 Saccharinus

1. Fucus longiſſimo, latiſſimo, craſſoque folio. *Ray's Syn.* 39. *Common*
2. Fucus folio ſingulari longiſſimo, lato, in medio rugoſo, qui balteiformis dici poteſt. *Ray's Syn.* 39. *Sea-belt.* Hud.
3. Fucus latiſſimus et longiſſimus, oris criſpis. *Ray's Syn.* 38. *Curled.* Hud.
On rocks and ſtones in the ſea.

Waſhed in ſpring water and then hung up in a warm place, a ſubſtance like ſugar exſudes from it. Some people eat it freſh out of the ſea.

OARWEED. Leaves membranaceous; oblong; inclining to egg-ſhaped; very entire; on foot-ſtalks. Stalk cylindrical; branched—*Reſembling the leaves of the* Bleeding Dock *in ſize as well as form.* Dock-leaved 32 Sanguineus

Fucus, ſeu alga, folio membranaceo purpureo, lapathi ſanguineo figura et magnitudine. *Ray's Syn.* 49.
On the ſhores of Yorkſhire and Cornwall.

Fringed 33 Ciliatus *Common* OARWEED. Leaves membranaceous; ſpear-ſhaped; proliferous; fringed—*of a reddiſh colour.*

1. Fucus membranaceus rubens anguſtifolius, marginibus ligulis armatis. *Ray's Syn.* 47.

Purple 2. Leaves broader, purple.

Fucus membranaceus purpureus latifolius pinnatus. *Ray's Syn.* 47.

Sea-ſhore.

Curled 34 Criſpus OARWEED. Leaves ſomewhat membranaceous; forked; ſegments growing broader towards the end—*which is edged with blunt teeth; curled, torn, according to Hudſon cloven, with a tinge of white or purple.*

Fucus membranaceus purpureus, varie ramoſus. *Ray's Syn.* 47.

Sea-ſhore.

Jagged 35 Laciniatus *Hud.* OARWEED. Leaves membranaceous; branched: branches growing broader towards the end; hand-ſhaped. *Hudſon* 475.

On the Yorkſhire coaſt.

Winged 36 Alatus OARWEED. Leaves membranaceous; ſomewhat forked; ſtringy; ſegments alternate, running along the rib; cloven—*purple; ſtrap-ſhaped; ſemi-tranſparent; the ends ſomewhat toothed.*

Fucus dichotomus parvus coſtatus et membranaceus. *Ray's Syn.* 44.

Sea-ſhore.

Red 37 Rubens OARWEED. Leaves membranaceous; oblong; waved; indented. Stem cylindrical, branched.

On the ſhores of Cornwall.

Louſewort 38 Pinnatifidus *Hud.* OARWEED. Leaves (*membranaceous*). branched; branches with winged clefts, inclining to toothed; border callous. *Hudſon* 473.

Fucus dealenſis pedicularis rubræ folio. *Ray's Syn.* 48.

On ſtones and rocks in the ſea.

Pinnated 39 Pinnatus *Hud.* OARWEED. Leaves membranaceous; triply winged. *Hudſon* 474.

On the Yorkſhire and Suſſex coaſt.

Horny 40 Corneus *Hud.* OARWEED. Leaves griſtly; branches ſomewhat winged; ſegments ſharp; bearing fruit. Stem ſomewhat cylindrical, very much branched. *Hudſon* 474.

Fucus flavicans teretifolius, ramulis plennatim enaſcentibus. *Ray's Syn.* 50.

On the Devonſhire coaſt.

OAR-

OARWEED. Leaves griſtly; ſpear-ſhaped; doubly winged; feather-like: ſtem thread-ſhaped: compreſſed; branched.—*Leaves purple. It reſembles the following ſpecies, but is winged like* Feathermoſs, *and ſmall.* Feathered 41 Plumoſus

Fucoides purpureum eleganter plumoſum. *Ray's Syn.* 37.

OARWEED. Leaves griſtly compreſſed, more than doubly compound; winged; ſegments blunt. *Hudſon* 473. Fern-leaved 42 Filicinus *Hud.*

On the Lancaſhire coaſt—not common.

OARWEED. Leaf griſtly; compreſſed; more than doubly compound; winged. Segments ſtrap-ſhaped—*coloured*; Stem *ſomewhat cylindrical: compreſſed; griſtly; very much branched: little branches alternately pointing one way. Fruit globular, on the ſides of the branches. Hudſon* 473. Scarlet 43 Cartilagineus

Fucoides rubens, varie diſſectum. *Ray's Syn.* 37.

Sea-ſhore.

Obs. *Hudſon does not admit of Linnæus's character but refers it as a Synonym of his* Conferva rubra, *though both agree in giving* Ray's Synonym.

OARWEED. Leaves membranaceous, ſtrap-ſhaped, compreſſed, branched; fruit globular, ſcattered; without footſtalks. Purple 44 Confervoides *Lin.*

Thread-ſhaped cylindrical, very much branched; branches alternate; little branches crowded, bearing fruit; fruit globular. *Hudſon* 471.

1. Fucus teres rubens ramoſiſſimus. *Ray's Syn.* 51. *Branched*

2. Not ſo much branched, but running out in length. *Long.* Hud.

Fucus teres rubens minus ramoſus in longum protenſus. *Ray's Syn.* 51.

Fucus purpureus. *Hudſon* 471.

On ſtones and rocks in the ſea.

Obs. *All the ſpecies of Oarweed may be uſed to manure land, or burnt to make Kelp, which is an impure foſſil alkaly.*

430 RIVERWEED. 1207 Conferva.

Fibres ſimple; uniform; like hair or thread. GEN. Plant. Unequal *Tubercles*, on very long, hair-like fibres. SYST. NATURAE.

* *Threads ſimple, equal, without joints.*

Common 1 Rivularis — RIVERWEED. Threads undivided; equal; extremely long—*Deep green.*

True — 1. Threads extremely fine and long, floating with the ſtream. Conferva fluviatilis ſericea vulgaris et fluitans. *Dill.* 12. t. 2. f. 1.

Conferva Plinii. *Ray's Syn.* 58.

Threads coarſer, not ſo long, winding in different directions, from one to two feet in length, about the thickneſs of human hair.

Hair-like. Hud. — 2. Conferva paluſtris ſericea filamentis craſſioribus et longioribus. *Ray's Syn.* 477.

Conferva paluſtris ſericea, craſſior et varie extenſa. *Dill.* 13. t. 2. f. 2.

(1.) Rivulets, ponds, and bogs; in all ſeaſons, but chiefly ſpring and autumn. (2.) Marſhy meadows, ſpring and ſummer.

OBS. *Are not theſe diſtinct ſpecies?*

Spring 2 Fontinalis — RIVERWEED. Threads undivided; equal; ſhorter than one's finger.

Small — 1. Threads extremely ſhort; ſometimes not more than half an inca in length: of a browniſh green.

Conferva minima Byſſi facie. *Dill.* 14. t. 2. f. 3.

Conferva fontalis fuſca omnium minima mollis. *Ray's Syn.* 58.

On ſtones, in ſlow ſtreams and near ſprings. Winter.

Larger. Dill? — 2. Threads longer, of a ſlippery conſiſtence, uniting into fleecy locks.

Conferva mucoſa, confragoſis rivulis innaſcens. *Dill.* t. 2. f. 4.

(1.) On ſtones in ſlow ſtreams, and near ſprings. Winter. (2.) In rocky rivulets. Llanberris, Wales. *Dill.*

* * *Threads*

* * *Threads branched, equal.*

RIVERWEED Threads equal; branched at the ends. Branches mostly undivided. *Hudson* 478. Threads pretty long, dividing into two or three short branches at the ends. Forked 3 Furcata *hud.*

1. Conferva capillacea, filamentis bifidis et trifidis. *Dill.* 16. t. 2. f. 6. *Cloven* Hud.

Conferva rivularum nostras bicornis, filamentis tenuissimis. *Ray's Syn.* 58.

2. Threads shorter; thicker; and more branched—*green when taken out of the water, but when dry it turns to a greyish dark green.* *Branched* Hud.

Conferva palustris, filamentis brevioribus et crassioribus. *Dill.* 17. t. 3. f. 10.

Conferva palustris subhirsuta filamentis brevioribus et crassioribus. *Ray's Syn.* 477.

Obs. *This variety appears to be a different species, at least not a variety of this. Dillenius speaks of it as having some affinity both to the* Common *and the* Mill Riverweed *but shews it to be distinct from both. The threads he says are shorter, thicker, and more branched than in the former; longer, thicker and less branched than those of the latter.*

(1) In rivulets. (2) In ditches. Spring—Summer—Aut.

RIVERWEED. Threads equal; somewhat branched; shining like white silk.—*Pretty much resembles in form the* Cloven Forked Riverweed *but it is not quite so long.* From Dillenius. White 4 Candida *Dill.*

Conferva capillacea, serici candidi instar nitens. *Dill.* t. 2. f. 7.

Penryn, Wales. Anglesey rocks. *Dill.*

RIVERWEED. Threads equal; branched; inclosing air bubbles. Cotton 5 Bullosa

Conferva palustris bombycina. *Ray's Syn.* 60. *Dill.* 18. t. 3. f. 11.

Alga bombycina. *Bauh. pin.* 363. *Park.* 1261.

Ditches and stagnant waters. Spring—Summer and Autumn, disappearing in the winter. On the sides of cisterns in all seasons.

RIVERWEED. Threads equal; more branched toward the base.—*Branches crowded; soft and herbaceous when taken out of the water, but when dry they acquire an almost stony hardness, which Dillenius attributes to a sediment deposited by the water. This may possibly be the case, but may it not be owing rather to an earth absorbed by the plant itself in the process of vegetation, as is probably the case in the* Stonewort? Mill 6 Canalicularis

Conferva rivulorum capillacea, densissime congestis ramulis. *Ray's Syn.* 58. *Dill.* 21. t. 4. f. 15.

Alga in tubulis aquam fontanam ducentibus. *Bauh. pin.* 364.

Clear streams, and mill pond troughs.

RIVER-

Amphibious 7 Amphibia

RIVERWEED. Threads equal; branched; when dry uniting into ſtiff ſharp points--

Conferva amphibia, fibrilloſa et ſpongioſa. *Dill.* 22. t. 4. f. 17.

Conferva terreſtris exilis fibrilloſa. *Ray's Syn.* 59.

Fungus vel potius ſpongia viridis doliolis adnaſcenti ſimilis. *Ray's Syn.* 57.

Banks of rivers; ditches; damp walls. Autumn and Winter, and in Summer in moiſt ſhady places.

Fennel-leaved 8 Fæniculacea *Hud.*

RIVERWEED. Threads equal; very much branched; branches and ſubdiviſions of the branches very long, ſcattered. *Hudſon* 479.

Conferva marina foeniculacea. *Dill.* 16. t. 2. f. 8.

On rocks on the ſea-ſhore.

Rough 9 Rigida *Hud.*

RIVERWEED. Threads equal; very much branched; little branches very ſhort, alternate. *Hudſon* 479.—*Diſtant; of a rigid conſiſtence.*

Conferva fluviatilis fibrilloſa, ſubrigida. *Dill.* 21. t. 4. f. 16.

In ſlow rivers.

Fibrous 10 Fibrilloſa *Dill.*

RIVERWEED. Threads equal; very much branched; little branches very ſhort, numerous; crowded; very finely divided;—*Soft; of a pale green. Very much reſembles the preceding ſpecies. Iu Dillenius's figure there appears little kuots at the diviſions of the branches, which are probably the flowers or fruit.* From Dillenius.

Conferva fluviatilis brevis, extremis ramulis creberrime et tenuiſſime diviſis. *Dill.* t. 4. f. 18.

On ſmall ſtones in Hackney River. *Dill.*

Matted 11 Littoralis

RIVERWEED. Threads equal; very much branched, pretty long; roughiſh. Larger threads jointed, branched; leſſer threads very much branched, matted together. *Hudſon* 484.

Conferva marina capillacea longa ramoſiſſima, mollis. *Ray's Syn.* 59. *Dill.* 23. t. 4. f. 19.

Conferva plicata. *Hudſon* 484.

On rocks and ſtones in the ſea.

RIVER-

RIVERWEED. Threads equal, very much branched; Wool
branches undivided; crowded together; brown, (*or whitiſh.*) 12
Hudſon 480. Tomentoſa
Hud.

1. Of a more tender ſubſtance, and of a whitiſh colour. *White*
Conferva marina tomentoſa, tenerior et albicans. *Ray's Syn.* 59. *Dill.* 19. t. 3. f. 12.
Muſcus maritimus goſſipio ſimilis. *Bauh. pin.* 663.
2. Of a firmer confiſtence, of the colour of ruſty iron. *Ruſt-coloured*
Conferva marina tomentoſa, minus tenera et ferruginea. *Ray's Syn.* 59. *Dill.* 19. t. 3. f. 13.
Salt ditches, and the ſea-ſhore.

RIVERWEED. Threads branched; ſoft; ſhorter than Sea-green
one's finger; of a beautiful green. 13
Conferva marina capillacea brevis, viridiſſima mollis. *Dill.* Æruginoſa
23. t. 4. f. 20.
On Oarweed.

RIVERWEED. Threads equal; branched; branches very Black
ſhort; in bundles. *Hudſon* 481. 14
On the Yorkſhire coaſt. Nigra
Hud.

RIVERWEED. Threads branched; equal. Little branches Sponge
undivided; tiled. *Hudſon.* 480. 15
Fucus teretifolius ſpongioſus piloſiſſimus. *Ray's Syn.* 46. Spongioſa
Hud.
Sea-ſhore.

RIVERWEED. Threads equal; forked— Briſtle
Conferva dichotoma, ſetis porcinis ſimilis. *Dill.* 17. t. 3. f. 9. 16
Conferva Plinii ſetis porcinis ſimilis. *Ray's Syn.* 58. Dichotoma
Salt marſh ditches. All the year.

RIVERWEED. Threads proliferous; of the ſame length; Broom
rough with hair— 17
Conferva marina pennata. *Dill.* 24. t. 4. f. 23. Scoparia
Sea-ſhore.

RIVERWEED. Larger threads branched; leſſer threads Latticed
alternate; ſhort; with many finger-like diviſions— 18
Conferva marina cancellata. *Ray's Syn.* 59. *Dill.* 24. t. 4. Cancellata
f. 22.
On ſhells and pebbles near the ſea-ſhore.

* * * *Threads*

* * * *Threads growing into one another.*

Net 19 Reticulata — RIVERWEED. Threads uniting ſo as to form a ſort of network.
Conſerva reticulata. *Ray's Syn.* 59. *Dill.* 20. t. 4. f. 14.
Ditches, rivulets, and banks of rivers. Summer.

* * * * *Threads knotted.*

Horſe-tail 20 Fluviatilis — RIVERWEED. Threads undivided, briſtle-ſhaped, ſtraight. Knots angular and thicker than the thread —
Conſerva fluviatilis lubrica ſetoſa, Equiſeti facie. *Dill.* 39. t. 7. f. 47.
On ſtones in rivers.

Flat-headed 21. Compreſſa *Dill.* — RIVERWEED. Threads undivided; ſomewhat necklace-ſhaped. Joints broad, flatted. *From Dillenius.*
Conſerva fluviatilis nodoſa, fucum æmulans. *Dill.* t. 7. f. 48.
Near Ludlow, Shropſhire; on the new bridge. *Dill.*

Frogſpawn 22 Gelatinoſa — RIVERWEED. Threads branched; necklace-ſhaped; joints globular, of a jelly-like ſubſtance,—*ſet cloſe together.*

Brown — 1. Larger; brown, or mouſe-colour.
Conferva fontana nodoſa, ſpermatis ranarum inſtar lubrica, major et fuſca. *Ray's Syn.* 62. *Dill.* 36. t. 7. f. 42.

Green — 2. Smaller, green.
Conſerva fontana nodoſa, ſpermatis ranarum inſtar lubrica, minor et viridis. *Dill.* 37. t. 7. f. 43.

Pale green — 3. Tender ſlimy ſubſtance; pale green.
Conſerva ſtagnatilis, globulis vireſcentibus mucoſis. *Dill.* 38. t. 7. f. 44.

4. Large; blue; pretty much branched.
Conſerva alpina lubrica, filamentis nodoſis cæruleis. *Dill.* 38. t. 7. f. 45.

5. Threads black, fine and ſlender; very much branched.
Conſerva fontana nodoſa lubrica, filamentis tenuiſſimis nigris. *Dill.* 39. t. 7. f. 46.
In clear ſprings and rivulets. (2) Spring—Summer. (4) Aug.

Pearl 23 Moniliformis *Dill.* — RIVERWEED. Threads branched; necklace-ſhaped; joints globular, of a jelly-like ſubſtance; little branches ſtraight, finely tapering to a point;—*Joints diſtinct from each other. A very elegant ſpecies.* Colour *of a pale red.* From Dillenius.
Conſerva marina nodoſa, lubrica, ramoſiſſima et elegantiſſima rubens. *Ray's Syn.* 62. t. 2. f. 3. *Dill.* t. 7. f. 40.

RIVER-

RIVERWEED. Threads very much branched; branches numerous, finely divided: little branches necklace-shaped. *From Dillenius.* Shrubby 24 Fruticulosa *Dill.*

* * * * * *Threads jointed.*

RIVERWEED. Threads jointed, undivided; joints alternately compressed.—*Threads winding in different directions.* Thread 25 Capillaris

Conferva filamentis longis geniculatis simplicibus. *Dill.* 25. t. 5. f. 25.

Conferva palustris, seu filum marinum anglicum. *Ray's Syn.* 60.

1. Threads fine; about the thickness of a horse-hair; sometimes a little branched. *Dill.* f. 25. B. *Fine.* Dill.

2. Threads considerably thicker; about the thickness of a very coarse thread. *Dill.* f. 25. 4. *Coarse.* Dill.

(1) In fresh water ditches. (2) In salt water ditches.

RIVERWEED. Threads jointed; forked; hairlike,—*Rather short; dividing into forks about four or five times in the whole length.* From *Dillenius.* Hair 26 Crinita *Dill.*

Conferva ramosa, filamentis geniculatis brevioribus. *Dill.* t. 5. f. 26.

On small stones in Kenchurch River, Herefordshire. *Dill.*

RIVERWEED. Branches forked; flowers surrounding the joints—*Ellis Phil. Trans.* vol. 57. p. 425. Scarlet 27 Geniculata *Ellis.*

On the sea-coast near Brighthelmstone.

RIVERWEED. Branched forked; fringed at the joints—*Ellis Phil. Trans.* vol. 57. p. 425. Fringed 28 Ciliata

RIVERWEED. Threads jointed, forked—*Of a jelly-like substance.* Coralline 29 Corallinoides

1. Threads finer, of a reddish colour. *Red*

Conferva marina gelatinosa, corallinæ instar geniculata tenuior. *Dill.* 33. t. 6. f. 37.

Corallina confervoides gelatinosa rubens, ramulis et geniculis per-angustis. *Ray's Syn.* 34.

2. Threads thicker, white.; joints transparent. *White*

Conferva marina gelatinosa, corallinæ instar geniculata, crassior. *Dill.* 33. t. 5. f. 36.

Corallina confervoides gelatinosa alba, geniculis crassiusculis pellucidis. *Ray's Syn.* 34.

On the sea-shore. (2.) Adhering to pebbles.

RIVERWEED. Threads jointed, very much branched; lesser branches very long, distant, sharp. *Hudson* 484. Pointed 30 Elongata. *Hud.*

Conferva marina geniculata ramosissima lubrica, longis sparsisve ramulis. *Ray's Syn.* 61. *Dill.* 35. t. 6. f. 38.

On the sea-shore.

RIVER-

Hand-shaped 31 Polymorpha

RIVERWEED. Threads jointed, branches in bundles.—*Pretty much divided.*

This plant as its Latin name implies, assumes a variety of appearances. At first it is red; it then turns of a brownish colour, and at last becomes almost black. At this stage of its growth, it changes too its form. Most of the middle, or lower branches drop off, and the upper ones alone remain, which when grown to nearly an equal length, have very much the appearance of an open hand, whence its English name.

Conferva marina geniculata nigra palmata. *Dillenius* 32. t. 6. f. 35.

Conferva marina geniculata ramosissima lubrica, brevibus et palmatim congestis ramulis. *Ray's Syn.* 61.

Muscus capillaceus multifidus niger. *Bauh. pin.* 363.

On the sea-shore.

In the *Philos. Trans.* vol. 57. p. 424. Mr. Ellis has demonstrated, that this species ought to be considered as bearing its Chives on one plant, and its Fruit upon another. The Capsules containing the Seeds are placed in the Forks at the termination of the branches. They are almost transparent, and when magnified the Seeds are visible within them if the plant is kept moist with water. The flowers furnished with Chives are collected into Catkins, which stand on the terminations of the branches, not in the Forks.

Larch-leaved 32 Laricifolia *Dill.*

RIVERWEED. Threads jointed; branched; branches in bundles; undivided: bundles alternate— *From Dillenius.*

Conferva fluviatilis, sericea, tenuis. *Dill.* t. 6. f. 31.

Munnow River near Kenchurch, Herefordshire. *Dill.*

Pale-red 33 Plumula. *Ellis*

RIVERWEED. Branches nearly horizontal; feathered on the upper side. *Ellis. Phil. Trans. Vol.* 57. p. 425.

Spreading 34 Vagabunda

RIVERWEED. Threads jointed; winding in different directions; branches, and divisions of the branches rather short.—*Joints hardly visible to the naked eye, especially in the finer branches; They may however be discovered with the assistance of a microscope.*

Conferva marina trichoides, lanæ instar expansa. *Ray's Syn.* 60. *Dill.* 30. t. 5. f. 32.

Salt marshes.

Short 35 Fulva. *Hud.*

RIVERWEED. Threads jointed, branched; branches and divisions of the branches very short, alternate. *Hudson* 484—*Of a tawny colour.*

Brown 36 Fusca *Hud.*

RIVERWEED. Threads jointed, very much branched; little branches alternate, undivided. *Hudson* 486.

On the Yorkshire coast.

RIVER-

RIVERWEED. Larger threads jointed: forked; branches mostly alternate; diſtinct; branching out into roundiſh tufts: little branches crowded. *From Dillenius.* — Bunched 37 Racemoſa *Dill.*

Conferva marina trichoides, ramoſiſſima ſparſa. *Ray's Syn.* 60. *Dill.* 28. t. 5. f. 30.

On ſmall ſtones, Shepey and Angleſey. *Dill.*

OBS. *Hudſon has noted it as a ſubject of enquiry whether it may not be a variety of the next ſpecies.*

RIVERWEED. Threads jointed, little branches rather ſhort, with many clefts—*Green.* — Cluſter 38 Glomerata

Larger threads jointed, forked; leſſer threads branched; little branches ſhort, rough, in bundles. *Hudſon* 483.

Conferva fontalis ramoſiſſima, glomeratim congeſta. *Ray's Syn.* 59. *Dill.* 28. t. 5. f. 31.

Springs and rivers. *Dill.* And on the ſea ſhore. *Hudſon.*

RIVERWEED. Threads jointed; very much branched; little branches with many diviſions, in bundles. *Hudſon* 485. — Branched 39 Fucoides *Hud.*

On the Yorkſhire coaſt.

RIVERWEED. Threads jointed; branched; branches in bundles; crowded; green. *Hudſon* 485. — Silk 40 Sericea *Hud.*

Conferva trichodes virgata, ſericea. *Dill.* 31. t. 5. f. 33.

Conferva marina trichodes, ramulis virgatis longioribus glabris. *Ray's Syn.* 60.

Rivers, and on ſtones on the ſea ſhore.

RIVERWEED. Threads jointed; very much branched; green— — Rock 41 Rupeſtris

Threads jointed, branched; branches alternate, ſubdivided, green. *Hudſon* 485.

1. Threads branched all the way. — *Sea*

Conferva marina trichodes ramoſior. *Dill.* 28. t. 5. f. 29.

Conferva marina trichodes, ſeu Muſcus marinus virens tenuifolius. *Ray's Syn.* 60.

2. Threads branched only at the ends.

Conferva fluviatilis trichodes, extremitatibus ramoſis. *Dill.* tab. 5. f. 28. — *River.* Hud.

(1) Sea Beard. (2) River Beard.

(1) On rocks on the ſea ſhore. (2) In rivers; on the bottoms of boats.

RIVERWEED. Threads jointed; very much branched; joints cylindrical; branches oppoſite. *Hudſon* 483. — Tranſparent 42 Pellucida *Hud.*

Sea ſhore.

Red 43 Rubra *Hud.*

RIVERWEED. Threads jointed; very much branched; compreſſed; branches more than doubly compound; winged; coloured. *Hudſon* 486.

Sea ſhore.

OBS. *Hudſon gives the* Fucus cartilagineus *of Linnæus as a ſynonym of this ſpecies of which it is probably only a variety. See Dillenius's opinion of this matter in* Ray's Syn. 37.

Feathered 44 Pennata *Hud.*

RIVERWEED. Threads jointed; branched; branches doubly winged, brown. *Hudſon* 186.

Conferva marina pennata. *Ray's Syn.* 59. *Compare this with No.* 17.

Conferva plumoſa. *Ellis.* Phil. Tranſ. vol. 57. p. 424.

Sea ſhore.

This ſpecies produces Chives upon one plant and Seeds upon another. The Capſules are ſeated in the diviſions of the branches near to the principal Stem. The flowers with Chives, only form ſpikes, which ſtand in the forks of the branches, partly at an equal diſtance betwixt the Stem and the ends of the branches. *Philoſ. Tranſ.* vol. 57. p. 424.

Winged 45 Paraſitica *Hud.*

RIVERWEED. Threads jointed; winged—*Hudſon* 486.

On Oarweed.

Carmine 46 Floſculoſa *Ellis*

RIVERWEED. Flowers on fruit-ſtalks; empalement leaſy. *Ellis. Phil. Tranſ.* vol 57. p. 425.

On the ſea coaſt near Yarmouth.

Moorball 47 Ægagropila

RIVERWEED. Threads jointed; very much branched; branches extremely crowded; proceeding from the center, and forming a round ball—*about the ſize of a walnut, very much reſembling thoſe hairy balls found in the ſtomachs of Cows and Oxen; green; joints of the threads brown. It is found looſe in the water, and not adhering to ſtones or any other ſubſtance, nor does there appear to be any ſolid body in the center from whence the threads might proceed.*

Conferva globoſa. *Phil. Tranſ.* vol. 41. p. 498.

In lakes on mountains.

431 MUSHROOM. 1209 Agaricus.

FUNGUS horizontal, with *Gills* on the under surface.

OBS. *See Pl.* 1. *fig.* H.

* *Hat circular ; standing on a pillar.*

MUSHROOM. Gills branched; running down the pillar. —*Hat turban-shaped, rather flat ; border angular, as if cut into segments.* Gills *yellow* ; *thick* ; *not distinct from the substance of the Mushroom.* Pillar *yellow short and naked. It varies both in size and colour. It is mostly of a pale yellow, but sometimes of a deep, or even Saffron colour. Gerard has called them the* Fungi lethales, *but very improperly, as they are not only an innocent but an excellent food, and of a fine flavour.* Champignon 1 Chantarellus

1. Fungus angulosus et velut in lacinias sectus. *Bauh. pin.* 371. *Common.*
Fungus luteus seu pallidus, Chantarelli dictus, se contorquens esculentus. *Ray's Syn.* 2.
Fungi lethales. *Gerard.* 1385.

2. Hat with the border not circular but running into angles ; reflected upwards ; in form of a drinking glass, or inverted cone ; yellow, and when full grown, with a tinge of red. *Stalk* very short, thick. *Cup.* Ray
Fungi esculenti 14 generis species 2 Clusii, aut ei similis. *Ray's Syn.* 2.
(1) Meadows and pastures, and in woods under trees. (2) in Woods.
The French and Italians eat them.

MUSHROOM. Gills all of an equal size.—*Hat not fleshy* ; *flat* ; *rather concave* ; *as broad as one's hand* ; *pale tawny, or flesh-coloured* ; *scored and dotted at the edge.* Gills *pale, or white, and betwixt them and the edge of the hat, little teeth.* Pillar *white* ; *nearly half as long again as the hat, about the thickness of one's little finger.* White 2 Integer

Fungus major pediculo longo, modice crasso, lamellis albis creberrimis, superne ad margines apparentibus. *Ray's Syn.* 4.
In woods.

MUSHROOM. Hat broad ; stalk very long ; variegated. Of a finer flavour than the *Common Mushroom.* From *Ray.* Variegated 3 Varigatus *Ray*

Fungus pileolo lato, longissimo pediculo variegato. *Bauh. pin.* 371. *Ray's Syn.* 3.
In Chesterton Close near Cambridge, and in the Wolds of Lincolnshire.

Teasel-headed 4 Dipsacoides *Ray*

MUSHROOM. Brown; root of the stalk bulbous.- *Has a near affinity to the preceding, but judged to be a distinct species by those who have examined both. From* Ray.

Fungus quercinus *Ray's Syn.* 3.

Fungus bulbosus fuscus duplici pileolo. *Bauh. pin.* 371.

Navel 5 Umbilicalis

MUSHROOM. White; growing together in clusters. From *Ray.*

Fungi plures simul albi, ad arborum radices, esculenti. *Ray's Syn.* 10.

Fungi umbilicum exprimentes, plures simul albi. *Bauh pin.* 370.

At the roots of Elms near Camberwell.

Reddish 6 Muscarius

MUSHROOM. The gills that extend bur half way single. Pillar furnished with a cap; thicker towards the top; egg-shaped at the base.—*Hat large; almost flat; either white, red, or crimson; sometimes beset with angular red warts.* Gills *white, flat, inversely spear-shaped.* Pillar *hollow.* Cap *fixed to the middle of the pillar, limber, hanging down.*

Fungus minor campestris rotundus, lamellatus, inferne albus, superne purpureus. *Ray's Syn.* 3?

Obs. *This species of Ray seems to correspond pretty exactly with the* Muscarius. *Purple above and white underneath. He describes it indeed as round, and seldom larger than a Pigeon's egg, but says he never saw it expanded.* Pillar *short and thick.*

In pastures.

It will effectually destroy bugs if the juice is rubbed upon the walls and bed posts. The inhabitants in the North of Europe whose houses are greatly infested with flies at the decline of summer, infuse it in milk and set it in their windows, and the flies upon tasting the least drop of it are instantly poisoned.

Toothed 7 Dentatus

MUSHROOM.- Hat convex; gills with a sharp pointed tooth at the base.—*Hat yellowish, smooth, somewhat clammy.* Gills *parting from the pillar.* Pillar *slender; hollow; cylindrical; without a cap; scored towards the top. It grows in clusters, several rising from the same point.*

Fungus fasciculosus, pileo orbiculari lutescente, pediculo fusco tenerrime villoso, lamellis ex flavo candicantibus. *Ray's Syn.* 9.

Hat small, from half an inch to an inch broad, of a brownish yellow. *Flesh* hardly any. *Gills* numerous; of a yellowish white. *Pillar* brown; covered with a fine wooly down; about the thickness of a straw, and about two inches high: Four or five grow together. *Ray.*

At the bottom of gate-posts. September.

On Hampstead heath. December. *Ray.*

MUSH-

MUSHROOM. Hat circular; of a yellowiſh brown; from Purple ſtalked
two to three inches broad. Gills light brown; pillar purple— 8
Several growing together. From Ray. Faſciculoſus
Fungus faſciculoſus pileo orbiculari luteſcente, pediculo pur- *Ray*
pureo, *Ray's Syn.* 10.
On rotten wood.

Obs. *This may perhaps prove to be only a variety of the* Toothed.

MUSHROOM. Yellow; ſeveral riſing from one point. Gold knotted
From *Ray*. 9
Fungi plures ex uno pede e Prunorum radicibus enati. *Ray's* Nodi Aurei
Syn. 9. *Ray*
Fungi dumetorum ex uno pede prodeuntes. *Bauh. pin.* 375.
In thickets at the roots of trees.

MUSHROOM. Hat flat; fleſh coloured, with a milky Milky
juice; gills browniſh-red. Pillar long, fleſh-coloured.— 10
Fungus lacteſcens non acris. *Ray's Syn.* 4. Lactifluus
In woods.
It is agreeable to the taſte and not pernicious.

MUSHROOM. Hat nearly flat, containing a milky juice. Pepper
Border bent downwards, gills pale fleſh-colour—*Milky juice very* 11
acrid. Piperatus
Fungus piperatus albus, lacteo ſucco turgens. *Ray's Syn.* 4.
Fungus albus acris. *Bauh. pin* 370.
In woods.

MUSHROOM. Hat pale red or tawny. Gills numerous; Tawny
whitiſh; pillar ſhort; thick; fleſh extremely white. Has a 12
peppery taſte. From *Ray*. Fulvus
Fungus major rubeſcens pediculo brevi craſſo, lamellis crebris *Ray*
albentibus. *Ray's Syn.* 4.
Under Oaks in Auguſt.

MUSHROOM. Hat convex; ſcaly; whitiſh. Gills browniſh Common
red.—*Hat hemiſpherical; with ragged ſcales; bent inwards at the* 13
edge Gills *concave; entire at the edge; unequal in length*. Pillar Campeſtris
cylindrical; above the cap ſmooth and white; beneath aſh-coloured.
Fungus campeſtris albus ſuperne, inferne rubens. *Ray's Syn.* 2.
Fungus Eſculentus. *Park.* 1317.

Its degree of convexity and the colour of its gills depend upon its age. When it firſt makes its appearance it is ſmooth and almoſt globular; the edges of the hat preſs upon the pillar, and the Gills which are then almoſt white are covered with a white membrane extending from the edge of the hat to the ſummit of the

the pillar. In this ſtate it is called a button. By degrees it expands; the membrane burſts, the edges of the hat recede from the pillar and the gills are expoſed to view, of a bright fleſh-colour: this however ſoon fades, and ſinks at length into a dark brown or chocolate colour. The hat now loſes its convexity and becomes almoſt flat, rough and ſcaly.

Yellowiſh 2. Hat yellowiſh rather ſmaller; gills fleſh-colour. *Ray* ib.
Ray

3. Hat yellow, convex; gills white.—*Pillar firm, ſolid; with*
St. George's *whitiſh down.*
Georgii

4. Hat broad, white above; gills very numerous, of a pale red
Wood? or fleſh-colour; ſtalk ſhort, pretty thick.—*Deſcribed as a diſtinct ſpecies, but ſuppoſed by Dillenius to be either the ſame or at leaſt only a variety of the common Muſhroom.*

Fungus minor pileolo lato, ſuperne candido, lamellis ſubtus creberrimis, pallide rubentibus, ſeu incarnati, ut vocant, coloris. *Ray's Syn.* 3.

In parks and lands that have been long unploughed, (1) commons and poor land. (2) In paſtures. (3--4.) In woods.

This ſpecies is one of the corner ſtones of modern luxury; either dreſſed in ſubſtance, or the juice boiled up with wine and ſpices, taking the name of Catchup.

The Seeds are contained in the ſubſtance of the Gills. Each Gill is compoſed of two layers, and betwixt theſe layers are the Seeds, which fall to the ground when ripe. Some of them in their fall catch upon the Cap, and are entangled in its woolly ſurface, where by the aſſiſtance of a Microſcope they may eaſily be found.

MUSHROOM. Large; of a pale red or fleſh colour. From
Fleſh coloured *Ray.*
14 Fungus magnus rubentis ſeu incarnati coloris. *Ray's Syn.* 3.
Incarnatus OBS. *Perhaps only a variety of the* common Muſhroom.
Ray

MUSHROOM. Hat and gills white. From *Ray.*
Eſculent Fungus eſculentus pileo et lamellis albis. *Ray's Syn.* 2.
15 Grows with the common Muſhroom, to which it is little
Eſculentus inferior in taſte.
Ray

MUSHROOM. Large; of a whitiſh green—*Fleſh of a fine*
Green *flavour.* From *Ray*
16 Fungus magnus viridis. *Ray's Syn.* 2.
Viridis Fungi umbilicum referentis variegati 3 ſpecies. *Bauh. pin.* 370.
Ray In Hornſey-wood. *Dill.*

MUSH-

MUSHROOM. Hat covered with a mucus viſcid ſubſtance Verdigris
of the colour of Verdigris. Of a moderate ſize. From *Ray*. 17
Fungus medius pileo muco æruginei coloris obducto. *Ray's* Æruginoſus
Syn. 6. *Ray*
In the garden belonging to the Company of Apothecaries of London, and in St. James's Park, and obſerved in a gravel pit in the middle of September.

MUSHROOM. Hat full of cracks or chinks; violet co- Violet
loured and downy at the border; pillar bluiſh, with wool the 18
colour of ruſty iron—*Hat hemiſpherical*. Gills *horizontally*; *co-* Violaceus
lour of dirt.
Fungus pileo magno, orbicularis, ſublivido, lamellis albis, pediculo brevi bulbiformi violaceo. *Ray's Syn*. 3.
In meadows. December.
Hat large; broad, or orbicular; ſomewhat of a livid colour; of a thick ſubſtance; at firſt convex, by degrees becoming ſomewhat hollow, with the border ſlightly reflected. *Gills* quite white, numerous, contracted towards the ſtalk. *Pillar* ſhort; thick; ſwelling into a kind of bulb toward the baſe. *Cap* at firſt pale, then turning to a livid colour, and at laſt brown. *Ray*.

MUSHROOM. Hat orbicular ſomewhat of a livid colour; Purple-edged
edge ragged and tinged with purple. Gills browniſh—*Fleſh little* 19
or none. Hat *about two or three inches in diameter*. From *Ray*. Purpuraſcens
Amanita orbicularis ſublivida, lamellis ſubfuſcis, pilei oris *Ray*
leviter purpuraſcentibus. *Ray's Syn*. 5.
Near Camberwell. Aug.

MUSHROOMS. Hat of a dirty yellow; gills a browniſh Brown
red, inclining to yellow—*Hat convex but flattiſh, with frequently* 20
a little ſudden riſing in the middle; *not milky*; *ſharp at the edge*; Cinnamomeus
of a Cinnamon colour. Fleſh *of a fine flavour*. Pillar *long*; *naked*; *yellowiſh*.
Fungus ovinus. *Ray's Syn*. 2.
Fungus planus orbiculatus aureus. *Bauh. pin*. 371.
In woods

MUSHROOM. Hat round; of a yellowiſh brown, with a Singed
tinge of red; about the middle of darker red, appearing as if 21
ſinged. Gills numerous, of a dirty green—*Hat about two inches* Uſtulatus
in diameter. Pillar *not very thick*; *about twice as long*. *Several* *Ray*
riſing from the ſame point. From *Ray*.
Fungus mediæ magnitudinis, pileolo ſuperne e ruſo flavicanti, lamellis ſubtus ſordide virentibus. *Ray's Syn*. 10.
September.

Cheſnut coloured 22 Caſtaneus *Ray*

MUSHROOM. Hat of a cheſnut colour. Gills white. Pillar ſpotted.—*Of a pretty large ſize.* From *Ray.*

Fungus pileatus major, ſuperne coloris caſtanei, lamellis candidis, caule maculato. *Ray's Syn.* 4.

County of Down in Ireland.

Clammy 23 Viſcidus

MUSHROOM. Hat of a purpliſh brown; clammy. Gills of a brown purpliſh colour.—*Hat at firſt convex; afterwards hemiſpherical; and when old turban-ſhaped, with the edge bent inwards.* Gills *diſtinct, remote; the ſides woolly, and duſted with a powder.*

Amanita dura, ex fuſco rubens quercina. *Ray's Syn.* 5.

Hat moſtly flat and even, but ſometimes curled and irregular; from two to four inches in diameter; of a firmer conſiſtence than moſt of the other ſpecies. *Fleſh* little or none. *Gills* pale reddiſh brown. *Pillar* about the thickneſs of one's finger, two or three inches high; of a pale reddiſh brown. *Ray.*

In woods, moſtly at the root of oaks, and ſingle, but ſometimes ſeveral together. October. *Ray.*

Excentric 24 Excentricus *Ray*

MUSHROOMS. Hat fixed to the pillar on one ſide—*not in the center. It is ſubject to great varieties in colour, form, and ſize. It has ſometimes no pillar or at leaſt a very ſhort one, and that, though very rarely, inſerted in the center of the hat. Colour whitiſh or as if daubed with ſoot. It always grows on trees, but is exactly of the ſame ſubſtance with the other Muſhrooms.* From *Ray.*

Fungus arboreus mollis multiformis. *Ray's Syn.* 5.

On trees, in St. James's Park.

Mealy 25 Farinaceus *Ray*

MUSHROOM. Hat flat; white, and as if ſprinkled with meal. From *Ray.*

Fungus ſimetarius in plano orbicularis, candidus. *Bauh. pin.* 372. *Ray's Syn.* 6.

On dung-hills.

Starry 26 Equeſtris

MUSHROOM. Hat pale, with a yellow ſtar in the center. Gills brimſtone colour—*not clammy*; Pillar *naked, ſmooth.* Hat *convex.*

Fungus parvus, parvi galeri formam exprimens, ruſus. *Ray's Syn.* 7. *Bauh. pin.* 373.

On heaths.

Sooty 27 Fuliginoſus *Ray*

MUSHROOM. Hat, in form nearly the ſame as the former; of a ſooty colour; not clammy. Gills black; extending from the hat to the cap.—*Of ſhort duration.* From *Ray.*

Fungus parvus lethalis galericulatus. *Ray's Syn.* 7.

Fungus parvus galericulatus alter parvus. *Bauh. pin.* 373.

In paſtures near paths and hedges. Autumn.

MUSH-

MUSHROOM. Hat convex, tapering to a point in the center; grey. Gills convex; grey; fcolloped. Pillar naked—*fcored, very long, cylindrical. Hat blunt at the top; rather inclining to one fide; of a dirty yellowifh afh-colour;* Gills *not numerous, only every fourth extending to the pillar. Pillar fometimes near fix or eight inches long. Slender though firm; fcored; hairy; yellowifh. A very beautiful mufhroom.* Grey 28 Mammofus

Fungus fordidi fulvus, capitulo in conum faftigiato, pediculo longiffimo firmo ftriato. *Ray's Syn.* 4.

In woods.

MUSHROOM. Hat convex, tapering into a fharp cone, of a dirty yellow,—*very much refembling the preceding, but the* Pillar *fhorter and flenderer, and the* Gills *of a blackifh livid colour.* From *Ray.* Sugar-loaf 29 Acuminatus *Ray*

Fungus fordide fulvus in acutum conum faftigiatus. *Ray's Syn.* 6.

Obs. *This is perhaps only a variety of the preceding.*

MUSHROOM. Shield-fhaped, of different colours. Several rifing from one point; border either fcolloped or entire. From *Ray.* Cottage 30 Ruralis *Ray*

Fungi multi ex uno pede clypeiformes variorum colorum, per oras crenati vel non. *Ray's Syn.* 10.

On old ruinous cottages.

MUSHROOMS. Hat hemifpherical, tapering to a point; clammy. Gills white. Pillar long; cylindrical; white—Gills *not concave; dufted with a fine powdery fubftance on each fide.* Root *bulbous, long, hooked at the end.* Long-ftalked 31 Clypeatus

Fungus parvus, pediculo oblongo, firmo, lento, pileolo in medio faftigiato, ftriis exterius apparentibus? *Ray's Syn.* 8. border of the *Hat* fcored.

In wood-land-paftures. September.

Dr. Percival in the laft vol. of his Effays page 267, relates the cafe of a man who was poifoned by eating a Mufhroom, which Mr. Hudfon thinks was one of this fpecies.

MUSHROOM. Hat fpherical; pale; fpotted. Pillar growing in form of a bulb. Rather fmall. From *Ray.* Bulbous 32 Bulbofus *Ray*

Fungus pediculo in bulbi formam excrefcente. *Bauh. pin.* 373. *Ray's Syn.* 3.

Amongft the brakes in Middleton park Warwickfhire.

MUSHROOM. Hat Bell-fhaped; whitifh; ragged. Gills exceedingly white; pillar with a fort of bulb; awl-fhaped; naked— Conic 33 Extinctorius

In paftures and on dunghills.

MUSH-

Party coloured MUSHROOM. Hat conical. Gills numerous; parti colour-
34 ed the outer half of a pale red, the inner half black.—*Pretty*
Versicolor *large.* From *Ray.*
Ray Fungus pascuorum majusculus, capitulo conico, lamellis subtus creberrimis, exteriore medietate rubentibus, interiore nigris. *Ray's Syn.* 5.

In pastures.

Mouse-coloured MUSHROOM. Hat circular at the margin; about three inches broad; of a mouse-colour; clammy; not very fleshy.
35 Gills whitish; pretty broad. Pillar thick; short; when full
Murinus grown the border of the hat somewhat reflected. From *Ray.*
Ray Fungus superficie murini coloris, lamellis albicantibus. *Ray's Syn.* 5.

In the fields near Chelsea.

rsh MUSHROOM. Hat of a yellowish brown. Gills whitish.
36 —*The Hat assumes a variety of forms. At first it is conical, by de-*
Palustris *grees it becomes flat, and at length hollow; appearing as if inverted.*
Ray Pillar *of a yellowish brown*; *small.* From *Ray.*

Fungus parvus ex luteo fuscus, pileo per maturitatem inverso. *Ray's Syn.* 6.

In damp moist places, amongst Moss in Chelsea garden and at Lambeth, and in wet marshy places on heaths, and on rotten Moss. Spring Autumn.

Scotch Bonnet MUSHROOM. Hat broad; thin; of a leather-like sub-
37 stance; compressed; in form of a drinking-glass. *From Ray.*
Mitratus Fungus lamellatus, pileo lato, tenui, coriaceo, compresso,
Ray umbilicato *Ray's Syn.* 6.

Frequent about Hyde-park the latter end of Oct. *Ray.*

Funnel-shaped MUSHROOM. Hat clammy; of a pale red; when full
38 grown the border reflected upwards in form of a funnel or invert-
Infundibuliformis ed cone. Gills few.—*Of the smaller size.* From *Ray.*
Ray Fungi pratenses minores, externe viscidi rubentes. *Ray's Syn.* 7.

In pastures, especially in wet weather. Aug.—Sept.

Meadow MUSHROOM. Hat white, or yellow; clammy. Pillar
39 short.—*Of the smaller size.* From Ray.
Pratensis Fungi pratenses minores, externe viscidi, albi et lutei, pedi-
Ray culis brevibus. *Ray's Syn.* 7.

With the former. Aug.—Sept.

MUSHROOM. Hat bell-ſhaped; ragged; white. Gills black; bending outwards. Pillar hollow.—*Hat before it unfolds egg-ſhaped; underneath exactly like the inſide of a bell.* Gills *white.* Egg 40 Fimectarius

Fungus albus ovum referens. *Ray's Syn.* 5.

On dung-hills.

MUSHROOM. Hat convex; aſh-coloured; with white gills and warts. Pillar thickiſh at the baſe. *Hudſon*, 494.—*Surrounded with a cap.* Warty 41 Verrucoſus

Fungi albi venenati viſcidi. *Ray's Syn.* 7.

In ſhady places.

Obs. *This ſpecies Linnæus inſerted in ſome of his former pieces, bu for ſome reaſon or other has omitted it in his later works. Ray' plant however is a native of England. He compares it to the* Common Muſhroom; *from which it differs in having a longer and more ſlender pillar, hat broader, not ſo thick and fleſhy, and of a poiſonous diſagreeable ſmell.*

MUSHROOM. Hat bell-ſhaped; ſcored; tranſparent. Gills aſcending. Pillar naked,—*very long, ſmooth.* Hat *clammy, of an aſh or mouſe-colour.* Gills *black, or aſh-coloured.* Bell 42 Campanulatus

Fungus perpuſillus, pediculo oblongo, pileo tenui utrinque ſtriato, ſeu flabelli in modum plicatili. *Ray's Syn.* 8.

Gardens, meadows, and paſtures. Autumn.

MUSHROOM. Hat conical; blunt at the top; ſcored. Pillar ſlender; hollow; brittle.—*Small; brown; not fleſhy.* From Ray. Spring 43 Vernalis *Ray*

Amanita parva, verna, utrinque ſtriata, fuſca, pileo obtuſe coniformi, Muſco paluſtri ramoſo majori, fol. membranaceis acutis, *Vern.* innaſcens. *Ray's Syn.* 8.

In Charlton Bogs, on moſs.

MUSHROOM. Hat conical; of a tawny reddiſh colour; gills few;—*very ſmall.* From Ray. Houſe 44 Tectorum *Ray*

Fungus minimus capitulo conico, rufeſcens, lamellis ſubtus paucis. *Ray's Syn.* 9.

Amongſt moſs on the roofs of houſes.

Obs *Ray remarks that he has obſerved one very like it, if not the ſame, in marſhy places.*

MUSHROOM. Hat ſmooth; livid colour. Gills blackiſh. Pillar bulbous; furniſhed with a cap.—Hat *expanding; fleſhy.* Gills *black, as if ſooted.* Pillar *long; cylindrical; pale.* Cap *pendant; membranaceous; fixed to the middle of the pillar.* Blackiſh 45 Separatus

Fungus parvus pediculo oblongo, galericulatus; ſtriis lividis aut nigris. *Ray's Syn.* 8.

On dunghills.

MUSH-

Hemiſpherical 46 Hemiſphæricus *Ray.*

MUSHROOM. Hat hemiſpherical; of a livid colour. Gills of the ſame colour; flat,—*not hollow, but extending ſtraight from the brim of the hat to the pillar, without having any cavity underneath; numerous.* Pillar *ſlender; flexible; white; three or four inches long; ſwelling into a kind of bulb towards the root. Several ſometimes riſing from the ſame point.* From Ray.

Fungus parvus pediculo oblongo, pileolo hemiſphærico, ex albido ſubluteus. *Ray's Syn.* 7.

In paſtures upon horſe and cow-dung. Sept.—Oct.

Cat's-Milk 47 Lacteſcens *Ray*

MUSHROOM. Pillar very fine; long, and ſlender, with a milky juice;—*grows ſingle and is very ſmall.* From Ray.

Fungus minimus, pediculo longo tenuiſſimo, lacteſcens. *Ray's Syn.* 9.

In paſtures amongſt graſs.

Brittle 48 Fragilis

MUSHROOM. Hat convex; clammy; tranſparent; yellow. Gills of the ſame colour. Pillar naked—

Fungus pratenſis minor, externe viſcidus, capitulo præcædentis, ſtriis ſubtus fulvus ſeu croceis. *Ray's Syn.* 8.

On heaths.

Turban 49 Turbinatus *Ray.*

MUSHROOM. Hat turban-ſhaped; ſcored; of an aſh or tawny colour above, with blackiſh furrowed ſcores underneath. —*Several growing together.* Pillar *white; ſlender.* From Ray. *Hat not fleſhy.*

Fungi minores plurimi ſimul naſcentes turbinati, exterius cinerei aut ſubfulvi, ſtriis nigricantibus. *Ray's Syn.* 10.

In hedges and thickets.

Bunch 50 Racemoſus *Ray*

MUSHROOM. Hat turban-ſhaped; of a clear white; ſmooth. Gills of the ſame colour,—*arched, extending to the ſkin of the hat; pillar very ſlender; white; ſeveral growing from the ſame point.* From Ray.

Fungi plures juxta ſe naſcentes, parvi turbinati, candidi ubivis coloris. *Ray's Syn.* 10. t. 1. f. 2 a. a.

On rotten wood and branches of trees. Nov.

Wood 51 Umbelliferus

MUSHROOM. Hat plaited; membranaceous. Gills broadeſt at the baſe.—*Very ſmall; of a clear white.* Stalk *long, ſlender.*

White

1. Fungus parvus candidiſſimus lamellatus, pediculo longo gracili. *Ray's Syn.* 9.

In woods amongſt rotten leaves, and on rotten wood. Latter end of Autumn.

Aſh-coloured

2. Of a whitiſh aſh-colour. Pillar very long and ſlender; gills few. *From Ray.*

Fungus minimus e cinereo albicans, tenui et prælongo pediculo paucis ſubtus ſtriis. *Ray's Syn.* 9.

On rotten wood, riſing out of the very ſubſtance.

MUSH-

MUSHROOM. Hat plaited; membranaceous; white. Pillar black,—*hair-like*. Gills *extremely thin*; *white*. Black ſtalked 52 Androſaccus

Fungus minimus Adianti aurei capitulis. An Mithridaticus Mentzelii? *Ray's Syn.* 9.

On rotten ſticks and leaves; and on rotten ſticks at the bottom of lakes.

MUSHROOM. Tubular, and nearly cylindrical. On the outſide of a purple-livid colour; clammy; white within. Gills whitiſh—*About half an inch high*, Hat *never obſerved expanded*, *but upon being broken*, Gills *of a whitiſh colour appeared within*. *From* Ray. Purple veined 53 Hæmorrhoidalis *Ray*

Fungus hæmorrhoidalis, purpureus, minimus, viſcidus. *Ray's Syn.* 7.

On trees that have been cut down, and on rotten wood.

† *Five ſtringed Pricklycap.*

* * *Without a pillar; hat but half a circle. Growing on other plants.*

MUSHROOM. Gills winding like a labyrinth.—*Hat woolly above, with ſtripes of different colours. Pretty much reſembling the* Striped Spunk. Oak Agaric 54 Quercinus

1. Agaricus quernus lamellatus coriaceus villoſus. *Ray's Syn.* 25. *Common*

2. Smaller; white above. *Smaller* *Ray*

Fungus parvus arboreus villoſus albus, inferne lamellatus. *Ray's Syn.* 25.

On trunks of trees.

MUSHROOM. Leathery. Gills indented, ſimple. *Hudſon* 495.—*Divided into teeth like a comb.* Hat *of the colour and ſubſtance of a Cow's-hide, but rather ſofter.* Teeth *about half an inch long, from a line to two lines in breadth, with a tinge of blue at the edge; diſtinct, but ſet cloſe together; very numerous.* Comb Agaric 55 Pectinatus *Hud.*

Agaricus coriaceus longiſſimus, pectinatim inferne diviſus. *Ray's Syn.* 25. t. 1. f. 5.

On rotten wood, in cellars, &c.

MUSH-

Birch Agaric 56 Betulinus — MUSHROOM. Leathery; woolly; blunt at the edge. Gills branching and growing into one another.—*Hat about four inches broad, and half an inch thick, white and woolly above, with* oblong holes *underneath.* Ray.

Common — 1. Agaricus villosus, lamellis sinuosis et invicem implexis. *Ray's Syn.* 24.

Smooth Ray — 2. White above, with *long and round holes* underneath. *Ray.* Agaricus quernus lamellatus, coriaceus albus. *Ray's Syn.* 24. On trunks of trees.

Obs. *A comparison of Linnæus's and Ray's characters will be sufficient to convince us, how the* Agaric Mushrooms *run insensibly into the* Spunks.

Alder Agaric 57 Alneus — MUSHROOM. Gills cloven, powdery,—*grey.*

Agaricus parvus lamellatus, pectunculi forma elegans. *Ray's Syn.* 25.

In woods on trunks of trees, but mostly on the *Alder.*

Crab Agaric 58 Mali sylvestris *Ray* — MUSHROOM. White: of a hard substance; with gills resembling the veins of the blood-stone. From *Ray.*

Fungus arboreus albus durus, lamellis instar lapidis Hæmatitis. *Ray' Syn.* 25.

On the crab.

432 SPUNK. 1210 Boletus.

Fungus horizontal; porous underneath.

* *Without a pillar; growing on other plants.*

Cork 1 Suberosus — SPUNK. Cushion-like; white; smooth. Pores angular; of different shapes—*Frequently divided into a number of lobes, disposed one over another.*

Agaricus intybaceus. *Ray's Syn.* 23.

The Fungus foraminosus arboreus lævis albissimus. *Ray's Syn.* Ed. 2. 340. differs only in age, becoming of a dry friable substance, smooth, exceedingly white within, and of a reddish tinge without.

On trunks of trees.

It is soft like Sponge and may be made into corks.

SPUNK. Cuſhion-like; uneven; blunt. Pores cylindrical; Spongy
equal; bluiſh—*In ſhape of a horſe's foot; white on the upper part.* 2
Pores ſet cloſe together. Fimentarius

Sometimes near a foot broad, and two inches thick, brown and marked with ſemicircular lines.

Agaricus poroſus igniarius Carpini. *Ray's Syn.* 24.

On the trunks of trees particularly on Elm.

It is uſed for tinder.

SPUNK. Cuſhion-like; ſmooth; pores exceedingly fine.— Touchwood
Hat ſcaly, convex; but depreſſed in the center. When young of a 3
light brown above and ſoft like velvet; white underneath and cover- Igniarius
ed with a ſlimy kind of matter, but when come to maturity, it turns to a dark brown approaching to black. From ſix to ten inches in diameter; ſometimes growing on a footſtalk an inch or two long.

Agaricus pedis equini facie. *Ray's Syn.* 22.

Fungus in candicibus naſcens, unguis equini figurea. *Bauh. pin.* 372.

Fungus durus ſive igniarius. *Park.* 1323.

Touchwood, or ſpunk.

On trunks of trees, chiefly on Aſh.—It is made uſe of in Germany and ſome parts of England for tinder. The Germans boil it in ſtrong lye, dry it, and boil it again in a ſolution of ſaltpetre. The Laplanders burn it about their habitations in order to keep off a ſpecies of the Gadfly which is fatal to the young Reindeer. It has been uſed to ſtop the bleeding from arteries after amputations. *Philoſ. Tranſ.* vol. 48. p. 2 49. p. 1. for this purpoſe the hard outer part is cut off, and the ſoft inner ſubſtance is beat with a hammer to make it ſtill ſofter. It is beſt when gathered in Auguſt or September.

SPUNK. White above, brown underneath. From *Ray.* Beech
Agaricus poroſus igniarius Fagi, ſuperne candicans, inferne 4
fuſcus. *Ray's Syn.* 24. Fagineus
On trees. *Ray*

This as well as the preceding ſpecies is made uſe of for tinder, of which it may perhaps be only a variety.

SPUNK. Cuſhion-like; of a fleſhy ſubſtance; either ſimple Eſculent
or divided into two or three lobes.—*Of a dark reddiſh colour with-* 5
out; white within; with a tinge of red. Contrary to the reſt of this Eſculentus
diviſion it is of a fleſhy ſubſtance and of no ungrateful flavour. It is *Ray*
eaten by the Italians, and is even ſaid by ſome to be richer than the Common Muſhroom.

Agaricus poroſus rubens carnoſus hepatis facie. *Ray's Syn.* 23.

On the trunks of trees.

Striped 6 Verſicolor

SPUNK. With ſtripes of different colours; white underneath;—*Of a leather-like ſubſtance, growing over each other like tiles. Covered above with a fine ſhort down. Stripes concentric; of different ſhades of red, orange, yellow, green, purple, grey, black, and white, and ſometimes extremely vivid. Whitiſh, or yellowiſh brown underneath.*

Agaricus varii coloris ſquamoſus. *Ray's Syn.* 24.

Fungus Ceraſorum imbricatim alter alteri innatus variegatus. *Bauh. pin.* 372.

On trunks, and ſtumps of trees.

Woolly 7 Villoſus *Ray*

SPUNK. Of a leather-like ſubſtance; woolly above, ſomewhat ſtriped—*Pretty much reſembles the preceding.* Pores *larger.* From *Ray.*

Agaricus villoſus et poroſus, ſubſtantiæ coriaceæ. *Ray's Syn.* 24.

Moſtly on young trees.

Obs. *I have a ſpecimen which I believe to be this of* Ray, *which contrary to moſt of the Agaric Tribe is* orbicular. *It did not appear to be merely an accidental variety, as there were ſeveral different ſtages of growth, all exactly agreeing in this reſpect. And what appeared equally remarkable, the porous part in all of them was uppermoſt.*

Semicircularis 8 Semicircular *Ray*

SPUNK. Semicircular: pores numerous; roundiſh; about three inches broad, and half an inch thick. From *Ray.*

Fungus arboreus poroſus minor, abſque pediculo ſemicircularis. *Ray's Syn.* 24.

Obs. *Probably only a variety of ſome of the preceding.*

Prolific 9 Proliferus *Ray*

SPUNK. Very large; ſpreading on the ground; growing together in large quantities.—*It is ſaid ſometimes to ſpread over a plat of ground upwards of thirty feet in diameter. From* Ray.

Agaricus multiplex poroſus. *Ray's Syn.* 23.

In mountainous paſtures.

White 10 Suaveolens?

SPUNK. Smooth above; growing on the Willow,—*Of an agreeable ſweetiſh ſmell. It ſometims grows to a conſiderable ſize, weighing upwards of three pounds.* Pores *larger than thoſe of the firſt ſpecies. Cuſhion-like, ſmooth; pores oblong.* Hudſon 496?

Agaricus officinali ſimilis. *Ray's Syn.* 23.

Agarico ſimilis Fungus diverſarum arborum caudicibus adhærens. *Bauh. pin.* 375.

Boletus albus. *Hudſon* 469.

On the willow.

SPUNK?

SPUNK? Citron
BOLETUS? 11 Citrinus Ray

Woolly above; dotted underneath;—*Of a fine red above, covered with a soft velvet down. Underneath of a citron colour, dotted, but not pierced with holes; about six inches broad, and two thick; of a softish substance which when cut through very much resembles Rhubarb. From* Ray.

Fungus arboreus major aureus, nulla membrana superne tectus. *Ray's Syn.* 22.

On trees.

Obs. *This and the following will probably prove to be true species of Spunk.*

SPUNK? Dotted
BOLETUS? 12 Punctatus Ray

Divided into lobes; dotted: lobes of a pale red. From *Ray.*

Fungus arboreus lobis rubellis, diversi modo figuratis et punctatis. *Ray's Syn.* 23.

On the stumps of old Elms.

SPUNK? Three-lobed
BOLETUS? 13 Trilobatus Ray

Divided into three lobes; white; small. From *Ray.*

Fungus albus minimus trilobatus, sine pediculo, foliis quercinis adnascens. *Ray's Syn.* 22.

On Oak leaves in Madingly Wood near Cambridge.

Obs. *Very uncertain to what genus it belongs. Perhaps it is a* Turbantop.

Obs. *Ray mentions a large wet spongy substance exsuding from the trunks of Ash, which he calls the*

Fungus spongiosus maximus aqueus, e Fraxinorum truncis exsudans. *Ray's Syn.* 17.

Observed at Rocliff near York, and in Surry. Probably only a species of the Spunk in an imperfect state.

He mentions too a hard fungous substance which grows on Birch and Elm called *Birch-ball* in the North, and *Swan-ball* in Surry, which the boys play with instead of balls. It is what John Bauhine calls the

Fungus solidus ex Betula & quandoque Ulmo. *Ray's Syn. Ind. Plant. Dub.*

Probably only a variety of the first species.

** * Standing on a pillar.*

Shell 14 Testaceus *Ray* — SPUNK. Hat roundish; pores oblong, shallow. Pillar short; fixed to the edge of the hat,—*thick and tough. Hat large: of a reddish brown. Two or three spring from the same root. From Ray.*

Round — 1. Fungus maximus arboreus porosus, pediculo limbo affixo. *Ray's Syn* 11. n. 4.

Fungus angulosus pediculo exiguo. *Bauh. pin.* 370. *Ray's Syn.* 11. n. 5.

Circular — 2. Hat nearly orbicular, *Ray. ib.*

At the roots of decayed trees, Elms, Willows, &c. (2) On a tree near Charlton.

Perennial? 15 Perennis? — SPUNK. Perennial: hat nearly flat on both sides. *Lin.* —*Thin and very flexible; broad, of a black reddish colour.* Pores *extremely fine, scarce perceptible.* Pillar *very short. Ray.*

Fungus coriaceus, pileolo latissimo atro-rubente, pediculo brevissimo. *Ray's Syn.* 11.

Yellow 16 Luteus — SPUNK. Hat cushion-like; somewhat clammy. Pores rounded at the angles. Convex; bright yellow. Pillar whitish;—*about two inches high.* Hat *convex; fleshy; from half an inch to two inches in diameter; sometimes of a yellowish brown, or livid colour.* Pores *larger than those of the following species.*

Boletus luteus. *Ray's Syn.* 10.

In woods.

Brown 17 Bovinus — SPUNK. Hat smooth; cushion-like; edged with a border. Pores compound, angular; little pores angular, shorter.—*Hat convex fleshy; dusky yellow; sometimes reddish brown.* Pores *grey, or greenish yellow.* Pillar *smooth, pale, or reddish brown.*

Fungus porosus crassus. *Ray's Syn.* 11.

In woods. August—September.

It is eaten by the Italians. Cows eat, but it is said to make their milk nauseous.

433 PRICKLYCAP. 1211 Hydnum.

Fungus horizontal, with awl-shaped fibres on the under surface, like a Hedgehog.

Common Imbricatum — PRICKLYCAP. Hat convex, tiled, standing on a pillar,—*Pale flesh-colour.* Prickles *white* Pillar *smooth.*

Fungus pæne candidus, prona parte erinaceus. *Ray's Syn.* 11.

In woods.

It is eaten in Italy, and is said to be of a very delicate taste.

PRICK-

PRICKLYCAP? Five ftringed
HYDNUM? Quinquenerve
Hat ftanding on a pillar, with five fibres extending from the pillar to the edge of the hat.—*Pillar tall; flender; of a pale red. From* Ray. *Ray.*

On dry Ivy leaves. *Ray's Syn.* 12. under No. 11.

Obs. *This fpecies feems to form the connecting link between the* Mufhroom *and the* Pricklycap. *If future obfervation fhould confirm Ray's defcription, will it not form a new genus, and might it not with propriety be ftiled the* Penteneuros *or* Stringcap?

434 MORELL. 1212. Phallus.

Fungus fmooth underneath, with a fort of network on the upper furface.

MORELL. Hat egg-fhaped; full of cells. Pillar naked, Efculent
wrinkled.— Efculentus

Fungus favaginofus. *Park.* 1317. *Ray's Syn.* 11.

Fungus efculentus 1, feu porofus. *Bauh. pin.* 370.

Woods, groves, meadows, paftures, and about hedges, and dry ditchbanks. March—April.

MORELL. Hat full of cells; pillar with a cap.—*White.* Stinking
The Roots *confift of very long white threads, fpreading under ground,* Impudicus
and matted together. On thefe grow little balls which Ray calls Volvæ.

Phallus hollandicus. *Park.* 1322.

Fungus phalloides. *Ray's Syn.* 12.

Fungus fœtidus penis imaginem referens. *Bauh. pin.* 374.

Fungus virilis penis arrecti facie. *Gerard.* 1385.

Stinkhorn. Stote.

Woods, hedges, and dry ditchbanks. July—Auguft.

It is fo putrid in autumn, when ripe, that the Common *Flefh Fly* devours it with great avidity.

435 TURBANTOP. 1214 Helvella.

Fungus fmooth both above and underneath. *Gen. Plant.*
Fungus turban-fhaped. *Syft. Nat.*

* *With a Pillar.*

TURBANTOP. Hat bent down; growing to the pillar; di- Curled
vided into lobes of different fhapes,—*Plaited; white; mem-* 1
branaceous; fcarce broader than the pillar. Pillar *white; nearly as* Mitra
thick as one's finger, and two or three inches long; fcored; hollow, and full of cracks.

Fungus terreftris pediculo ftriato et cavernofo, capitello plicatili fubtus plano. *Ray's Syn.* 8.

On rotten trunks of trees. October.

Purple 2 Purpurea *Ray*

TURBANTOP?
HELVELLA?

Pillar cylindrical; hat small; surface uniform; without either gills or pores. *Purple. The whole plant appears to consist only of a single pillar. From* Ray.

Fungus fontanus, purpureus elegans. *Ray's Syn.* 12.

Grows on grass floating in the water, in springs. A circumstance which seems peculiar. There is however another *aquatic* Fungus; See the *Black-stalked Mushroom.*

White 3 Candida *Ray*

TURBANTOP?
HELVELLA?

Hat very small, about a quarter of an inch in diameter; of a clear white; without the least appearance of gills—*About an inch high. Said to have some affinity to the* Fivestringed Pricklycap. *From* Ray.

Fungus minimus candidus absque lamellis. *Ray's Syn.* 12.

In the inside of hollow Oaks, near the bottom where they are moist.

* * *Without a pillar.*

Flaky 4 Scindalina *Ray*

TURBANTOP?
HELVETTA?

Convex, smooth without; substance within consisting of a number of concentric layers composed of a number of minute tubes or threads—*pointing from the center. Substance hard; covered with a thin bark of a brownish black, somewhat wrinkled and rather glossy, grey within. Not growing horizontal, but adhering close to the substance on which it grows.*

Fungus fraxineus niger, durus, orbiculatus. *Ray's Syn.* 16?

On Ash trees when rotten or in a decaying state, and observed on no other tree. *Ray.* In Worcestershire. S.

Fingered 5 Digitata *Ray.*

TURBANTOP. Long and narrow; smooth; elegantly variegated with yellow and scarlet interspersed with black at the sides,—*Ten or more grow together, as if springing from the same root, from three to six inches in length, and from* 1 *to* 3 *inches broad. Pretty thick, with an upper and a lower surface, which distinguishes it from the* Clubtop *to which from its form one might be inclined to refer it. Smooth, without any appearance of pores or gills. From* Ray.

Agaricus digitatus maximus, ex luteo coccinneo et nigro colore eleganter variegatus. *Ray's Syn.* 21.

Fungus 4. *Park.* 1321?

Found at the root of a Yew tree near Boxhill in May.

TURBANTOP. Woolly above; ſmooth underneath.—*Thin; of a leatherlike ſubſtance; horizontal; from half an inch to an inch in breadth, growing one over another. Margin lobed; bent inwards; moſtly of a browniſh yellow, but ſometimes with a tinge of purple or variegated above with white and yellow, or purple and white. Under ſurface perfectly ſmooth; without the leaſt appearance of either gills or pores. From* Ray. Agaric 6 Agaricus *Ray*

1. Agaricus villoſus, tenuis, inferne lævis, *Ray's Syn.* 21. *Plain*

2. Margin jagged, curled inwards.—*In this variety I have obſerved the under ſurface covered with a fine ſoft purpliſh bloom like that upon the Plumb, which it is not improbable may be the flowers or ſeed.* *Curled*

Obs. *This ſpecies bears the ſame affinity to the* Turbantop *as the other Agarics to the* Muſhroom *and* Spunk.

(1) On trees and wood. (2) In woods at the roots of trees.

TURBANTOP. Wrinkled; indented; ſmooth underneath; of a violet colour,—*adhering flat to the ſubſtance on which it grows. Subſtance leathery but gelatinous. Smell not diſagreeable, ſomewhat reſembling that of the* Morell. Violet 7 Violacea *Ray*

Agaricus meſentericus violacei coloris. *Ray's Syn.* 22.

On rotten wood and trunks of trees.

TURBANTOP. Wrinkled; membranaceous; expanding; margin edged all round with a ſoft wool underneath;—*Of a ſubſtance ſimilar to that of the* Jew's Ear Funneltop, *but the leaf is expanded and not formed into little wrinkled cups like that. From* Ray. Wrinkled 8 Corrugata *Ray.*

Fungus membranaceus expanſus. *Ray's Syn.* 18.

In the garden belonging to the ſociety of apothecaries of London. *Ray.*

TURBANTOP?
HELVELLA?

A fungous ſubſtance, compared to the loaves which the modern Jews eat at the paſſover. From *Ray.* Jew's loaf 9 Collyrica *Ray.*

Fungus collyricus in putreſcento Salice natus. *Ray's Syn.* 19.

On a rotten Willow.

TURBANTOP?
HELVELLA?

Saffron Crocei *Ray* Waved; membranaceous; of a jelly-like ſubſtance; pale yellow, or ſaffron colour.—*Form irregular, ſomewhat reſembling the kernel of a Walnut; about a line in thickneſs. Stains the hands of a ſaffron colour.*

Agaricus membranaceus ſinuoſus, ſubſtantia gelatinæ. *Ray's Syn.* 21.

Agaricus putridus arborum ramis inherens, plurimis ſimul cohærentibus. *Bauh. pin.* 372.

Agaricus parvus lamellatus croceus, e Corylorum ramulis dependens. *Ray's Syn.* 25.

On rotten wood, and on dead ſticks and branches of trees, particularly Hazel.

Obs. *It ſeems to have an affinity to the* Jew's Ear Funneltop. *Ray deſcribes it in one place as having gills, but they are probably no more than wrinkles in the ſubſtance of the plant.*

Obs. *On the* Turbantop.

Ought not the limits of this genus to be extended ſo as to comprehend all thoſe Funguſſes which have an upper and a lower ſurface, and the lower ſurface ſmooth? This would comprehend that part of the Agarics which are excluded from the agarics of Linnæus and the Boletus. In a word cught not the Turbantop to be defined "Fungus horizontal; ſmooth underneath"——? The network ſurface of the Phallus and the open network of the Clathrus will ſufficiently diſtinguiſh them.

436 FUNNELTOP. 1215 Peziza.

Fungus bell-ſhaped. *Seeds* round, convex, or flat.

* *Purple Starjelly. Lentil Starjelly.*

Black I Lentiſera FUNNELTOP. Bell-ſhaped; containing a number of ſmall flat ſeed-like ſubſtances—*aſh coloured. In its younger ſtate the mouth of the cup is contracted, incloſing the ſeeds ſurrounded with a white mucilage. When it opens the ſeeds appear diſengaged from their mucilage, each connected by a fine thread to the bottom of the cup.*

Common 1. Fungi calyciformes ſeminiferi. *Ray's Syn.* 20.

Fungus minimus ligneis tabellis areolanum hortorum adnaſcens. *Bauh. pin* 374.

Corn-bells. Worceſterſhire.

Scored Ray. 2. Cup ſcored within; rough without; longer and narrower.

Fungus ſeminifer externe hirſutus, interne ſtriatus. *Ray's Syn.* 20.

(1) On old wood, in plowed fields, and gardens; in barley ſtubble growing on the rotten ſtalks, whence called Corn-bells. (2) In Gardens.

FUN-

FUNNELTOP. Betwixt globular and bell-ſhaped; con- Veiled
taining a number of ſmall, flat ſeed-like ſubſtances. Mouth in 2
its younger ſtate covered with a thin membrane.—*Seeds adhering* Calyptrata
to the cup by a fine thread. Smaller than the former ſpecies. From *Ray.*
Ray.

Fungus feminifer minor, fere hemiſphæricus. *Ray's Syn.* 2. t. 1. f. 2. b. and c.

In gardens, and on rotten wood.

FUNNELTOP. Bell-ſhaped; without any ſeed-like ſub- Seedleſs
ſtances;—*at leaſt none could be diſcovered in ſeveral ſpecimens exa-* 3
mined for that purpoſe. Pretty thick, of a blackiſh colour, ſcored Aſpermia
when dry. About three parts of an inch long, ſupported by a ſhort pillar. From *Ray.*

Fungus minor calyciformis, vernus, craſſior, nigricans. *Ray's Syn.* 20. Spring.

FUNNELTOP. Turban-ſhaped; lopped; ſurface dotted; Spotted
Center clear white, with black riſing dots. Pillar *very ſhort. The* 4
Seeds *are contained in the pores, and are thrown out with a ſpring.* Punctata

Fungus minimus infundibuliformis, ſuperne nigris punctis notatis. *Ray's Syn.* 12.

On horſe and cow-dung, and obſerved no where elſe.

FUNNELTOP. Funnel-ſhaped: ſurface expanding; indent- Cornucopia
ed; dotted—*ſtanding on a pillar; dark grey.* 5
Peziza tubæ fallopianæ æmula. *Ray's Syn.* 20. Cornucopoides
In woods.

FUNNELTOP. Glaſs-ſhaped: angular on the out-ſide; Cup
with branching veins—*of a dirty aſh colour without; browniſh* 6
within; ſtanding on a pillar. Subſtance brittle. Acetabulum

Peziza ſubfuſca major. *Ray's Syn.* 17.

On rotten wood, *Lin.* and on hot-beds, *Ray.*

OBS. *It ſeems doubtful whether this ſpecies of Ray's be the true Acetabulum of Linnæus. The Acetabulum of Linnæus has a pillar; Ray deſcribes his ſpecies as having none. Poſſibly it is only a variety of the* Agaricus Chantarellus *or Champignon* MUSHROOM.

FUNNELTOP. Glaſs-ſhaped; of a cheſnut or amber colour; Amber
about an inch broad; of a thickiſh ſubſtance, brittle; ſome- 7
what tranſparent.—*Pillar about an inch, or an inch and a half* Succinea
high, ſolid. From *Ray.* *Ray.*

1. Peziza acetabuliformis ſubfuſca. *Ray's Syn.* 19. *Brown*

2 ? Cup of a ruſt colour within. *Ruſty*

Fungus arboreus pyxidatus coloris intus ferruginei. *Ray's Syn.* 20.

(1.) On the ground and on rotten wood. Spring. (2.) On trees.

Goblet 8 Amphoralis *Ray* FUNNELTOP. Glaſs-ſhaped ; contracted upwards : about four inches broad. *Pillar* ſometimes near half a foot high. Some of the larger ones it is ſaid will hold more than half a pint. From *Ray*.

Fungus maximus pileolo pyxidato. *Ray's Syn.* 19.

Near Cambridge.

Obs. *Is not this the Fungoides maximum pyxidatum of* Vaillant *which Linnæus gives as a ſynonym of the* Acetabulum, *and conſequently is it not the real* Acetabulum ?

Glaſs-ſhaped 9 Pyxidata *Ray* FUNNELTOP. Glaſs-ſhaped ; of a pale livid colour. From *Ray*.

Fungi Pezizæ Plinii. *Ray's Syn.* 17.

Fungus noxius 5. ſeu acetabulorum modo cavus, radice carens. *Bauh. pin.* 372.

On the ground in gardens, and ſhady lanes.

Scarlet 10 Coccinea *Ray* FUNNELTOP. Glaſs-ſhaped ; border upright ; ſmooth. Beautiful crimſon or ſcarlet within, pale red without—*ſometimes purple*.

Common 1. Cup entire ; ſitting.

Fungus membranaceus ſeu coriaceus, acetabuli modo concavus, colore intus coccineo ſeu cremeſinos ſaturo. *Ray's Syn.* 18.

Stalked ? 2. Cup entire ; ſupported by a pillar—*from half an inch to an inch in diameter, margin ſmooth and even.* Pillar *from half an inch to an inch high.*

Peziza acetabuliformis, coccinei intus coloris. *Ray's Syn.* 19.

Curled ? 3. Cup ſupported by a pillar ; curled ; wrinkled, and ſometimes jagged ;—*pretty much reſembling the* Common, *but a good deal larger.*

Fungus membranaceus acetabuli modo cavus, coccineus criſpatus. *Ray's Syn.* 19.

(1.) On rotten ſticks in woods and hedge banks. Spring. (2.) On rotten wood. Spring.

Sweet ? 4. Round ; ſcarlet ; ſweet-ſcented.

Fungus rotundus ſcarlatinus odoratus. *Ray's Syn. Ind. Plan. dub.*

(1.) On rotten ſticks, in woods and in hedge banks. Spring. (2.) On rotten wood. Spring. (3.) At Church Lench in Worceſterſhire. *Ray*.

Hairbrimmed 11 Piloſa *Ray* FUNNELTOP. Glaſs-ſhaped ; margin edged with hairs. —*Scarlet.* Cup *ſupported by a pillar ; about half an inch in diameter, with black ſtiff hairs on the brim.* From *Ray*.

Peziza acetabuliformis coccinea marginibus piloſis. *Ray's Syn.* 19.

On rotten oaks in Kilwarlin near Hilſborough in Ireland. June.

FUN-

FUNNELTOP. Glafs-fhaped ; jagged ; fegments refembling leaves ; edges curled inwards—*of a deep orange colour. Subftance of the* Cup *pretty thick, but brittle.* From *Ray.* Leafy 12 Foliacea Ray

Peziza miniata major. *Ray's Syn.* 17.

At the roots of trees, lying on the ground, in Middleton Park, Warwickfhire.

FUNNELTOP. Glafs-fhaped ; border blunt ; upright—*Small, yellow.* Pillar *very fhort.* Smooth 13 Cyathoides

Peziza lutea parva, marginibus lævibus. *Ray's Syn.* 18. t. 24. f. 4.

In woods. Dec.

FUNNELTOP. Flat ; border convex ; hairy.—*Yellow.* Hairy 14 Scutellata

1. Peziza lutea parva, marginibus pilofis. *Ray's Syn.* 18. t. 24. f. 3. *Larger*
2. Smaller, hollowed like a faucer at the top : of an orange colour. *Smaller?* *Ray*

Peziza miniata minor. *Ray's Syn.* 18.

Obs. *Dillenius (in Ray's Syn.) makes this variety a different fpecies from the former, probably from not having obferved any hairs on the margin, but had he examined them with a microfcope, it is likely he would have found them.—There is however a fmall fpecies of the* Funneltop *of a bright yellow, or orange, feldom larger than a pin's head ; that appears on cow-dung in the fpring, which if examined through the microfcope, appear to be furrounded at the margin with a number of ftraight ftiff upright yellow hairs, and correfponding exactly with the* Scutellata *of Linnæus. This I take to be the latter variety, and that it differs from the former only in fize.*

( ) On rotten wood. (2) On old cow-dung. Spr. Aut.

FUNNELTOP. Flat ; oval ; fmooth—*Not hollowed in the leaft ; of a faffron colour ; fmall, like Lentil feeds.* From *Ray.* Saffron 15 Crocea *Ray.*

Peziza lenticularis parva miniata. *Ray's Syn.* 18.

On cow-dung, and on Hampftead-heath in a fandy foil. Aut.

FUNNELTOP. Concave ; wrinkled ; fhaped like an ear, —*greenifh afh colour ; covered with fhort down ; wrinkled into little cups* Jews ear 16 Auricula

Peziza auriculam referens. *Ray's Syn.* 18.

Fungus membranaceus auriculam referens, five fambucinus. *Baub. pin.* 372.

Fungus fambucinus, feu auricula judæ. *Gerard.* 1385. *Park.* 1320.

Tremella auricula. *Hud.* 464.

On rotten wood.

437 CLUBTOP. 1216 Clavaria.

FUNGUS perpendicular; smooth; oblong; of one uniform surface.

Simple — ** Undivided.*

Pistillaris — CLUBTOP. Club-shaped; undivided;—*yellowish or whitish.*

Common — 1. Fungoides clavatum minus. *Ray's Syn.* 14.

Smaller ? Ray — 2. Slenderer, and somewhat longer.
Fungus clavatus minimus *Ray's Syn.* 14.

Sharp pointed ? *Ray* — 3. Curved and bent inwards; terminating in a sharp point; of a citron colour.

Fungoides clavatum incurvum inacutum mucronum productum. *Ray's Syn.* 14.

(1) Woods, heaths, and pastures. 2. In woods near Badmington in Gloucestershire. Oct. (3) Hamstead-heath.

Black
2
Ophioglossoides — CLUBTOP. Club-shaped; undivided; compressed; blunt—*black.*

Fungus ophioglossoides. *Ray's Syn.* 14. In woods

Pile
2
Sublicia
Ray — CLUBTOP. Nearly cylindrical, ends blunt and finely notched—*Rising from a flat smooth crust; from a quarter to half an inch high, like piles or little pillars, some single, others growing several together.* Crust *as well as pillars* of a *whitish livid colour; of a fungus substance.* From *Ray.*

Fungoides humile ex albo livescens, apicibus tenuissime crenatis. *Ray's Syn.* 14. t. 1. f. 4.

On rotten boards. Nov. Dec.

Toothed
4
Denticulata
Ray — CLUBTOP? CLAVARIA?

Small toothed Mushroom of *Parkinson.* From *Ray.*

Fungus parvus denticulatus. *Park.* 1321. *Ray's Syn.* 14.

In pastures, springing out of dry horse or cow-dung in Warwickshire.

** * Branched.*

Fingered
5
Digitata — CLUBTOP. Branched; woody: black.—

Fungus piperi æthiopico similis, vel digitatum niger. *Ray's Syn.* 14.

On old planks, and half rotten wood.

Flat
6
Hypoxylon — CLUBTOP. Between branched, and horned; compressed—*black, covered with a kind of wool, and the ends with a white mealy dust, which is probably the* Flowers *or* Seed.

Common — 1. Fungus ramosus, niger, compressus, parvus, apicibus albidis. *Ray's Syn.* 15.

Reindeer — 2. Branched like the horns of a Rein-deer.
Fungus cornu dorcadis facie. *Hudson* 501.

Awl-shaped *Ray* — 3. Ends of the branches somewhat cylindrical, and awl-shaped; sometimes cloven. From *Ray.*

Fungus niger subularis, apicibus albidis. *Ray's Syn.* 15.

On rotten wood; in cellars, ships, &c. and at the roots of trees.

CLUB-

CLUBTOP. Branched; winding in various directions,—*and insinuating itself between the bark and wood of trees, sometimes spreading to at least two feet in length, and as much in breath; black without; white within, near half an inch broad.* Branches *flatted, thin, sometimes running into one another.* From *Ray*. Winding 7 Insinuans *Ray*

1. Fungus niger compressus, varie divaricatus et implexus inter lignum et corticem. *Ray's Syn.* 15. *Flat*
2. Branches round and brittle; not running into one another, but winding in the same manner amongst stalks and dead leaves. From *Ray*. *Round ?*

(1) On a dead Elm in St. James's-Park. (2) From Ireland.

CLUBTOP. Branched; thread-like—*about the thickness of small pack-thread; black; of the same substance with the two preceding, sometimes growing three or four inches long.* From *Ray*. Thread 8 Filiformis *Ray*

1. Fungus tenuis niger ramosus. *Ray's Syn.* 15. *Large*
2 ? Hair-like; very small. *Small*

Fungus niger minimus ramosus capillaceus. *Ray's Syn.* 15.

In vaults. Obs. *Both this and the preceding species may possibly prove only varieties of the 6th species.*

CLUBTOP. Branches crowded; very much divided; unequal.— Yellow 9 Coralloides

Fungus ramosus flavus et albidus. *Ray's Syn.* 16.

Fungus ramosus et imperati. *Bauh. pin.* 371.

Fungus corallinus ad antiquarum arborum radices. *Ray's Syn. Ind. Plant, dub.*

Heaths and woods. The latter in the woods near Petersfield in Hampshire.

CLUBTOP. Very much branched; ends of the branches mostly forked;—*about three or four inches high.* Stem *not forked, but subdividing into a number of branches at every division; about the breath of one's hand high. Of a dirty yellow.* From *Ray*. Forked 10 Furcata *Ray*

Fungus ramosus minor, colore sordide flavicante. *Ray's Syn.* 16.

CLUBTOP. Very much branched; branches terminating in round irregular heads, uniting together, and forming little knobs of a greenish yellow—*resembling those of Cauli-flower. It sometimes weighs two or three pounds.* From *Ray*. Cauliflower 11 Caulifloræ *Ray*

Fungoides ramosum maximum, Brassicæ cauliflloræ facie et magnitudine. *Ray's Syn.* 16.

In meadows.

CLUBTOP. Branches crowded; very much divided; of an equal height blunt; yellow— Stinking 12 Fastigiata

Fungoides coralliforme luteum fætidum et minus ramosum. *Ray's Syn.* 479. t. 24. f. 5.

In pastures and woods. Aut.

CLUB-

Pointed 13 Muſcoides

CLUBTOP. Branches divided; tapering to a point at the ends; unequal; yellow.—

Fungus parvus luteus ramoſus. *Ray's Syn.* 16. t. 24. f. 7.

Paſtures, woods and heaths.

Beech 14 Faginea *Ray*

CLUBTOP. Branches divided; of an orange colour.—*Very ſmall.* From *Ray.*

Fungus ramoſus minimus coloris aurantii. *Ray's Syn.* 16.

On the bark of Beech; very common in England.

White 15 Candida *Ray*

CLUBTOP. Branches divided, of a clear white.—*Small.* From Ray.

Fungus ramoſus candidiſſimus ceranoides, ſeu digitatus minimus. *Ray's Syn.* 16.

Fungus digitatus. *Park.* 1318?

In paſtures, Staffordſhire. Aut.

* * * *Obſcure ſpecies.*

Oak leather 16 Hæmatodes *Ray*

CLUBTOP? CLAVARIA?

Broad; flat, of a ſoft leather-like ſubſtance,—*exactly reſembling tanned leather, except that it is thinner and ſofter. Of no determinate form.*

Fungus coriaceus quercinus hæmatodes. *Ray's Syn.* 25.

Grows in the clefts and hollows of old oaks in Ireland, and in ſeveral parts of England; ſometimes on aſh.—It is made uſe of in Ireland to dreſs ulcers, and in Virginia to ſpread plaiſters on inſtead of leather.

OBS. *This remarkable ſubſtance ſeems to bear ſome reſemblance to the* Winding Clubtop *in its mode of growth, and to the Agarics in colour and ſubſtance. Its uniform ſurface however ſeems to refer it to the former.—There are ſometimes little round hard tubercles within the ſubſtance of the plant, which may be felt by the hand, but do not appear on the ſurface. Theſe may poſſibly be the* Fruit *or* Seed-veſſels. *If the Naturaliſts would examine them they may poſſibly lead to the true arrangement of this very remarkable vegetable.*

438 PUFFBALL. 1217 Lycoperdon.

FUNGUS roundiſh; opening at the top; full of powdery impalpable *Seeds.*

* *Solid; growing under ground; without a root.*

Truffle 1 Tuber

PUFFBALL. Globular; ſolid; beſet with a number of pointed prominences; without a root.—*about the ſize of a prune; white, full of a number of veins or winding cells within, with a ſmall quantity of brown powder in the center. Does not open.*

1. Tubera

1. Tubera. *Bauh. pin.* 376. *Ray's Syn.* 28. *Common*
Tubera terræ. *Gerard.* 1385.
Tubera terræ edulia. *Park.* 1319.
Trubs, or Truffles.
2. ? Small; about the ſize of a nut: purple. *Ray's Syn.* 28. *Purple*
3. ? Large; fringed; ſomewhat in form of a cup. *Ray's Syn.* Ray
28. *Cup*

(1.) Beneath the ſurface of the ground in parks, &c. (2) In Ray
Hampton-court park. (3) From Lancaſhire.

This is one of the eſculent Funguſſes, and one of the beſt of them. Dogs are taught to hunt it, and when they ſcent it they bark a little and begin to ſcratch up the earth. Pigs likewiſe in Italy root it up, and an attendant takes it from them.

PUFFBALL. Globular; rather ſolid, powdery in the center. Without a root.—*Skin thick; hard; warty. The inner ſubſtance of a purpliſh white. Powder black. Does not open.* Branny 2 Cervinum

Tubera perniciosa terreſtria, ſeu cervina. *Ray's Syn.* 28.
Tubera cervina. *Bauh. pin.* 376. *Park.* 1319, 1320.
In woods. Oct.

Obs. *Linnæus arranges it as* growing under ground, *and yet Dillenius in Ray's Syn. has not placed it amongſt the* Fungi ſubterranei.

* * *Powdery; growing above ground; with roots.*

PUFFBALL. Nearly globular; opening with a rent; powder greeniſh—*ſomewhat flatted at the top; about the ſize of a pigeon egg; white; ſprinkled with a kind of yellowiſh meal, and when ripe full of a very fine powder of a dirty yellow or greeniſh colour. Skin livid when ripe.* Common 3 Vulgare *Ray*

Crepitus Lupi ſive Fungus ovatus. *Park.* 1323. *Ray's Syn.* 26.
Fungus rotundus orbicularis. *Bauh. pin.* 374.
Fungus tertius ſeu orbicularis. *Gerard.* 1385.
Puff-balls, Duſty Muſhrooms, Puckefiſts, Bullfiſts, Puſſeballs and Fuſſeballs.
In barren dry paſtures. Aut. It is recommended by ſome as a ſtyptic.

PUFFBALL. Nearly globular, opening with a rent,—*powder black. Sometimes as large, and even larger than a man's head; of a pretty firm ſubſtance; whilſt growing, of a greyiſh white, and weighing ſeveral pounds, but when ripe it turns of a brown colour, and becomes exceedingly light; ſo that one larger than a man's head will not weigh an ounce.* Bullfiſt 4 Boviſta

Fungus maximus rotundus pulverulentus, dictus Germanis Boſiſt. *Ray's Syn.* 26.
Fungi rotundis orbicularis 2 ſpecies. *Bauh, pin.* 371.
Bunt, Puckefiſt, Frogcheeſe.
In rich paſtures and on dunghills.

The

The fumes of this when burnt have a narcotic quality, and on this account it is ſometimes made uſe of to take a hive without deſtroying the bees. This too as well as the former is ſometimes uſed as a ſtyptic.

Orbicular 5 Orbiculare *Ray*

PUFFBALL. Round and flatted, quite full of duſt. *From Ray.*

Fungus orbicularis per totum pulvere repletus. *Ray's Syn.* 26.

Cup 6 Calyciforme *Ray*

PUFFBALL. Flatted; ſomewhat in form of a cup; large; black; duſty only towards the top *From Ray.*

Fungus Lupi crepitus dictus in ſummitate ſolum pulverulentus. *Ray's Syn.* 26.

Fungus niger calcyis figuram referens. *Bauh. pin.* 375.

Pear-headed 7 Pyriforme *Ray*

PUFFBALL. Head in form of a pear, aſh-coloured. *From Ray.*

Fungus pyriformis. *Ray's Syn.* 27.

In old paſtures not far from Yarmouth, and on the north of the Iſle of Wight.

Short ſtalked 8 Compreſſum *Ray*

PUFFBALL. Head ſupported by a pillar; ſomewhat flatted; quite full of duſt; ſkin roughiſh. thick and tough; pillar not more than half an inch long. *From Ray.*

Fungus pulverulentus compreſſus pediculatus, cortice craſſiore. *Ray's Syn.* 27.

Spongy 9 Spongioſum *Ray*

PUFFBALL. Head of a ſpongy ſubſtance within; covered with a membranaceous ſkin: ſupported by a pillar. Pillar ſhort, thick, extending to the edge of the head. *From Ray.*

Fungus pulverulentus cute membranacea, ſubſtantia intus ſpongioſa, pediculo brevi craſſiore in oras ſere ducto. *Ray's Syn.* 27.

Obſerved near Packington, and Alrewas Hays in Staffordſhire, near the deep ſpring. *Ray.*

Thick-ſtalked 10 Stipitatum *Ray*

PUFFBALL. Head ſupported by a pillar; pillar pretty long; diſtended. *From Ray.* Pillar thickeſt in the middle; about ſix inches long.

Fungus pulverulentus, Crepitus Lupi dictus major, pediculo longiore ventricoſo. *Ray's Syn.* 27.

In the north of Ireland, in orchards. *Ray.*

PUFFBALL. Head ſupported by a pillar. Pillar pretty Rough
long; rough —*While in its growing ſtate, it will ſtick to cloaths* 11
like the flowering heads of Burdock. From Ray. Scabrum
Fungus pulverulentus, Crepitus Lupi dictus, pediculo longi- *Ray*
ori ſcabro. *Ray's Syn.* 27.
Near Waring's Town in the county of Down in Ireland, under pine trees. *Ray.*

PUFFBALL. Pillar long. Head globular; ſmooth: mouth Long-ſtalked
cylindrical; very entire.—*ſmall.* 12
Fungus pulverulentus minimus, pediculo longo inſidens. *Ray's* Pedunculatum
Syn. 27.
In paſtures. March.

PUFFBALL. Cap with many clefts; expanding. Head Star
ſmooth; mouth tapering, plaited.—*Clefts of the cap expanded flat* 13
on the ground in form of a ſtar. Head *in the middle, ſupported by a* Stellatum
ſhort pillar; mouth opening when ripe, dividing into very fine ſegments
Fungus pulverulentus, Crepitus Lupi dictus, coronatus et inferne ſtellatus. *Ray's Syn.* 27. t. 1. f. 1.
In meadows, and paſtures.

PUFFBALL. Cap with many clefts; expanding. Head Cullander
orbicular; pierced with a number of holes like a cullander, or 14
ſtr iner,—*which give a paſſage to the duſt. Clefts of the* cap *of a* Coliforme
thicker ſubſtance than the former, expanding on the ground in form of *Ray*
a ſtar. From Ray.
Fungus pulverulentus coli inſtar perforatus, cum volva ſtellata. *Ray's Syn.* 28.
Found in a lane leading from Crayford to Bexley-Common in Kent, and near Hampton-Court. *Ray.* Sept.

PUFFBALL. *Cap* with four clefts; upright; arched. Turret
Head ſmooth: mouth blunt; fringed. *Hudſon,* 502. 15
Geaſter volvæ radiis et operculo elevatis. *Phil. Tranſ.* No. 474. Fornicatum
p. 234. *Hud.*
In meadows and paſtures.

439 MOULD. 1218 Mucor.

FUNGUS. A roundiſh bladder, containing a number of *Seeds* fixed to croſs-ſhaped *Receptacles.*

* *Permanent.*

MOULD. Pillar thread-ſhaped; black. Head globular, Round-headed
aſh-coloured.—*About a quarter of an inch high.* 1
Fungus pileatus minimus, pediculo tenui capillaceo. *Ray's* Sphærocephalus
Syn. 12.
On rotten wood, and on dead ſticks in hedges.

MOULD.

Dew-drop MOULD. Permanent? pillar thread-ſhaped: *Head* like a
2 dew-drop, marked with a black dot at the point. *From Ray.*
Roridulus Fungus (ex ſtercore equino) capillaceus, capitulo rorido,
Ray nigro punctulo in ſummitate notato. *Ray's Syn.* 13.
On horſe-dung about London.

Grey-headed MOULD. Pillar awl-ſhaped; black. Head lentil-ſhaped,
3 aſh-coloured,—*about as large as a poppy-ſeed.*
Lichenoides Coralloides fungiforme arboreum nigrum vix cruſtoſum. *Dill.* 78. t. 14. f. 3.
On rotten wood, on the bark of old oaks. Aut.—Wint.

Black MOULD. Like a black briſtle, beſet with ſoft brown hairs.
4 —On rotten wood.
Embolus

* * *Of ſhort duration.*

Cotton MOULD. Capſule globular, ſupported by a pillar.
5 Fungus bombycinus murini coloris, e fimo felino, tenuiſſimis
Mucedo capillis. *Ray's Syn.* 13.
On various putrid ſubſtances; ſtale bread, plants, &c.

Yellowiſh MOULD. Like briſtles; ſeeds at the roots—*At firſt white,*
6 *afterwards yellow.* Seeds *egg-ſhaped.*
Leproſus In vaults and caverns.

Greeniſh MOULD. Heads nearly globular, ſupported by a pillar;
7 incorporated.—
Glaucus On apples, melons and other decaying ſubſtances.

Fingered MOULD. On pillars in fingered ſpikes;—*White.* Seeds
8 *round.*
Cruſtaceus On ſeveral ſorts of food when in a decaying ſtate.

Branching? MOULD. Pillar branched, with ſpikes growing by threes.
9 *Lin.*
Cæſpitoſus? Of a ſnowy white, very much branched, ſoft. *Ray.*
Fungus fimoſus, niveus, ramoſiſſimus, mollis. *Ray's Syn.* 13.
In cellars on cat's-dung. Aut. *Ray.*

OMISSIONS

APPENDIX.

OMISSIONS.

SNOWDROP. 401 Galanthus.

EMPAL. *Sheath* oblong, blunt, compreſſed, ſhrivelling; opening at the flat ſide.

BLOSS. *Petals* three, oblong, blunt, concave, looſe, equal; ſtanding open.

Honeycup cylindrical; half as long as the petals; compoſed of three leaves reſembling petals, parallel, blunt, broken at the margin.

CHIVES, *Threads* ſix; hair-like, very ſhort. *Tips* oblong, approaching, tapering to a briſtle-ſhaped point.

POINT. *Seedbud* globular: beneath. *Shaft* thread-ſhaped; longer than the chives. *Summits* ſimple.

S. VESS. *Capſule* nearly globular, with three blunt corners, three cells and three valves.

SEEDS. Many; globular.

SNOWDROP. As there is only one ſpecies known, Linnæus gives no deſcription of it—*Roots bulbous.* Garden Nivalis

Leucojum bulboſum trifolium minus. *Baub. pin.* 56.

Fair Maids of February.

Near Cirenceſter in Glouceſterſhire, and plentifully at the foot of Malvern Hills Worceſterſhire. P. February—April. It is common in gardens and becomes double by cultivation.

PIPEWORT. 100 Eriocaulon.

EMPAL. *Common*, formed by ſeveral circular, concave, membranaceous ſcales; fringed on the upper part.

B. *Barren Florets* in the center; numerous.

Outer; an egg-ſhaped ſcale, fringed at the top, and tapering into a claw at the baſe; placed on the outer ſide of each floret.

Proper Cup of two concave, wedge-ſhaped; fringed leaves.

BLOSS. *Individuals* of one funnel-ſhaped petal; divided into two lips at the mouth, and fringed.

CHIVES. *Threads* four; thread-ſhaped; as long or longer than the bloſſom. *Tips* oblong.

F. *Fertile florets* in a double row round the circumference.

EMPAL. *Outer*; an egg-ſhaped ſcale, fringed at the top and tapering into a claw at the baſe; placed on the outer ſide of each floret.

Proper Cup of two egg-ſhaped concave leaves; fringed at the top and tapering into a claw at the baſe.

BLOSS. *Individuals* of two oblong, concave petals; fringed at the top and on the back; gradually tapering at the baſe into ſlender claws.

POINT. *Seedbud* globular, but compreſſed. *Shaft* ſhort. *Summits* two; thread-ſhaped.

S. VESS. *Capſule* globular, but compreſſed; *Cells* two.

SEEDS. Smooth; dimpled at the end; one in each cell.

Wreathed Septangulare

PIPEWORT. Stalks upright; twiſted; compoſed of ſeven tubes. Leaves nearly upright—*Petals white; with a black ſpot in the middle*. Tips *black*.

In a ſmall lake in the Iſland of Skye. P. September.

This plant was found by Mr. Robertſon in the year 1768; and Doctor Hope, Botanical Profeſſor in the Univerſity of *Edinburgh*, hath given us a deſcription and an engraving of it in the Philoſ. Tranſ. vol. 59. p. 243. The Doctor inclines to think it the *Eriocaulon decangulare* of Linnæus; but it neither agrees with that in the generic or ſpecific characters; and as Linnæus never ſaw a recent ſpecimen of the plant, I ſhould rather truſt to the well known accuracy of Doctor Hope, who had an opportunity of examining it in its recent ſtate. If it is an *Eriocaulon* it can never be the *Decangulare*; for the ſtalks of that are beſet with black ſhining wool, and the leaves which are few in number ſpread upon the ground; nor can it be any other of the ſpecies enumerated by Linnæus, for not to mention that

they

they are all natives of India, they differ from this in the number of angles or tubes in the stalk.

The root is slightly acrimonious when chewed.

At the end of the specific character of the Royal MOONWORT (p. 648.) add—*The first leaves consist only of three or five roundish little leaves, resembling some species of the* HEMIONITIS, *in which state of growth it was figured and described in the octavo edition of Ray's Synopsis as a distinct species, under the name of* Hemionitis pumila trifolia vel quinquefolia maritima. *Ray's Syn.* Ed. 1. p. 26. t. 1. f. 2.

Add the following Synonym from Ray, to the Starjelly FEATHERMOSS (*p.* 691.)

Conferva gelatinosa tenerrima et viridissima, muscum quendam filicifolium repræsentans. *Ray's Syn.* 60.

After the Crab's-eye CUPTHONG (*p.* 708.) add the following species.

CUPTHONG?
LICHEN?

Of a black fungous substance, with saucers. From *Ray*. Fungous *Ray* Fungosus

Fungoides quercinum peltatum nigrum. *Ray's Syn.* 16.

On the trunks and branches of Oaks that have been cut down.

THE

GLOSSARY.

ABORTIVE. See BARREN.

ABRUPT. When a winged leaf ends without a tendril or a little leaf.

ACORN. The seed of the *Oak*.

AIRBAGS. Vessels distended with air; being a sort of seed-vessel: as in some species of the *Oarweed*, and the *Bladder Sena*.

ALTERNATE. As the leaves of *Borrage*. Or *Chequered Daffodil*. *Pl*. 9. *f*. 3. (*d. d. d. d. d.*) *Pl*. 8, *f*. 54.

ANNUAL. Living only one year; as the *Larkspur*.

ANGULAR. *Stem*, &c. having edges or corners: opposed to cylindrical. A stem or stalk may have 1, 2, 3, 4, or more angles or corners. The *White Archangel* hath 4.

——— *Capsule*, *Flower de Luce*, or *Flag*.

APPLE. A species of seed-vessel in which the seeds are contained in a capsule, and this again is enclosed in a fleshy substance; as the *Common Apple* or *Pear*. *Pl*. 5. *fig*. 20.

APPROACHING-*Leaves*: bent inwards towards the stem. *Pl*. 9. *f*. 5. (*a. a.*)

——— *Petals* converging to the center of the flower, as in the *Pæony*. *Globe Flower*.

——— *Threads*; as in *Borrage*.

——— *Tips*; leaning towards each other; as in the *White Archangel* and *ivy-leaved Gill*.

ARROW-SHAPED-*Leaves*; shaped like the head of an Arrow, as the leaves of *Sorrel*; the *Small* or *Great Bindweed*. *Pl*. 7. *f*. 13.

——— *Tips*; as in the *Crocus*. *Elder*.

——— *Props*. *Pea*.

ASCENDING. Growing first horizontally and then bowed upwards. It is applicable either to *Leaves*, to *Stalks*, to *Stems*, as in spiked *Speedwell*, or to Chives as in all the *Speedwells*. *See the Chive next below* (*a*) *in Pl*. 1. *f*. 8.

AWL-SHAPED. Slender, and becoming finer towards the end, like an awl. *Pl*. 7. *f*. 8. *Pl*. 5. *f*. 15. (*a*) *Leaves* of *Rock Stone-crop*.

——— *Threads* of *Crocus*. *Borrage*. *Daffodil*. *Hawthorn*.

——— *Seeds* of *Shepherds Needle*.

AWN. The slender sharp substance growing to the valves of corn or grass, and frequently called a beard. It is remarkable enough in *Oats* and *Barley* It is sometimes used to signify a sharp point terminating a leaf, &c. *Pl*. 2. *f*. 21. (*b. b.*) *f*. 23. (*b. b.*)

BARK. The universal covering of the stems, roots and branches of vegetables. It is generally spoken of as *inner* and *outer*. Blossoms are an expansion of the *inner*, and empalements are a continuation of the *outer* bark.

BARREN *Flowers*; those that produce no perfect seeds. The barren flowers are generally those that have chives, but no pointals. Flowers that have only pointals are sometimes barren, owing to the absence of other flowers that have chives. In the Rundled flowers (Class V. Order II.) it is not uncommon to have several of the florets barren, though they are furnished both with chives and pointals; perhaps owing to some imperfection in the pointals; but future observation must determine this matter.

BASE. That part of a leaf, &c. nearest to the branch or stem.

- - - - *of the leaves or branches.* Flowers or fruit-stalks are often said to grow at the *base* of the leaves, or the branches; that is when they are placed at the bottom of a leaf or branch, and on the inner side, where it joins to the stem. Pl. 9. f. 5. (*m.*) the fruit-stalks of the *Common Pimpernel*; the *Great Periwinkle* and the *Flowers* of the *Common Calamint* are examples.

BEADED. Consisting of many little knobs connected by small strings. As the roots of the *White Saxifrage*.

BEARDED. Beset with straight parallel hairs.

BELL-SHAPED. The idea this term is intended to convey cannot well be mistaken; examples of it occur in the *Cup* of the *Cherry*; in the *Blossoms* of the *Convolvulus* or *Lilly of the Valley*; and in the *Honey-cup* of the *Wild Daffodil.* Pl. 5. f. 1. (*a.*) Pl. 4. f. 2. 3. 4. 5.

BENEATH. A BLOSSOM is said to be *beneath* when it includes the seed-bud and is attached to the part immediately below it, as the blossom of *Sage*; *Borrage*; *Convolvulus. Polyanthus*.

- - - - A SEEDBUD is said to be *beneath* when it is placed below the attachment of the blossom and therefore not included within it; as in the *Honey-suckle*; *Currant*; *Haw-thorn.*

BERRY. A pulpy seed-vessel without valves; in which the seeds are naked, as in the *Goosberry* or *Elderberry*. Pl. 5. f. 19.

BIENNIAL plants or roots; are those which continue alive two years.

BIRDSFOOT. Bearing some resemblance to the feet of land-fowl; as the leaves of the *Passion Flower*, or the seed-vessel of the *Birds-foot Trefoil.* Pl. 7. f. 49.

BITTEN. Not tapering to a point, or ending in any even regular form, but appearing as if bitten off; as in the root of *Devil's-bit*; and the petals of common and *Marshmallows*. Pl. 7. f. 18.

BILL. A long awl-shaped substance resembling the bill of a Wood-cock; as in *Shepherd's-Needle* or *Crane's-bill.* Pl. 5. f. 15.

BLADDER-SHAPED. Distended like a blown bladder; as is the cup of the *Bladder Campion*, and the blossom of the *Figwort.*

BLISTERED. When the surface of a leaf rises high above the veins, so as to appear like blisters.

BLOSSOM. One of the parts of a flower. It may consist of one or more *Petals*; and is formed by an expansion of the inner bark of the plant. Pl. 4.

BLUNT.

BLUNT. Oppoſed to *ſharp*; as the leaves of the *Spiked Speedwell*; the cup of the *Convolvulus* and the Capſule of the *Yellow Rattle*.

BOATSHAPED. Like a little keel-bottomed boat; as are the valves of the ſeed-veſſels of the *Woad* and the *Mithridate*. Pl. 5. f. 13.

BORDER. The upper ſpreading part of a bloſſom of one Petal; as in the *Primroſe* and *Auricula*. It is ſometimes uſed to ſignify the thin membranaceous part of a ſeed or ſeedveſſel. Pl. 4. f. 1. (*b. b.*)

BRISTLES. Strong ſtiff cylindrical hairs.

BRISTLE-SHAPED. Slender, and nearly cylindrical; as the ſtraw of the *leaſt Bullruſh*; the leaves and props of the *Aſparagus*.

BROAD-TOPPED-SPIKE. A ſpike wherein each of the flowers ſtand elevated upon its proper fruit-ſtalk ſo that they all riſe to nearly the ſame height.

Exemplified in the *Pear* and the *Common Bethlem Star*. Pl. 6. f. 7.

BUD. A protuberance upon the ſtem or branches, generally ſcaly and gummy or reſinous. It contains the rudiments of the leaves, or flowers, or both, that are to be expanded the following year.

BULB. May be conſidered as a *Bud* placed upon the root. It contains the rudiments or embryo of a future plant. Bulbs ſometimes are found upon the ſtem, as in ſome ſpecies of Garlic.

Bulbous Roots are either.

Solid as in the *Tulip*; Pl. 11. f. 3.

Scaly as in the *Lily*; Pl. 11. f. 4. or

Coated as in the *Onion*. Pl. 11. f. 2.

BUNCH. A fruit-ſtalk furniſhed with ſhort lateral branches. The *Grape*, the *Currant* and the *Barberry* are inſtances. Pl. 6. f. 8.

BUNDLE. When ſeveral flowers ſtand on their reſpective fruit-ſtalks which grow nearly from the ſame point and riſe to the ſame height; as in the *Sweet William*.

BUNDLED. *Leaves*; when they ariſe nearly from the ſame point and are crowded together; as in the *Larch*. Pl. 9. f. 3. (*f.*)

- - - - *Roots*; a ſort of tuberous roots in which the knobs are connected without the intervention of threads, as in the *Pæony*.

BUSH. A collection of *Floral Leaves* terminating the flowering ſtem: as in *Sage* and *Crown Imperial*; remarkable in the *Pine Apple*.

BUTTERFLY SHAPED. From an imaginary reſemblance that ſome bloſſoms bear to that inſect. The *Pea* and the *Broom* furniſh examples. See the introduction to the 17th claſs; and alſo pl. 4. f. 13. 14. 15. 16. 17.

CAP. The membranaceous empalement of Funguſſes; ſurrounding the pillar. Pl. 1. fig. H. *a*.

CAPSULE. A dry hollow ſeedveſſel, that opens naturally in ſome determinate manner; as at the *Side* by a ſmall hole in *Orchis* and *Campanula*; *horizontally* in *Pimpernel*: *longways* in *Convolvulus*: at the bottom in *Arrowgraſs*; or at the top as in moſt plants. See pl. 5.

CATKIN; is a compoſition of flowers and chaff on a long receptacle, reſembling a Cats tail. The *Willow* the *Hazel* and the *Reedmace* are inſtances. Pl. 6. f. 12.

CELL. A vacuity in a capſule for lodging the ſeed. Capſules have either one cell as in *Primroſe*: two as in *Thornapple*: three as in *Lilly*, four as in *Spindletree*: five as in *Rue*: ſix as in *Aſarabacca*,&c.

It alſo ſignifies the vacuity in the tips that contains the duſt.

CENTRAL *Florets*; thoſe that occupy the middle part of compound flower; as the yellow ones in the middle of a common Daiſey; pl. 4. f. 24. (*b.*) and it likewiſe is uſed to ſignify the flowers in the middle part of a rundle.

—— *Leaf-ſtalk* is fixed not to the baſe but to the middle part of a leaf, as in the garden *Naſturtium* and *Marſh Pennywort*. Pl. 9. f. 4. (*a.*)

CHAFF. A thin membranaceous ſubſtance growing from a common receptacle to ſeparate the florets from each other, as in *Teaſel*; *Scabious*; *Willow*; *Burdock*.

CHAFFY *Leaves* the leaves of the *Fir*, the *Yew*, the *Pine*, and the *Cedar* are ſo called. Pl. 9. f. 3. (*e.*)

CHANNELLED *Leaves* having a deep furrow or channel extending from the baſe to the end.

CHIVE. Open the bloſſom of a Tulip or Lily and you will ſee ſix long threads placed round the central pillar, with a tip on the top of each thread. One of theſe threads together with its tip is called a chive. Pl. 3. f. 2. (*b. b. b. b. b.b.* f. 3. (*b. i.*) f. 6.

CIRCULAR. Nearly in the form of a circle, as are the leaves of the *Alder*, or the petals of the *Strawberry* and *Hawthorn*. Pl. 7. f. 2.

CIRCUMFERENCE. The part of a circle moſt diſtant from the center. Thus in a ſhilling or half crown the inſcription is round the circumference. It is uſed in botany to expreſs the florets that are furtheſt from the center of a compound flower; as the white ones that ſurround the yellow ones in the *Common Daiſie*, or the florets in the outer part of a rundle. Pl. 4. f. 24. (*a. a. a. a.*)

CLAMMY. Adheſive like birdlime; as are the leaves of the *Alder*; or the ſtalks of *Fraxinella*; and *Gum Ciſtus*.

CLASS. See the introduction.

CLAW. Bloſſoms that are compoſed of ſeveral petals have frequently thoſe petals ſo formed as to admit of two diſtinct names; the claw and the limb. The claw is the lower part, or that next the baſe: thus if you take a *Pink*, a *Campion*, or a *Wallflower* and draw out one of the petals, the lower and the ſlender part by which it was connected and which was included within the cup, is the part which is called the *Claw*. Pl. 4. f. 11. (*a. a.*)

CLIMBING. A term applied to plants that take the advantage of twining round ſome other body to ſupport and raiſe themſelves; as the *Ivy* and *Honeyſuckle*.

CLOATHING. Every ſpecies of hairineſs on the ſurface of plants. See DOWN; HAIR; WOOL; BRISTLES.

CLOSE. When a branching fruit-ſtalk bears its flowers cloſely compacted together but without regularity.

CLOVEN. Divided half way down, as are the ſummits of *Ground Ivy* and *Jacob's Ladder*; the petals of *Campion*, and the leaves of wormwood.

—— *Tips* ſee pl. 1. f. 3. (*a. a. a.*)

CLUBSHAPED. Thin at the baſe and thicker upwards, as is the fruit fruitſtalk of the *Cuckowpint*, and of the *African Marigold*.

CLUSTER. A collection of flowers ſomewhat in an egg-ſhaped form, as thoſe of the *Lilac* and *Butterbur*.

COLOURED. When a leaf or cup is any other colour than green; as the floral leaves of *Golden Saxifrage*.

COMMON *Empalement*; including ſeveral flowers: ſee the introduction to the 19th claſs. We have a well known inſtance in the *Dandelion* and in all the *Thiſtles*. Pl. 4. f. 20.

——— *Receptacle*. A ſeat for ſeveral flowers or florets included within one common empalement; as is the caſe with moſt of the plants in the nineteenth claſs. The *Dandelion* is an example. Pl. 4. f. 23. (*a*.)

COMPACT. Growing cloſe and as it were preſſed together.

COMPOUND *Flowers*; conſiſt of many florets upon one receptacle or ſeat, and included within one common empalement; as moſt of thoſe in the 19th claſs; a *Thiſtle* is a familiar example. Pl. 4. f. 19. 24. 25. Sometimes; but with leſs propriety, the flowers that grow in rundles are called compound, as thoſe in the ſecond order of the 5th claſs; of which the Carrot is a well known inſtance.

——— *Rundle*; when each rundle, is divided into other little rundles or rundlets. Pl. 6. f. 9.

——— *Leaf*; when each leafſtalk ſupports more than one leaf; or when one leaf is inſerted into a nother, as in *Wood Horſetail*. Pl. 7. f. 47. 49. Pl. 8. f. 52. 53. 54. 55. 56. Pl. 9. f. 3. (*a*.) See alſo *Doubly compound*; *Triply compound*.

——— *Berry*; when one large berry is compoſed of ſeveral ſmall ones as for inſtance the *Raſpberry*.

COMPRESSED. A term applied to a cylindrical ſubſtance more or leſs flatted. Thus ſuppoſe a ſtraw to be the cylindrical ſubſtance; if this is preſſed betwixt the thumb and finger ſo as to flatten it, we ſhould then ſay it was compreſſed. The cup of the *Gilliflower* or the *Wallflower* is compreſſed, and ſo is the bloſſom of the *Rattle* and the pod of the Ladyſmock.

——— *Leaf*; one that is thicker than it is broad.

CONCAVE. Hollowed out like a bowl; as are the petals of the *Cherry* or the *Hawthorn*; and the the leaves of *Broad leaved Plantain*.

CONE. A ſpecies of ſeed-veſſel exemplified in the *Pine* and *Fir*. Pl. 5. f. 18.

CONGREGATED. When ſeveral little ſpikes or panicles are crowded together ſomewhat in a globular form. Examples are not uncommon amongſt the Graſſes, *Rough Cocksfoot* is one.

CONICAL. The ſhape of the *Alpine Strawberry*, nearly reſembling the form of a ſugar loaf.

CONNECTED *Leaves* or *Props*; thoſe that have their upper ſurface at the baſe growing to the ſtem or branch.

CONTIGUOUS. When a leaf, branch, or ſeed veſſel riſes up ſo perpendicularly as to ſtand almoſt parallel and cloſe to the ſtem. The pods of the *Common Muſtard* furniſh an example; and the leaves of the *Creſs Mithridate*. Pl. 9. f. 6.

CONVEX. Oppoſed to concave. Riſing like the ſurface of a globe.

globe. The receptacle of the garden *Tanſey* is convex.

CREEPING *Stem*, creeping along the ground and ſending forth little roots; the *Violet* and *Ivy* are inſtances. Pl. 10. f. 8.

——— *Root*; as in the Spearmint. Pl. 10. f. 7.

CRESCENT-SHAPED. Shaped like a new moon; as are the tips of the *Strawberry*.

——— *Leaf*. Pl. 7. f. 11.

CROOKED *Fruitſtalk*; ſo much bent that the flower faces the earth, and ſo ſtiff that it cannot be ſtraightened without breaking, as in *Crown Imperial*.

CROSS-PAIRS. When leaves grow in pairs, and each pair points in a different direction to the pair next above or below it. Thus if one pair point Eaſt and Weſt, the pair next below it point North and South; the third pair croſſes the ſecond, and the fourth the third, and ſo on. Pl. 9. f. 1.

CROSS-SHAPED; *Flowers*, are thoſe which have four petals diſpoſed in the form of a croſs. The *Gilliflower Candytuft*, and *Cabbage* are familiar inſtances. Pl. 4. f. 11. f. 12.

CROWNED *Seed*; is a ſeed to which the cup of the flower adheres, as in *Teaſel*; or it is a ſeed furniſhed with a feather as in *Dandelion*. Pl. 4. f. 22. f. 27.

——— *Berry*; is a berry with the flower cup adhering; as in the *Honeyſuckle*.

CUP. A ſpecies of empalement contiguous to the other parts of the flower. It either includes one flower as in the *Convolvulus* and *Gilliflower*; or ſeveral florets as in the *Sunflower* and *Daiſie*. Pl. 3. f. 1. f. 10. f. 5. (*a*.) Pl. 4. f. 7. (*c*.) f. 12. (*b*) f. 13. 14. 18. (*a*. *a*. *a*.)

CURLED *Leaves*; as in *Endive* and *Curled Mint*. Pl. 8. f. 67.

CURVED. Bent back, ſo that the extremity is lower than the baſe. Pl. 9. f. 5. (*e*. *e*.)

CUT-ROUND. When a ſeed veſſel does not open longways as is generally the caſe, but in a circle ſurrounding it, like a ſnuff box. as in *Pimpernell*. Pl. 5. f. 9.

CYLINDRICAL. The form of the trunk of a tree. Pl. 8. f. 68.

——— Straw; *Bullruſh*.

——— Stalk; *Great Plantain*.

——— Stem; *Aſparagus*.

——— Leaf; *Wild Garlic*; *Onion*.

——— Cup; *Pink*.

——— Catkin; *Reedmace*.

DECIDUOUS *Leaves*; thoſe that fall off at the approach of winter.

——— *Cup*; falling off before the bloſſom; as does that of the *Thorn-apple*, the *Cabbage*, the *Lady-ſmock* and the *Muſtard*.

——— *Seedveſſel*: falling off before it opens, as in the *Sea Rocket* and *Woad*.

DECLINING. Bent like a bow, with the arch downwards; as the ſeedveſſel of the *Watercreſſet*, the threads of the *Bugloſs*. See the lower chive in pl. 1. f. 11. f. 12.

DENTED. A blunt leaf &c. with a dent or blunt notch at the end; as in the *Broad-leaved Sea Heath*.

DEPRESSED. When the ſurface of a leaf, &c. is in a ſmall degree concave—preſſed down—flatted.

DIAMOND-SHAPED. Applied to leaves that reſemble the figure

figure of a diamond as painted on cards.

DIMPLE. A little hollow dot; as in the seed of the *Berberry*.

DISTANT. Far asunder; as the chives of the *Mint*; or the whorls of the flowers in the *Corn Mint*.

DISTENDED. As the cup of the *Rose*, or the under part of the blossom of the *Foxglove*. Pl. 4. f. 4.

DISTINCT. A term applied to chives and pointals when they are not only found in different flowers, but these flowers must likewise grow upon different plants. See the 22d class. Thus in the *Yew Tree*, if you find it in flower and one of the flowers is furnished with chives; all the flowers upon that particular tree have *only* chives and no pointals; but if you find a flower with a pointal and no chives, then all the flowers upon that tree will be found equally destitute of chives. Pl 1. f. 22.

DIVERGING. Spreading wide from the stem, almost horizontally. This term is opposed to *Compact*.

DIVISIONS. See the next article.

DIVIDED. Applied to a leaf, a cup, or a petal, it signifies that it is parted more than half way down; as the petals of *Chickweed*; the cup of *Comfrey* or *Borrage*. Pl. 7. f. 28.

DOTTED. Marked with little hollow dots; as are the leaves of the *Sea Chamomile*.

Or the receptacles of some of the compound flowers. Pl. 4. f. 23.

DOUBLED *together*. As are the leaves of the *Black Cherry* before they unfold.

DOUBLE. Applied to the tips of several flowers when upon one thread there are two tips united, like a double nut; as in the *Ranunculus*, *Anemone*, *Celandine Plumb*, *Cherry*. Pl. 3. f. 6. *b*.)

——— *Cup*; when the cup of a flower hath another outer cup surrounding it, as in the *Marshmallow*, and *Hollyhock*.

——— *Seedbud*; when two seed-buds are united together, as in *Goose-grass* or *Cleavers*.

DOUBLY-COMPOUND *Leaves*, are of three different kinds,

1. *Twinfork*; when a forked leaf-stalk bears two little leaves at the end or each division or fork. Pl. 10 f. 4.
2. *Doubly threefold*; when a leaf-stalk with three divisions bears three little leaves upon the end of each division. Pl. 8. f. 57.
3. *Doubly winged*; a leaf-stalk having lateral ribs and each of these ribs being a winged leaf; example *Tansey*. *Yarrow*. Pl. 8. f. 56.

For leaves *more than doubly compound*, see *Triply compound*.

DOWNY. Covered with a whitish down; as the leaves of the *Marshmallow* and *Great Mullein*.

——— *Seeds*; the feather of seeds is sometimes downy, as in *Dandelion*. Pl. 4. f. 22. (*l*.) Pl. 6. f. 2. (*b*.)

DUST. A fine powder contained in the tips of flowers: it is too minute for the naked eye to examine, but by the assistance of a microscope, it appears very different in different plants: Thus in the *Bloody Geranium* it is a perforated globule; in the *Marshmallow* like the wheel of a watch; in the *Pansie*

Pansie it is triangular; in the *Narcissus* kidney-shaped, and in *Comfrey* the globules are double. Pl. 3. f. 5. (*f.*) a tip discharging its dust. f. 8. A particle of dust greatly magnified.

DUSTED. Some plants appear as if covered with a kind of dust or powder. e. g. the *English Mercury*.

EAR-SHAPED. Somewhat resembling a human ear.

ELLIPTICAL. The same as *oval*.

EGG-SHAPED. Signifies a shape resembling the solid substance of an egg as in the seedbud of *Jacob's Ladder* and the seeds of *Fennel*; or it implies only the form of an egg if divided longways, as in the leaves of the *Beech tree* or *Peppermint*. Pl. 7. f. 3.

EMBRACING *the Stem*. When the base of a leaf half surrounds a stem, as in *Solomon's Seal*, *Poppy* and *Borrage*. Pl. 9. f. 4. (*f.*)

EMPALEMENT. Is a continuation of the outer bark of a plant constituting a part of the flower. It is either a

——— Cup; as in *Primrose*; pl. 3. f. 10. a.

——— Fence; as in *Carrot*; pl. 6. f. 9. (*c. c.*) a

——— Catkin; as in *Hazel*; pl. 6. f. 17. a

——— Veil as in several *Mosses*; pl. 1. f. D. (*a*) a

——— Husk; as in *Oats*; pl. 2. f. 21. (*a. a.*) a

——— Sheath; as in *Narcissus*; pl. 3. f. 9. (*a. a.*) or a

——— Cap; as in several Funguses. Pl. 1. f. H. (*c*) See those terms.

ENTIRE *Leaf* or *Petal*. This term is opposed to cloven, gashed, indented, &c. but it does not signify that it is not serrated or scolloped. When a leaf is said to be very entire we understand that it is not so much scolloped or serrated. The leaves of a Nettle are *entire*, but those of a Lilac are *very entire*. pl. 7. f. 31. 35. entire leaves. f. 39. 40. very entire lerves.

EQUAL. Sometimes signifies regular; all alike; as the blossoms of *Angelica*. The florets forming the compound flowers in the first order of the 18th class, are said to be equal; that is all are alike in being equally furnished both with chives and pointals.

ESSENTIAL *Character*; Is a single circumstance serving to distinguish a species or a genus from every other species or genus. Thus the genus *Crowfoot* (Ranunculus) is distinguished from other genera by the honeycup at the base of each petal; and the *Colewort* is known from all the other genera in the same natural order, by the four longer threads being forked at the top.

EXCRESCENCE. A substance growing from the seat of the flower of some of the Mosses.

EXPANDING. Standing in a direction betwixt upright and horizontal; as the petals of the *Strawberry*, the branches of most plants, and the leaves of the Brooklime *Speedwell*. Pl. 9. f. 5. (*c. c.*)

EYE. The external scar upon a seed by which it was fixed to the seedvessel; it is very remarkable in a *Bean*. Pl. 6. f. 3. (*e.*)

FEATHER. The downy or hairy substance affixed to the seeds of some plants, enabling the wind to scatter them abroad. The feather of the *Dandelion* is downy: that of the *Sowthistle* hairy. Pl. 4. f. 22. (*l.*) pl. 6. f. 2. (*a. b.*)

FENCE

FENCE. The Empalement of a rundle: it is placed at ſome diſtance from the flowers. It is either *General* or *Partial.* The *Carrot* furniſhes inſtances of both. The *General* Fence is placed under the rundle; the *Partial* under the rundlets. Pl. 6. f. 9. (*c. c.*) (*d. d. d. d.*)

FERTILE Flowers; thoſe that produce ſeed capable of vegetation: as is very generally the caſe in thoſe flowers that have both chives and pointals. Flowers that have only chives never can produce ſeeds; and flowers that have only pointals muſt be barren, if they are ſo ſituated as to be out of the duſt from the tips of the barren flowers: In ſome inſtances they will indeed produce ſeeds to all appearance perfect, but theſe ſeeds will never vegetate.

FIBROUS *Roots*; compoſed of ſmall threads or fibres. Pl. 10. f. 7.

FIDDLESHAPED. The ſhape of a leaf of one ſpecies of Dock, that is ſuppoſed to reſemble a violin.

FINGERED *Leaves*; reſembling the expanded fingers of a man's hand. e. g. thoſe of the *Wild black Hellebore*, *Lupine* and *Horſe Cheſnut*. Pl. 7. f. 48.

FLESHY. More ſolid than pulpy; as the fruit of the *Apple*; the root of the *Turnep* and the leaf of the *Round leaved Stonecrop*.

FLEXIBLE *Branches*; eaſily bending; as thoſe of the *Barren Strawberry*.

FLOATING. Applied to aquatic plants whoſe leaves or flowers float upon the ſurface of the water: e. g. *Waterlily*.

FLORAL LEAVES differ in ſhape or colour from the other leaves of the plant; they are generally placed on the fruit-ſtalk, and often ſo near the flower as in ſome inſtances to be eaſily miſtaken for the cup; but the cup dries or withers when the fruit is ripe, whereas the floral leaves endure as long as the other leaves of the plant. Examples of floral leaves may be ſeen in the *Panſie*, the *Limetree*, the *Hellebore*, the *Paſſion flower*, the *Sage*, the *wild Marjoram*; and many others. Pl. 9. f. 8. (*a. a.*)

FLORET (a little flower) one of the ſmall flowers compoſing a compound or incorporated flower. See the introduction to the 19th claſs. They are *Tubular*; that is, formed of a tube cloven into five parts at the border; as in *Tanſey*; or *Narrow* when the bloſſom is long and ſtrap-ſhaped as in *Dandelion*. In the *Daiſie* and *Sunflower* the florets in the center are *Tubular*, and thoſe in the circumference *Narrow*, or *Radiate*. Pl. 4. f. 21. f. 24. f. 26. In the ſecond order of the fifth claſs the florets compoſing the rundles are compoſed of five petals. When the petals are all of the ſame ſize and ſhape the florets are ſaid to be *equal*; as in *Angelica* and *Celery*; but when the outer petals are larger than the others, the florets are ſaid to be *Radiate*; as in *Shepherds Needle* and *Carrot*.

FLOWER. A temporary part of a plant appropriated to the production of ſeeds. It is compoſed of ſeven parts; the *Empalement*; the *Bloſſom*; the *Chives*; the *Pointals*; the *Seed-veſſel*; the *Seeds*; and the *Receptacle*. To theſe perhaps we may add an eighth *viz.* the *Honey-cup*. It is not neceſſary that all theſe parts ſhould be preſent

ſent to conſtitute a flower. *Imperfect* flowers are deficient in one or more of the parts. The *Hyacinth* and *Tulip* have no cup. The *Misletoe*, *Gale*, *Hop*, *Yew*, *Dog's Mercury*, *Nettle*; and the flowers of the plants bearing catkins, have no bloſſoms. The *Ground Ivy*, the *white* and *red Deadnettle*, and all the plants in the firſt order of the fourteenth claſs have no ſeed-veſſels.

FORKED. Dividing into forks as the branches of moſt of the *Spurges*; the fruit-ſtalks of the *Common Calamint* and the *Pink*; the *Shaft* of the *Currant*. Pl. 10. f. 4.

FOURCORNERED. As the ſtem of the *Deadnettle*.

FRINGED. As the bloſſom of the *Buckbean* and the *Garden Naſturtium*; or the leaves of the *Croſs-leaved Heath*.

FRUIT. A part of a flower, conſiſting of the *Seedveſſel*, the *Seed* and the *Receptacle*.

FRUITSTALK. A part of a ſtem or branch, bearing flowers but not leaves. Pl. 9. f. 8. (*c.*) pl. 6. f. 7. (*a. a. a. a. a. a.*)

FUNNELSHAPED; A bloſſom of one petal; the lower part of which is tubular, the upper part conical, as in *Hounds-tongue*, *Bugloſs*, *Cowſlips*. Pl. 4. f. 7.

——— *Cup*; as in *Thrift*.

FURROWED. Marked with deep lines running length-ways.

GAPING *Bloſſoms*; So called from their reſemblance to a gaping mouth. *Toadflax and ſnapdragon* are inſtances. Pl. 4. f. 8. 9. 10.

GASHED. Divided nearly half way down, into lobes that are convex at the edges and diſtant from each other; as the leaves of *Ladies Mantle* and *Water-elder*. Pl. 7. f. 19.

GENERAL FENCE. A ſpecies of empalement ſurrounding a general rundle, as in a *Carrot*, *Angelica*, or *Lovage*. It conſiſts of one, or more leaves. Pl. 6. f. 9. (*c. c.*)

GENERIC *Character*. conſiſts of an accurate deſcription of the different parts compoſing a flower, and all thoſe plants whoſe flowers agree with this deſcription are ſpecies of the ſame genus. (See the introduction.)

GILLS. The thin plates on the under ſide of the Hats of the Funguſſes, remarkable in Muſhrooms.

GLANDS. Secretory veſſels, differently ſituated in different plants. In the *Willow* they are placed at the margins of the leaves; in the *Bird's Cherry* and *Almond Tree* at the baſe of the leaves; in *Butterwort* and the *Sundew* upon the leaves, and in the *Plumb* on the inner ſide of the cups. Pl. 10. f. 6. (*c. c.*) pl. 11. f. 1. (*a. a. a. a*)

GLASS-SHAPED. Tubular but dilated towards the top like a drinking glaſs; as the cup of *Jacob's-ladder*; the ſummits of the *Field Southern-wood*, the honey-cup of the *Nettle*.

GLOBULAR. Like a round ball; as the cup of the *Burdock*; the ſeedveſſel of the *Flax*; the ſeed of the *Pea*; the capſule of the *Poppy* Pl. 5. f. 5.

GLOSSY. Smooth and ſhining; as the ſeedbud of the *Sweet-briar*; the leaves of the *Holly*, *Ivy*, and *Box*.

GNAWED. As when a leaf is indented, and appears as if it had been

been gnawed or bitten at the edges. Pl. 7. f. 21.

GRANULATIONS. The small berries which join together and compose a large one, as in the *Mulberry*, *Blackberry* or *Raspberry*.

HAIRLIKE. Slender, undivided and cylindrical; as the threads in *Plantain*, *Raygrass*, *Reed*, and most of the Grasses.

HAIRS. Are supposed to be secretory ducts.

HALBERD-SHAPED As the floral leaves of the *Pansie*; the leaves of *Sheeps Sorrel*, and *Cuckowpint*. Pl. 7. f. 15.

HAND-SHAPED. resembling a human hand with the fingers expanded; as the leaves of *White Briony*; *Passionflower*; and the roots of *Spotted Orchis*. Pl. 7. f. 22.

HAT. The upper broad expanding part of Funguses. In Mushrooms the hat is often called the flap. Pl. 1. f. H. *c*.)

HEADED *Stalk*; When a stalk supports oné compact knob or head of flowers upon its extremity, as in *Thrift*.

HEADS Of Flowers; when flowers grow together in compact knobs; as in *Peppermint*, *Watermint*; *Common Thyme*.

HEART. That part of a seed which is the future plant in miniature. Pl. 6. f. 3. (*b*.)

HEART-SHAPED. A term used to express the form of a petal, a leaf &c. the leaves of *Waterlily*, *Deadnettle*, *Burdock* and *Violet*, are heart-shaped. Pl. 7. f. 10.

HELMET. A term to express the upper part of a gaping blossom, which bears some resemblance to a helmet. See the introduction to the 20th class.

HEMISPHERICAL. In the shape of half a globe; as the cup of the *Tansey*.

HERBACEOUS *Stem*. One that is succulent and tender, in opposition to one that is woody: it perishes annually down to the root. The *Pea* and the *Nettle* are instances. The stem of the *Gilliflower* is somewhat woody.

HONEYCUP. A part of a flower designed to secrete and contain honey. In flowers that have only one petal the tube of the blossom contains the honey; or else it is contained in a sort of horn-shaped appendage as in the *Butterwort*. In the *Violet*, the *Larkspur*, the *Columbine*, the *Fumitory*, &c. it is a sort of spur, or horn. In the *Ranunculus*, the *Lily*, the *Crown Imperial*, &c. it is a hollow cavity in the substance of the petals. In the *Daffodil* and *Hellebore* it is tubular. In the *Fraxinella* and *Campanula* it is fixed to chives; in the *Gilliflower*, *Turnip*, &c. it is placed on the seedbud in form of a gland. Its structure is no where more singular or beautiful than in the *Grass of Parnassus*. Pl. 5. f. 1. (*a*.) f. 2. (*a. a. a. a. a*.) f. 3. (*a. a*.) f. 4. (*a*.)

HOODED. A term applied to leaves that are rolled up, as the grocers roll paper to put sugar or spices in, like a cone; as the leaves of *Wall Pennywort*.

HORIZONTAL. A leaf or branch that grows from the stem pointing to the horizon, and parallel to the surface of the earth. Pl. 9. f. 5. (*d. d*.)

HORNSHAPED. Like the honeycup of the *Larkspur*. Pl. 5. f. 4. (*a*.)

HUNCHED

HUNCHED. Swelled out, as the under part of blossom of the *Foxglove*, the *Blossom* of the *Honey-suckle*; the cup of the *Turnep* and the *Wall-flower*. Pl. 4. f. 12. (*b*)

HUSK. The empalements and the blossoms of grasses are called the husks, they are thin, dry, and semi-transparent like chaff: a husk consists of one or more leaves called *Valves*, and when contiguous to the other parts of the flower, inclosing the chives and pointals, answers the purpose of a *Blossom:* but when placed on the outer-side and inclosing the inner valves, as well as the chives and pointals, it is called the empalement. The empalement frequently contains several florets. See the plate of *Grasses*.

JAGGED-*Leaves*; those that are variously divided into lobes and these lobes again divided in an irregular manner. The *Pansie* is an instance. Pl. 7. f. 24.

IMPERFECT-*Flowers*; those that want either the cup or the blossom. The *Tulip* wants a cup, and the *Nettle* is without a blossom.

INCORPORATED. When a number of little flowers or florets, are so disposed as to form one compound flower; all of them either inclosed within one common empalement, or situated upon one common receptacle; so that none of them can be taken away without destroying the uniformity of the whole. Thus the flowers of *Thrift*, *Parsley*, *Teasel*, *Scabious*, *Daisie*, are incorporated; several small flowers, or florets, combining to form one large flower.

INDENTED-*Leaf*; the sides of the leaf are hollowed or deeply scolloped, the lobes standing asunder as if part of the leaf had been cut out. The leaf of the *Turnep* is a familiar example. See also pl. 7. f. 25.

INDIVIDUAL-*Blossom*; the blossom belonging to a single floret in a compound flower. Thus in a *Carrot* each floret is composed of five petals which constitute the blossom of that individual floret. The individual blossoms in *Tansey* are all tubular; in *Dandelion* they are all long and strap-shaped. In the *Sun-flower* they are tubular in the center and strap-shaped in the circumference. Pl. 4. f. 21. f. 28.

INFLEXIBLE. Stiff; not easily bending: opposed to *Limber*.

INTERRUPTED. Broken in its regular form; as the spike of *Woodbetony*; the leaves of some species of the *Ladies-finger*. A spike may be interrupted by the intervention of leaves or smaller sets of flowers; a winged leaf is interrupted by the intervention of smaller pairs of little leaves. Pl. 8. f. 55.

INVERSELY-HEART-SHAPED. With the point of the heart next to the stem; as the seed-vessel of the *Shepherds Purse*; the petals of *Geranium* or *Marsh-mallow*; and the leaves of some of the *Trefoils*. Pl. 8. f. 60.

JOINTED-*Stem*. A Wheat-straw is an instance familiar to every one. Pl. 10. f. 3.

- - - - - - *Leaves*; as in the *Indian Fig*. pl. 9. fig. 3. (*a*.)

IRREGULAR. A term applied to compound flowers wherein the florets are not uniform; as in the *Carrot* and *Coriander*.

- - - - - *Blossom*. See *Regular*.

KEEL. A name given to the lowermost petal in a butterfly shaped blossom, from its supposed resemblance to the keel of a ship: see the introduction to the 17th class.

clafs. See alfo Pl. 4. f. 17. and f. 13. (*d*)

KEEL-SHAPED. Bent like the keel of a fhip or boat; as the fhaft of the *Pea*; the empalement of *Canary Grafs*. Pl. 2. f. 10. (*a. a.*)

KIDNEY-SHAPED. As the feed of the *French Bean*, the *Tips* of the Mallow; the leaves of *Ground Ivy*; *Golden Saxifrage*, and *Meadowbout*, Pl. 7. f. 9.

KNEE-JOINTED. When a ftraw or ftem is a little bent at the joints.

KNOB See *Head*.

LACTESCENT. Abounding with a milky juice.

LAMINATED. When the flat furfaces of leaves lye clofe one upon another.

LATERAL *Branches*; growing from the fides of the ftem: oppofed to *terminating*.

——*Flowers*; thofe that grow from the fides of the ftems or ftalks; thus the fpikes of flowers in the common *Speedwell* grow on lateral fruitftalks; or on fruitftalks proceeding from the fides of the ftem

LEAFSTALK. The footftalk of a leaf. It fupports the leaves but not the flowers In the *Great Periwinkle* the leaf-ftalks are very long. Pl. 9. f. 4. (*a. b. c.*)

LEAFY. Furnifhed with leaves.

——*Cup*; when the bafe of a cup is furrounded by a feries of leaves different from thofe which form the cup.

——*Seed*; a feed that is furrounded by a thin leafy edge, as in *Cow's Madnep*.

LEATHERY. Tough and pliable like leather; e g. the cup of the *Corn Cockle*, and moft of the plants in the third divifion of the 24th clafs.

LEVEL. When feveral branches or fruitftalks grow to equal heights, fo as to form a flat furface at the top; as in the flowers of the *Sweet William*.

LID. A cover to the tips of feveral of the Moffes; as in the *Bogmofs*. Pl. 1. f. D. (*b.*

LIMB. The upper part of a petal, in bloffoms compofed of more than one regular petal. Thus in the *Wall-flower*, the upper flat broad part of the petals is called the limb. The lower flender part included within the cup is called the claw. Pl. 1. f. 11.

LIMBER *Fruit-ftalk*; bending with the weight of its own flowers.

LIP. The upper or under divifion of a gaping bloffom. The *Dead-nettle* and the greater part of the plants in the 14th clafs furnifh examples. See the introduction to that clafs. See alfo pl. 4. f. 8. f. 9. and f. 10.

LITTLE *Fruit-ftalks*; the little foot-ftalks that fupport an individual flower, when there are feveral flowers upon one common fruitftalk.

LOBES. The divifions of a *Gafhed* leaf; fee gafhed. The lobes are rounded at the edges and ftand diftant from each other. The leaves of the *Hop*, *Anemone*, *Hepatica*, and *Sycamore* furnifh examples. Pl. 7. f. 17. f. 19.

LONG. A cup is faid to be long, when it is equal in length to the tube of the bloffom.

LOPPED. Appearing as if cut off with a pair of fciffars: the leaves of the *Great Bindweed* are lopped at the bafe, the petals of the *Periwinkle* are lopped at the end. Pl. 8. f. 63.

LYRE-SHAPED As the leaves of *Herb-Fennet*. Pl. 8. f. 62.

MATTED. Thickly interwoven

ven together as the fibres in turf-bogs.

MEMBRANACEOUS. Thin, ſkinny, and tough.

—— *Stem*; when the edges of the ſtem are bordered with a thin leafy ſubſtance, as in *Water Figwort* and *Broad Leaved Peaſe Everlaſting*.

MOUTH. The upper part of the tube of bloſſoms, conſiſting of a ſingle petal; as *Borrage*, *Houndstongue*, *Deadnettle*. Pl. 4. f. 9. (*d. d.*)

NARROW. The florets in ſome ſpecies of compound flowers are tubular at the bottom, but flat and narrow like a ſtrap or fillet at the top. In *Dandelion* the florets are all narrow: in the common *Daiſie* the florets in the circumference only, are narrow. Pl. 4. f. 10. f. 21. f. 24.

NAKED *Stalk*; deſtitute of leaves; as the ſtalk of the *Tulip* or *Lilly of the Valley*.

———— *Mouth*; when the mouth of the tube of a bloſſom is not cloſed by valves or hairs. The mouth of the bloſſom of *Borrage* is cloſed by five valves, or teeth: but that of *Gromwell* is open and *Naked*.

———— *Receptacle*; neither chaffy nor hairy; as that of the *Daiſie*.

NODDING *Flower*; when the fruitſtalk is bent near the end, as in the *Chequered Daffodil*, *Narciſſus and Jenquil*. Pl. 3. f. 9.

NOTCHED *at the end*; as the petals of the *Small campion*; and *Dove's-foot Crane's-bill*; the little leaves of *Vetch*; the leaves of the common *Maple*. Pl. 7. f. 16.

NOTCHED *Leaves*. The edges cut ſomething like the teeth of a large timber ſaw. *Dandelion*, *Broad leaved Watercreſs*, *Long rooted Hawke's-eye* and *Smooth Succory Hawkweed* are examples.

NUT. A ſeed covered by a hard woody ſhell; e. g. the *Hazel Nut*. This woody ſhell is ſometimes covered by a ſoft pulpy or fleſhy ſubſtance as in a Peach, or Apricot and then it is called a ſtone. Pl. 5. f. 21.

OBLONG Longer than broad, and rounded at the ends; as the leaves of the *Daiſie*; the tips of the *Honeyſuckle*. Pl. 7. f. 5.

OPEN. Standing wide.

OPPOSITE. Growing on the oppoſite ſides of the ſtem, but at the ſame height from the ground; as the leaves of the *Nettle*. In pl. 9. f. 5. all the leaves are oppoſite.

OVAL. As the leaves of box. Pl. 7. f. 4.

PALATE. The inner part of the mouth of gaping bloſſoms. Pl. 4. f. 10. (*e.*)

PANICLE. An aſſemblage of flowers growing without any very regular order, upon fruit-ſtalks that are variouſly ſubdivided; e. g. *Oats*. Pl. 6. f. 6. It is ſaid to be

———— *Spreading*; when the partial fruitſtalks diverge and ſtand wide aſunder, as in the *Common* and *Reed Meadowgraſs*.

———— *Compact*, when they ſtand near together as in the *Sheep's Feſcue*, and *Purple Hairgraſs*.

PANICLED *Bunch*; an aſſemblage of flowers partaking the properties of a panicle and a bunch. See thoſe terms. *Golden Rod* may ſerve for an example.

———— *Spike*; an aſſemblage of flowers partaking the properties of a panicle and a ſpike; as the *Wall Feſcue* and the *Manured Canary Graſs*, in which the collections of florets reſemble a ſpike in their general appearance, but the florets are furniſhed with fruitſtalks, ſhorter than themſelves.

PARA-

PARASITICAL *Vegetables*; not taking root in the earth but growing upon other vegetables. Thus *Misletoe* is found to grow upon the Apple Tree, the Pear, the Lime, the Elm, the Poplar, the Hawthorn and the *Buckthorn*, but never upon the ground.

PARTIAL. Belonging to a part, not to the whole.

—— FENCE. Pl. 6. f. 9. (*d. d. d. d.*)

PARTITIONS. The substances dividing seed-vessels into different cells. Thus the Seed-vessel of *Jacob's Ladder* is divided into three cells; and if you cut a Lemon across you will plainly see the partitions that divide it into nine cells. See also pl. 5. f. 14. (*b. b. b. b.*)

PENCILSHAPED. Like a camel-hair pencil; as the summits of *Millet*, or the appendages to the blossoms of the *Meadow Milkwort*. Pl. 2. f. 11. (*c. c*)

PENDANT. Hanging down; as the bunches of the *Red Currant*; the cones of the *Scotch Fir*; the flowers of the *Columbine*.

PERENNIAL. Continuing for several years.

PERFORATED *Leaves*; when the stem seems to go through the leaves: as in the *Round-leaved Thoroughwax*. Pl. 9. f. 4. (*g.*)

PERMANENT *Cup*, remaining till the fruit is ripe; as in *Borrage*; *Currant*; *Pink*; and *Deadnettle*.

PETALS. The leaves which constitute the blossom are called *Petals*, to distinguish them from the other leaves of the plant. See Pl. 3. f. 2. (*a. a a. a. a. a.*) Pl. 4. f. 12. (*a. a. a. a.*)

PILLAR. The cylindrical substance that supports the hat of a fungus, *e.g.* the *Common Mushroom*: also the little shaft upon which the feather of downy seeds is placed, as in *Dandelion* Pl. 1. f. H. (*b.*) Pl. 4. f. 22. (*i.*) Pl. 6. f. 2. (*d.*)

PITH. A soft spongy substance filling up the cavity in the stems of some plants; as in the *Rush* and the *Elder*.

PLAITED. Folded in plaits; as the blossom of *Convolvulus*; the cup of *Thrift* and the leaves of *Ladies-mantle*. Pl. 7. f. 37.

POD. A seed-vessel of two valves, within which the seeds are fixed alternately to each seam. When long it is called a pod, as in *Gilliflower*; when broad and short it is called a *Pouch*, as in *Honesty* and *Shepherds Purse*. Pl. 5. f. 10. f. 11. f. 12. f. 17.

POINTAL. A part of a flower, composed of the *Seedbud*, the *Shaft* and the *Summit*. Look into the blossom of a *Plumb* or *Cherry*, and in the center you will see the pointal surrounded by chives. In the blossom of the *Apple* or *Pear*, you will perceive five pointals in the center. In the *Deadnettle* you will find the pointal covered by the upper lip, and forked at the top. In the center of the blossom of the *White Lily*, the pointal stands surrounded by six chives. The *Seedbud*, which is the lower part of the pointal is long, cylindrical, and marked with six furrows. Next above this part is the *Shaft* which is long and cylindrical, and at the top of the pointal is the *Summit*, which is thick and triangular. See pl. 3. f. 2. (*d. e. f.*) f. 7. (*i. k. l.*) f. 5. (*c. d. e.*)

POINTING *from two opposite lines*; like the teeth in a double box or ivory comb. The leaves of the *Common Fir* and the flowers of *Sweet Cyperus* are examples.

——— *one way*; as the flowers of the *Foxglove*; the *Cocksfoot*, and the *Sheeps Fescue Grass*. Pl. 2. f. 13. *(d.)*

PORES. Little holes. At the inner side of the base of the petals, in all the species of *Ranunculus* or *Crowfoot*, are little pores filled with honey. See also pl. 3. f. 3. *(k.)*

POUCH. See POD.

PRICKLES. Sharp-pointed weapons of defence formed from the bark, and not from the woody part of a plant. The prickles of the *Rose* are a familiar example. Pl. 10. f. 2.

PROLIFEROUS *Blossoms*; when one grows out of another, as is not uncommon in the *Polianthus*.

———SHOOTS when one shoot springs out of another, as in the Proliferous *Feather-moss*.

PROMINENT. The partition of a seed-vessel is said to be prominent, when it projects beyond the valves, as in *Cabbage*, and many other plants of the fifteenth class.

PROPS. Small leaves or scales situated on each side the base of a leaf-stalk or fruit-stalk, for the purpose of supporting them at their first appearance. They are sufficiently evident in the *Garden Pea*. Pl. 10. f. 6. *(b. b.)*

PROTUBERANCES in seed-vessels; occasioned by the swelling of the inclosed seeds. They are sufficiently evident in the pods of *Mustard*, and in some sorts of *Beans*.

PULPY. Soft and tenacious. A *Strawberry* and *Cherry* is pulpy; but an *Apple* is fleshy.

PULPY *Seed-vessel*; consists of a nut or stone encompassed by a pulpy substance, exemplified in the *Plumb*, the *Cherry*, and the *Peach*. Pl. 5. f. 21.

PURSE-SHAPED. Like a purse that draws together with strings at the top; as the seed-vessel of the *Purple Marshlocks*, or the honey cup of the *Satyrion*.

RADIATE. A species of compound flowers in which the florets of the center differ in form from those in the circumference. Thus the *Daisie* and *Sunflower* are radiate flowers; the florets in the center are all tubular, but those in the circumference are narrow or strap-shaped. Pl. 4. f. 24.

——— *Summits*; placed in a circle; as in the *Poppy*. Pl. 5. f. 5. *(b.)*

RECEPTACLE. One of the parts necessary to compose a flower. It is the base or seat upon which the other parts of the flower are placed. Pl. 4. f. 11. *(c.)* f. 23. *(a.)*

REFLECTED. Bent back, as the segments of the cup of the *Currant*; the petals of the *Flower de Luce*; the blossoms of the *Hyacinth* and *White Lily*. Pl. 4. f. 5.

REGULAR *Blossom*; one that is regular in the figure, size, and proportion of its parts; as the *Jessamine* and *Syringo*.

REMOTE *Whorls*; when there is a considerable length of stem betwixt each whorl. Pl. 6. f. 11. *(a. a. a.)*

RIGID. Inflexible; opposed to limber and flexible.

ROLLED BACK. As the lateral edges of the leaves of *Rosemary*; the ends of the leaves of *Sweet Williams*, the summits of the *Pink*. Pl 1. f. 19. *(c. c.)*

ROOT may be *Fibrous*, *Bulbous*, *Tuberous*, *Bundled*, *Beaded*, *Spindle-shaped*, or *creeping*; see those terms. See also pl. 11.

ROOT-

ROOT-LEAVES. The leaves that proceed immediately from the root without the intervention of a ſtem. They often differ in ſhape and ſize from the other leaves. The *Field Bellflower* furniſhes an example. Pl. 9. f. 7.

RUNDLE. A compoſition of flowers in which a number of ſlender fruit-ſtalks proceed from the ſame center, and riſe nearly to the ſame height, ſo as to form a regular ſurface at the top. *Hemlock*, *Carrot*, and *Cowparſnep* are examples. Rundles of flowers are frequently called Umbels; and the plants producing them are ſaid to be *umbelliferous* plants. Pl. 6, f. 9.

RUNDLET. A little rundle. The fruit-ſtalks which compoſe a rundle are often divided at the top into ſeveral ſmaller fruit-ſtalks, and theſe ſmaller ſets of fruit-ſtalks are called *Rundlets*. *Hemlock*, *Carrot*, and *Angelica* furniſh inſtances. The fruit-ſtalks of a rundle and of a rundlet, are called *Spokes*. Pl. 6. f. 9. (*b. b. b. b*)

RUNNING *along the Stem*; means that a leaf has no leaf-ſtalk, and that its baſe is attached to the ſtem for a conſiderable length. The *Great White Mullein* and the *Muſk thiſtle* are examples. Pl. 9. f. 4. (*e.*)

SALVER-SHAPED. The ſhape of a bloſſom of one petal, the lower part of which is tubular, the upper part flat and expanded; as the bloſſom of the *Perriwinkle*, and the *Mouſe-ear ſcorpion Graſs*: Pl. 4. f. 1.

SAUCER. A ſort of flower of the *Cupthong* that is circular and concave like a china ſaucer. Pl. 1. f. F.

SCALY. Like the ſkin of a Fiſh; as the cups of *Burdock*. Pl. 4. f. 25 (*a.*)

SCATTERED. Diſpoſed without any regular order.

SCOLLOPED. Inſpect the edges of the leaves of *Bird's-eye* and *Gill*, and you will have a true idea of this term. Pl. 7. f. 35. 34. 33.

SCORED. Marked with ſuperficial lines, as the cup of a *Pink*, or the ſtems of *Butchers-broom*.

SEAM. The line formed by the union of the valves of a ſeed-veſſel. Thus the pod of a *Pea* is a ſeed-veſſel of two valves, and the two ſeams where the valves join are ſufficiently conſpicuous. As alſo in pl. 5. f. 6.

SEED. A deciduous part of a vegetable containing the rudiments of a new plant. It conſiſts of the *Heart*; the *Seed-lobes*; the *Eye* and the *Seed-coat*; See thoſe terms. Sometimes it is crowned with the cup of the flower, and ſometimes it is winged with a feather, or with a thin expanded membrane, which enables the wind to waft it abroad. See pl. 4. f. 22, and pl. 6 f. 3.

SEEDBUD. The lower part of a pointal. It is the rudiment of the embryo fruit. See *Pointal* Pl. 3. f. 2. (*d.*) f. 5. (*c.*) f. 7. (*i.*)

SEEDCOAT. The proper coat of a ſeed which falls off ſpontaneouſly. It is remarkable in *Spindle-tree*, *Hound's-tongue*, the *Cucumber*, the *Fraxinella* and the *Mallow*. Some ſeeds have only a dry covering or ſkin, as the *Bean*. Pl. 6. f. 1 *c. c.*)

SEEDLOBES. The periſhable parts of a ſeed, deſigned to afford nouriſhment to the young plant when it firſt begins to expand. A bean

bean after being soaked in water or moist earth, easily parts with its external skin, and divides into two parts called the *Seed-lobes*. Pl. 6. f. 3. (*a. a.*)

SEEDVESSEL. A vessel to contain the seed. It is of several kinds; as a *Capsule*; a *Pod*: a *Shell*; an *Airbag*; *Pulpy* including a nut or stone; an *Apple*; a *Berry*; and a *Cone*: see those terms. See also pl. 5.

SEGMENT. The small parts of a leaf, a cup, or a petal included betwixt the incisions.

SEMI-CYLINDRICAL; If the trunk of a tree was sawed lengthways through the middle, each part would be semi-cylindrical. The stalk of *Ramsons* is in this shape.

SEMINAL-LEAVES; those that arise immediately from a seed.

SEPARATE. Chives and pointals are said to be separate when they are found upon the the same plant, but within different empalements. Thus in the *Box*; the *Birch*: the *Cucumber*; and the *Melon*, some of the empalements contain chives and others contain pointals; but none of them contain both together. Pl. 1. f. 21.

SERPENTINE. The edge of some leaves is formed like a serpentine line; without any angles or corners. Pl. 7. f. 29.

SERRATED. Like the teeth of a common saw; as are the edges of the leaves of the *Apple*; the *Pear*, the *Spearmint*, the *Dead-nettle*, the *Sneezewort* or *Goose-tongue*, &c. Some leaves are *Doubly serrated*; that is the teeth are again cut into other little teeth; the *Common Elm* is an example. Pl. 7. f. 31. 32.

SHAFT. A part of a pointal; standing upon the seed-bud and supporting the summit See *Pointal*. Pl 3. f. 2. (*c.*) f. 5. (*d.*) f. 7. (*k.*)

SHARP. As the leaves of the *Jessamine* or the segments of the cup of the *Primrose*. Pl. 3. f. 10. Pl. 7. f. 40.

SHEATH. A species of empalement, exemplified in the *Crocus*, the *Iris*, and the *Daffodill*. Pl. 2. f. 9. (*a. a.*)

SHEATHED *Fruit-stalk*; one that is furnished with a sheath. Pl. 3. f. 9 (*d.*)

SHEATHING *Leaves*; when the base of a leaf enfolds the stem; as in most of the Grasses. Pl. 9. f. 4. *i.*)

SHEDDING. Continuing but a short time. Applied to a cup it signifies that it falls off before the the blossom, as in *Poppy*.

SHELL. A seed-vessel of two valves, wherein the seeds are fixed to one seam only; as in the *Pea*, and most of the plants in the fourth order of the seventeenth class. Pl. 5. f. 16.

SHORT. A cup is said to be short when it is shorter than the tube of the blossom, as in pl. 4. f. 7. (*c.*)

SHRIVELLING. Fading and withered, but not falling off. e. g. The blossoms of *Plantain* and *Stitchwort*

SHRUBBY. Somewhat woody, as the stems of the *Rose*.

SIMPLE; Undivided.

—— *Stem*; one that is undivided; only sending out small branches.

—— *Leaf*; when there is only one upon a leaf-stalk.

—— *Cup*; one that consists of a single series of segments: e. g. *Goatsbeard*.

— *Stalk*;

—— *Stalk*; undivided; as the ſtalks of *Tulips* and *Thrift*.

SINGLE. One flower only upa ſtalk; as the *Tulip*.

SITTING *Leaves*; have no leaf-ſtalk; as *Spearmint* and *Hound's-tongue*.

—— *Flowers*; are thoſe that have no *Fruit-ſtalk*, as the flowers of *Mezereon*.

SKINNY. Tough, thin, and ſemi-tranſparent; as the cup of *Thrift*.

SOLITARY. Only one in a place; as but one flower on a fruit-ſtalk, and only one fruit-ſtalk proceeding from the ſame part of a plant.

SPATULA-SHAPED. The form of a leaf. Pl. 8. f. 64.

SPEAR-SHAPED. As the leaf of *Ribwort Plantain* and *Spearmint*. Pl. 7. f. 6.

SPECIFIC-CHARACTER. One or more circumſtances of a plant ſufficient to diſtinguiſh it from every other plant of the ſame genus. The ſpecific characters are generally taken from the leaves or ſtem; ſometimes from the flowers; but ſeldom from the roots.

SPIKE. A compoſition of flowers placed alternately on each ſide of a common ſimple fruitſtalk, and not ſtanding upon little fruitſtalks. *Great Mullein*, *Agrimony*, and many of the *Graſſes* have their flowers collected into ſpikes. Pl. 6. f. 5.

SPIKE-STALK. A long rough receptacle upon which the flowers compoſing a ſpike are placed. Take a ſpike (or as it is frequently called, an ear) of Wheat; pull off all the ſeeds and chaff: what remains is a *Spike-ſtalk*. Pl. 2. f. 24, (*c. c.*)

SPINDLE-SHAPED *Root*. e. g a *Carrot*. Pl. 11. f. 6.

SPIRAL. Twiſted like a cork ſcrew. Pl. 10. f. 6. (*a. a.*)

SPOKES. The fruit ſtalks of flowers collected into *Rundles* or *Rundlets*: ſee thoſe terms. They ſpring from one point and diverge like the ſpokes of a cart wheel. Pl. 6. f. 9. *c. c. c. c.*)

SPREADING. Not riſing high but ſpreading wide upon the ground; as the ſtems of *Fumitory* and *Panſie*.

SPUR. Shaped like the ſpur of a cock, as the honey-cups of the *Larkſpur*.

STALK. That ſpecies of trunk which elevates and ſupports the flowers, but not the leaves of a plant. It differs from the *Fruit-ſtalk*, for that ſprings from the ſtem, or branches; but this riſes immediately from the root; as in *Narciſſus*; *Lilly of the Valley*, and *Hyacinth*. Pl. 6. f. 4.

STANDARD. The upright petal of a butterfly-ſhaped bloſſom; as in the *Pea*. See the introduction to the 17th claſs. See alſo pl. 4. f. 12. (*b.*) f. 14. (*b.*) f. 15.

STARRY. Plants whoſe leaves grow in whorls round the ſtem; as the *Gooſegraſs*, *Cheeſe-rennet*, and ſeveral other plants in the fourth claſs. Pl. 9. f. 3. (*b. b.*)

STEM. The proper trunk of a plant ſupporting the leaves, branches and flowers. It riſes immediately from the root.

STEM *Leaves*. Such as grow immediately upon the ſtem without the intervention of branches.

STINGS. Sharp pointed ſubſtances conveying poiſon into the part they penetrate. Few people are ignorant of the ſting of a *Nettle*.

STONE. See Nut.

STRADDLING; Branches ſtanding wide from each other.

STRAP-SHAPED. Long and narrow like a ſtrap or a fillet; as the leaves of *Thrift*; *Crocus* and *Roſemary*. Pl. 7. f. 7.

STRAW. A kind of trunk proper to Graſſes. Pl. 10. f. 3.

STRINGS. Woody fibres in leaves running undivided from the baſe to the extremity; as in the broad and narrow leaved *Plantain*. pl. 7. f. 46.

SUCKERS. Shoots that riſe from the root, ſpread along the ground and then take root themſelves; as in the *Sweet Violet*.

SUMMIT. The upper part of a pointal. See *Pointal*, pl. 3. f. 2. (*f*.) f. 5. (*e*.) f. 7. (*l*.)

SUPERIOR *Cup* or *Bloſſom*: when the cup or bloſſom is ſituated above the ſeedbud it is ſaid to be ſuperior as in the *Honeyſuckle*; *Currant* and *Campanula*.

SWORD-SHAPED. As the Leaves of the *Iris* or *Flower de Luce*.

TARGET. A kind of flower in the genus *Cupthong*, that is circular and convex. See *Saucer*.

TENDRIL. A ſpiral ſhoot or ſtring, by means of which ſome plants ſupport themſelves againſt the adjacent bodies. It is well known in the *Vine* and *Pea*. Pl. 10. f. 6. Pl. 8. f. 58.

TERMINATING. (Oppoſed to lateral) ſtanding at the ends of the ſtem or branches; as the fruitſtalks of *Borrage*, the *Bloſſoms* of *Groundſel*.

THORN. A ſharp pointed projection growing from the woody ſubſtance of a plant; as in *Gorze* and *Blackthorn*. Pl. 10. f. 1.

THREAD. A part of a chive ſupporting the tip. See *Chive*. Pl. 3. f. 3 (*b*.) f. 6. (*g*.) Pl. 1. f. 19. (*a a*.)

THREAD-SHAPED. Of the ſame thickneſs from top to bottom, like a piece of packthread. Take for example the leaves of *Fennel*, or the ſhaft of the *Crocus*, or *Honeyſuckle*.

THREE-EDGED. A ſtem having three corners or angles and the ſides not flat.

THREE-CORNERED. Having three corners or angles with flat ſides: as the ſtem of the *Panſie*.

TILED. One leaf or ſcale partly covering another like the tiles on a houſe. e. g. The cup of *Dandelion* or of *Burdock*. Pl. 4. f. 25. (*a*.)

TIP. A part of a chive fixed upon the thread, and containing the duſt. In *Dogs Mercury* it hath one cell; in *Hellebore* two; in *Orchis* three; in *Fritillary* four; ſee *Chive*. Pl. 3. f. 2. (*c.c.c.c.c.c*.) f. 5. (*b.b.b.b.b.b*.) f. 6. (*b*.)

TOOTHED. When the edges of a leaf are ſet with little teeth, not pointing towards the end as in the ſerrated leaves, nor towards the baſe as in the inverſely ſerrated leaves. *Common Eyebright*; *Primroſe*; *Cowſlips*; and *Mountain Willowherb*, have toothed leaves. Pl. 7. f. 30.

TRAILING *Stems*; Lying along upon the ground, and not ſending out roots. e. g. *Common Speedwell*; *Red Pimpernel*; *ſmall Sea Bindweed*.

TREBLY *Compound Leaves* are of three different kinds.

1. *Double Twinfork*; the leafſtalks twice forked and two little leaves at the end of each point; or three times divided, and three little leaves upon each diviſion. Pl. 8. f. 57.

2. *Triply*

2. *Triply three-fold*; the divisions of a triple leaf-stalk again subdivided into threes, and three little leaves at the end of each subdivision. Pl. 8 f. 59.

3. *Triply winged*; when the lateral ribs of a doubly winged leaf, have themselves other ribs with winged leaves. Pl. 8. f. 60 61.

TRIANGULAR. Expressing the form of a leaf that hath three sides and three angles, or corners. Pl. 7. 1. 12.

TRIANGULARLY *Spear-shaped*: Leaves in this form are broad at the base and nearly triangular, but spear-shaped at the point. e. g. *Black Poplar*. Pl. 7. f. 45.

TRIPLY WINGED. See *Trebly Compound*.

TRIPLY *Three-fold*. See *Trebly Compound*.

TRUNK. The main body of a plant: it is either a *Stem*, a *Stalk*, a *Straw*, or a *Pillar*: see those terms.

TUBE. The lower part of a blossom of one petal is frequently lengthened out into a tube, as in *Crocus* and *Polianthus*.

TUBERCLE. A little solid pimple.

TUBEROUS *Root*; consisting of many roundish knobs collected into a bundle, as the root of *Peony* and *Dropwort*. Pl. 11. f. 7.

TUBULAR. In the shape of a hollow tube, as the cup of *Privet*; the blossom of the *Honeysuckle*, or the honey-cup of the *Hellebore*.

—— *Florets* in compound flowers, are shaped like a hollow tube, and the top is cloven into five segments. In the *Tansey* all the florets are tubular, but in the *Sunflower* and the *Daisie* only those in the center. Pl. 4. f. 26.

TUFT. A composition of flowers in which a number of fruit-stalks proceeding from one common center rise to the same height; and these again shoot out other little fruit-stalks, which do not proceed from one central point. The *Elder*, the *Gelder Rose*, and the *Laurustinus* are instances. Pl. 6. f. 10.

TURBANSHAPED. Like a Turkish turban; exemplified in the cup of the *Elm*, or *French Wheat*; some *Pears* are in this form.

TWINING. Twisting round other bodies and ascending in a spiral line. Some plants twine from the left to the right ☾ in the direction of the sun's apparent motion, as *Hop*, *Honeysuckle* and *Black Briony*. Others twine from the right to the left ☽ contrary to the sun's apparent motion, as *Bindweed* and *Scarlet Kidney Bean*. Pl. 10. f. 5.

TWINFORK. See *Doubly compound Leaves*.

TWO EDGED. As the stem of *Tutsan* and the *Sweet-smelling Solomon's Seal*.

VALVE. The different pieces that compose a capsule are called valves. Thus in the *Thornapple* there are four valves. Pl. 5. f. 14. (*c. c c. c.*) In the *Loose strife* ten; in *Jacob's Ladder*, *Daffodil* and *Hyacinth* three. Pl. 5. f. 6. f. 12. (*a. a.*)

The petals and empalements that constitute the flowers of Grasses are called valves; thus in the *Common Meadow-Grass*, the empalement is a dry chaffy husk, composed of two valves, and the blossom is formed of two other valves See pl. 2. f. 1 . (*a. a.*) (*b. b*) and

most

most of the other figures in that plate.

The mouth of the tube of a blossom is frequently closed by several projecting substances; thus in the blossoms of *Borrage* and *Jacob's Ladder* the tube is closed by five of these substances; and they are called valves.

VANELIKE. Turning about like a vane or weathercock, as is the case with the tips of *Geranium*, and *Crown Imperial*.

VAULTED. Like the roof of one's mouth. The upper lip of many of the gaping blossoms is vaulted; *e.g.* red and white *Deadnettle*.

VEIL. The empalement of mosses, covering the tips. It is generally in a conical form like an extinguisher. Pl. 1. f. D. (*a.*)

VIVIPAROUS. A term applied to stems or stalks producing bulbs that are capable of vegetation. In *Toothwort* and *Star of Bethlem*, they are found at the base of the leaves; in *small Bistort*, on the lower part of the spike; in some species of *Garlic* at the origin of the rundle of flowers; and on the spikes of some of the grasses, as the *Cats-tail Canary*.

UMBEL. See RUNDLE.

UNARMED. Without weapons of defence. See WEAPONS.

UNEQUAL-*Florets*; when a rundle is not composed of equal florets, but those in the circumference are larger than those in the center; and the outer petals are larger and different in shape from the inner petals. As in the *Carrot* and *Cowparsnep*.

UNIFORM. A term applied to compound flowers when the florets that compose them are all alike; as those of *Fennel*, *Lettuce*, and *Burdock*.

UPRIGHT. Standing upright, as the cups of *Periwinkle*; the tips of *Polyanthus*; the stalks of *Tulips*; the stems of *Sparagus*. It is also applied to leaves. Pl. 9. f. 5. (*b.b.*)

WARTY. Having little hard lumps or warts upon the surface.

WAVED. When the surface of a leaf towards the edge does not lie flat, but appears waved, and full, like a man's ruffle. The leaf of the *Water Caltrops* is an example, Pl. 8. f. 66.

WEAPONS are either *Prickles*, *Thorns* or *Stings*. See those terms.

WEDGE-SHAPED. As the leaves of the *Garden Spurge*; and the *Garden Purslain*. Pl. 8. f. 65.

WHEELSHAPED. A term used to express a blossom of one petal, with a flat border and a very short tube. *Borrage* and *Speedwell* are familiar examples. Pl. 4. f. 6.

WHORLS of branches, leaves, or flowers. The branches of the *Fir*, the leaves of *Ladies Bedstraw*, and the flowers of the red and white *Deadnettle* grow in whorls round their respective stems. They somewhat resemble the spokes round the nave of a wheel. pl. 6. f. 11.

WINGS. The lateral petals of a butterfly-shaped blossom; e. g. in the *Pea*. See the Introduction to the seventeenth Class. See also pl. 4. f. 13. (*c. c.*) and f. 16.

WINGED-*Leaves*; when an undivided leaf stalk hath many little leaves growing from each side; as in *Jacob's Ladder*; *Bladder Sena*; *Ash* and *Pea*. Pl. 8. f. 52, 53, 54, &c.

WINGED-*Clefts*; applied to a leaf that is cut and divided so deeply on each side, down towards the middle

middle rib, as almost to resemble a winged leaf. The *Corn Poppy* and the *Polypody* are examples; and so are the root-leaves of the *Shepherds Purse*.

WINGED-*Leaf-stalk*; one that is not cylindrical, but flattish, with a thin leafy border at each edge.

WINGED-*Shoots*. When the shoots strike out from the sides, like the plumage along the sides of a quill. Instances will be found in several species of the *Feathermoss*.

WIRES. Barren twigs or shoots lying upon the ground, as in the *Garden Strawberry* and *Stone Bramble*.

WOODY. Opposed to herbaceous The stems of the *Wall-flower* or *Gilliflower* are woody.

WOOL. A kind of downy cloathing upon the surface of some plants. The leaves of *Horehound*, *Great Mullin* and *Gorze* are woolly.

WRINKLED. As are the leaves of *Sage*, *Primrose*, *Wood Strawberry* and *Hazel*.

ZIGZAG. Having many turnings and bendings, as the stems of *Rough Bindweed*, and *Woody Nightshade*; or the branches of *Golden Rod*.

THE

LATIN TERMS of *LINNÆUS*;

With the correfponding *Englifh* WORDS,

ABBREVIATUM, *fhort.*
Abortiens, *abortive.*
Abortivi (flofculi) *barren.*
Abruptum, *abrupt.* Pl. 8. f. 53.
Acaulis, *ftem-lefs.*
Acerofum, *chaffy.* Pl. 9. f. 3. (*e.*)
Acicularis, *Needle fhaped.*
Acinaciformis, *Scymeter-fhaped.*
Acini, *Granulations.*
Acotyledones, *Seeds without Lobes.*
Aculei, *Prickles.* Pl. 10. f. 2.
Aculeata, *prickly.*
Acuminatus, *taper.*
Acuminatum (fol.) *tapering to a Point.* Pl. 7. f. 41.
Acutus, *fharp.* Pl. 7. f. 40.
Adnatum, *connected.*
Adpreffa, *contiguous.* Pl. 9. f. 6.
Adfcendens, *afcending.*
Adverfum, *turned towards the South.*
Æqualis, *equal.*
Aggregatus, *incorporated.*
Ala, *wing.* Pl. 4. f. 16.
Alatus, *winged.*
Alburnum, *a foft white fubftance betwixt the inner bark and the wood.*
Algæ, *Thongs.*
Alternus, *alternate.* Pl. 9. fig. 3. (*d. d. d. d. d*)
Amentum, *Catkins.*
Amplexicaule, *embracing the Stem.* Pl. 9. f. 4. (*f.*)
Anceps, *Two-edged.*
Androgynia, *Chives and Pointals feparate.* Pl. 1. f. 21.
Angulatus, *angular.*
Anguftifolia, *narrow-leaved.*
Angiofpermia, *covered Seeds.*
Annua, *annual.*
Anomala, *without Order.*
Anthera, *Tip.* Pl. 3. f. 2. 3. 5. 6.
Apetalus, *without Petals.*

Apex,

Apex, *Point.*
Apophyſis, *Excreſcence.*
Appendiculatus, *with an Appendage.*
Approximata, *near together.*
Arborea, *Woody.*
Arcuatum, *bowed.*
Arillus, *ſee Caliculus.*
Ariſta, *Awn.* Pl. 2. f. 21. (*b. b.*) 23. (*b. b.*)
Arma, *Weapons.* Pl. 10.
Articulatum, *jointed.* Pl. 10. f. 3. Pl. 9. f. 3. (*a*)
Aſperifolia, *rough-leaved.*
Aſſurgentia, *riſing.*
Attenuata, *tapering.*
Auctus, *leafy.*
Auriculatum, *Ear-ſhaped.*
Avenis, *without Veins.*
Axillare, *at the baſe of the Leaves.* Pl. 9. f. 5. (*m.*)
Bacca, *berry.* Pl. 5. f. 19.
Barba, *Beard.*
Barbata, *bearded.*
Bicapſularis, *two Capſules.* Pl. 1. f. 8.
Biennis, *biennial.*
Bifarius, *oppoſite.*
Bifida, *Cleft or cloven.*
Biflorus, *two flowered.*
Bigeminum, *twinfork.*
Bijugum, *two Couple.*
Bilabiatus, *Two lipped.* Pl. 4. f. 8. 9. 10.
Bilobum, *two lobes.*
Binata, *in Pairs.* Pl. 7. f. 50.
Bilocularis, *two Cells.* Pl. 5. f. 12.
Bipartita, *two Diviſions.*
Bipinnatum, *doubly winged.*
Biternatum, *doubly threefold.* Pl. 8. f. 57.
Bivalve, *two Valves.* Pl. 5. f. 9. 16.
Brachiatus, *ſee Decuſſatus.*
Bractea, *floral Leaf.* Pl. 9. f. 8.
Bulboſa, *bulbous.* Pl. 11. f. 2. 3. 4.
Bulbus, *Bulb.*
Bullata, *bliſtered.*
Caduca, *ſhedding.*
Cæſpitoſus, *matted together.*
Calcariatum, *having a Spur.* Pl. 5. f. 4. (*a.*)
Caliculatum, *Cup double.*
Caliculas, *Seed-coat.* Pl. 6. f. 1. (*c. c.*)
Calyptra, *Veil.* Pl. 1. f. D. (*a.*)
Calyx, *Empalement.* Pl. 4. f. 7. (*c.*) f. 20. Pl. 3. f. 10. f. 5. (*a.*)
Campanulata, *Bell-ſhaped.* Pl. 4. f. 2. 3. 4.
Canaliculata, *channelled.*
Capillaris, *hair-like.*
Capitatus, *growing in Heads.*
Capitulum. *Knob.*
Capitulus, *flowering Head.*
Capreolus, *ſee Cirrhus.*
Capſula, *Capſule.* Pl. 5. f. 6. 9. 14.
Carina, *Keel.* Pl. 4. f. 7.
Carinatum, *Keel-ſhaped* Pl. 5. f. 13.
Carnoſa, *fleſhy.* Pl. 5. f. 20.
Cartilaginea *griſtly.*
Catenulata, *chained.*
Catulus, *ſee Amentum.*
Cauda, *Tail.*
Caudex, *Body.*
Cauleſcens, *having a Stem.*
Caulinum, *belonging to the Stem.*
Caulis, *Stem.*
Cernuus, *crooked, when applied to Fruit-ſtalks.*
Ciliatus, *fringed.*
Cingens, *binding round.*
Circumciſſa, *cut round.* Pl. 5. f. 9.
Cirrho um fol. *terminating in a Tendril.*
Cirrhus, *Tendril.* Pl. 10. f. 6.
Claſſis. *Claſs.*
Clauſa, *cloſed.*
Clavata, *Clubſhaped.*
Clavicula, *ſee Cirrhus.*
Coadunatum *joined.*
Coarctata, *compact.*

Cochleatum,

Cochleatum, *Snail-shell.*
Colorata, *coloured.*
Columnella, *Column.* Pl.5.f.14.(*d*)
Columnaris, *square Pillars.*
Coma, *Bush.*
Communis, *common.*
Compactum, *firm.*
Completi, *perfect.*
Compositus, *compound.*
Compressa, *compressed.*
Concava, *concave.*
Conduplicata, *doubled together.*
Confertus, *crowded.*
Confluentia, *thronging.*
Congesta, *collected into a ball.*
Conglomorati, *close.*
Conica, *conical.*
Connatum, *Twinleaves.* Pl. 9. f. 4. (*b. b.*)
Connivens, *approaching.*
Connivens calyx, *closing.*
Contraria, *contrary.*
Convexa, *convex.*
Convolutus, *from left to right.*
Corculum, *Heart.* Pl. 6. f. 3. (*b*)
Cordata, *Heart-shaped.* Pl. 7. f. 10.
Cornutum, *Horn-shaped.* Pl. 5. f. 3. (*a. a.*) Pl. 5. f. 4. (*a.*)
Coralla, *Blossom.* Pl. 4.
Corona, *Crown.* Pl. 4. f. 27. (*b*) Pl. 5. f. 5. (*a. b.*)
Cortex, *Bark.*
Corymbus, *broad topped Spike.* Pl. 6. f. 7.
Cotyledones, *Seed-lobes.* Pl. 6. f. 3. (*a. a.*)
Crenatus, *scolloped.* Pl.7.f.34.35.
Crispa, *curled.* Pl. 8. f. 67.
Cristatus, *crested.*
Cruciata, *Cross-shaped.* Pl. 4. f. 11. 12.
Cryptogamia, *Flowers inconspicuous.* Pl. 1. A. B. C. D. E. F. G. H.
Cucullata, *hooded.*
Culmus, *Straw.* Pl. 10. f. 3.
Cuneiformis, *Wedge-shaped.* Pl. 8. f. 65.
Cuspidatus, *Spit-pointed.*
Cyathiformis, *Glass-shaped.*
Cylindricus, *cylindrical* Pl.8.f.68.
Cyma, *Tuft.* Pl. 6. f. 10.
Dædaleum, *beautiful Texture.*
Debilis, *feeble.*
Decagynia, *ten Pointals.*
Decandria, *ten Chives.* Pl. 1. f. 10.
Decaphyllus, *ten leaved.*
Decidua, *deciduous.*
Declinata, *declining.*
Decompositus, *doubly compound.* Pl. 8. f. 56. 57.
Decumbens, *drooping.*
Decurrens, *running along the Stem.* Pl. 9. f. 4. (*c.*)
Decursivis, *running along the Leaf-stalk.*
Decussatus, *cross Pairs.* Pl. 9. f. 1.
Deflexus, *a little bent outwards.*
Deflorata, *having discharged.*
Deltoidea, *triangularly Spear-shaped.* Pl. 7 f. 45.
Demersum, *see Submersum.*
Dendroidis, *shrubby.*
Dentato-sinuatum, *toothed and indented.* Pl. 7. f. 26.
Dentatus, *toothed.* Pl. 7. f. 30.
Denticulatum, *little Teeth.*
Dependens, *hanging down.*
Depressa, *depressed.*
Diadelphia, *Threads in two Sets.* Pl. 1. f. 17. (*a. b.*)
Diandria, *two Chives.* Pl. 1. f. 2.
Dichotoma, *forked.* Pl. 10. f. 4.
Didyma, *double.* Pl. 3. f. 6. (*b.*)
Didynamia, *two Chives longer,* pl. 1. f. 14. (*a a*)
Difformia, *irregular and uncertain shaped.*
Diffusa, *spreading.*
Digitatum, *fingered.* Pl. 7. f. 48.
Digynia, *two Pointals.*
Dimidiatum, *going half way round.*
Dioecia, *Chives and Pointals distinct.* Pl. 1. f. 22.
Diphyllus, *two-leaved.*
Discus, *Center.*

Dissectum,

Diffectum, *fee* Laciniatum.
Difperma *two feeds.*
Diffepimentum, *Partition,* Pl. 5. f. 12. (*b. b.*) f. 14. (*b. b. b. b.*)
Diffiliens, *burfting.*
Diftans, *diftant.*
Difticha, *pointing from two oppofite lines.*
Divaricata, *ftraddling.*
Divergens, *diverging.*
Dodecandria, *twelve Chives.* Pl. 1. f. 11.
Dolabriformis, *battledore-fhaped.*
Dorfalis, *fixed to the Back.*
Drupa, *pulpy Seed-veffel.* Pl. 5. f. 21.
Duplicata, *doubly.*
Echinatum, *befet with Prickles.*
Elliptica, *fee* Ovale.
Emarginatus, *notched at the End.* pl. 7. f. 16. 36.
Enervis, *without Strings.*
Enneandria, *nine Chives,* pl. 1. f. 9.
Enodis, *without Joints.*
Enfiformis, *Sword-fhaped.*
Equitans, *laminated.*
Erecta, *upright.*
Erofum, *gnawed,* pl. 7. f. 21.
Exferta, *ftanding out.* pl. 2. f. 18. (*e. e. e.*)
Extrafoliacea, *beneath the Leaves.*
Farctum, *full.*
Fafcicularis, } *bundled,* pl. 9. f. 3.
Fafciculatus, } *f.*)
Fafciculus, *a bundle.*
Faftigiatus, *level.*
Faux, *Mouth,* pl. 4. f. 9. (*d. d.*)
Femineus Flos. *fertile Flower,* pl. 1. f. 21. (*b.*) 22. (*b.*) 23. (*c.*)
Fertiles, *fertile.*
Fibrofa, *fibrous.*
Filamentum, *Thread,* pl. 3. f. 6. (*g.*)
Filices, *Ferns,* pl. 1. A. B.
Filiformis, *Thread-fhaped.*
Fimbricata, *tattered.*
Fiffum, *cloven.*
Fiftulofa, *hollow.*
Flaccidi, *limber.*
Flagellis, *Wires.*
Flexuofa, *zigzag.*
Florale, *floral Leaf,* pl. 9. f. 8. (*a. a.*)
Flos, *Flower.*
Flofculus, *Floret,* pl. 4. f 21. f. 26.
Flofculofi, *tubular Florets,* pl. 4. f. 26.
Foliatus, *covered with Leaves.*
Foliaceum, *leafy.*
Folium, *Leaf.*
Folliculus, *Airbag.*
Fornicatum, *vaulted,* pl. 4. f. 8. (*a*)
Fruticofus, *woody.*
Fructificatia, *Flower.*
Fructus, *Fruit.*
Fulcra, *Supporters.*
Fungi *Funguffes,* pl. 1. H.
Furca, *Fork.*
Furcata, *forked.*
Fufiformis, *fpindle-fhaped.* pl. 11. f. 6.
Galea, *Helmet.*
Geminis, *in Pairs.* pl. 7. f. 50.
Gemma, *Bud.*
Genus, } *fee the Introduction.*
Genera, }
Geniculatus, *Knee-jointed.* pl. 2. f. 21. *the Awns.*
Geniculum, *Knee-joint.*
Germen, *Seed-bud.* pl. 3. f. 2. (*d*) f. 7. (*l.*)
Gibba, *bunched.* pl. 4. f. 12. (*b.*)
Glabra, *fmooth.*
Glandula, *Gland.* pl. 10. f. 6. (*c. c.*) pl. 11. f. 1. (*a. a. a. a.*)
Globofa, *globular.* pl. 5. f. 5.
Glochis, *Hook with many Points.*
Glomerata, *congregated.*
Gluma, *Hufk,* pl. 2. f. 1. (*a.a.*) f. 18. (*a. a*)
Glutinofitas, *gummy.*
Gramina, *Graffes.*
Granulata, *beaded.*
Gymnofpermia, *Seeds naked.*

Gynandria,

Gynandria, *Chives on the Pointal.* pl. 1. f. 20.
Hamus, *Hook.*
Hastata,. *Halberd-shaped.* pl. 7. f. 15.
Hemisphericus, *hemispherical.*
Heptandria, *seven Chives.* pl. 1. f. 7.
Herbacea, *herbaceous.*
Hermaphroditus, *Flowers containing both Chives and Pointals.*
Hexagonus, *six-sided.*
Hexagynia, *six Pointals.*
Hexandria, *six Chives.* pl. 1. f. 6.
Hians, *gaping.* pl. 4. f. 8, 9.
Hilum, *Eye.* pl. 6. f. 3. (*e*)
Hirsutus, *rough with Hair.*
Hispidus, *covered with strong Hair.*
Horizontalis, *horizontal.* pl. 9. f. 5. (*d. d*)
Hypocrateriformis, *Salver shaped,* pl. 4. f. 1.
Icosandria, *twenty Chives.* pl. 1. f. 12.
Imbricata, *tiled,* pl. 9. f. 2.
Inanis, *pithy.*
Incanum. *See* Tomentosum.
Incisum. *See* Laciniatum.
Inclinatus, *leaning.*
Inclusa, *inclosed.*
Incompleti, *imperfect.*
Incrassatus, *thicker towards the top.*
Incumbentes, *fixed Side-ways.*
Incurvata, *bowed inwards.* pl. 9. f. 5. (*a. a.*)
Inerme, *unarmed.*
Inferus, *beneath.*
Inflata, *bladder-shaped.*
Inflexa, *bent inwards.*
Inflorescentia, *Mode of flowering.*
Infundibuliformis, *funnel-shaped,* pl. 4. f. 7.
Integer, *entire.*
Integerrimus, *very entire.*
Interrupta, *interrupted,* pl. 8. f. 55.
Intrafoliacea, *upon the Leaves.*
Involucellum, *partial Fence.* pl. 6. (*d. d. d. d.*) f. c.
Involucrum, *Fence.* pl. 6. f. 9. (*c. c.*)
Involuta, *rolled inwards.*
Irregularis, *irregular.*
Labiatus, *with Lips.*
Labium, *Lip.* pl. 4. f. 8. 9. (*a. a.*) (*b. b.*) f. 10. (*a. b.*)
Lacerus, *ragged.*
Lacinia, *Segments.*
Laciniatus, *jagged.* pl. 17. f. 24.
Lactescentia, *milky Juices.*
Lacunosa, *pitted.*
Lævis, *even.*
Lamella, *Gills.*
Lamina, *Limb.* pl. 4. f. 11. (*b. b. b. b.*) f. 12. (*a. a. a. a.*)
Lana, *Wool.*
Lanata, *Cobwebbed.*
Lanceolata, *Spear-shaped,* pl. 7. f. 6.
Laterales, *lateral.*
Laxus, *flexible.*
Legumen, *Shell.* pl. 5. f. 16.
Leprosus, *spotted like a Leper.*
Liber, *the inner Bark.*
Ligulatus, *narrow.*
Limbus, *Border.* pl. 4. f. 1. (*b. b.*)
Linearis, *Strap-shaped.* pl, 7. f. 7. pl. 4. f. 21.
Lineata, *streaked.*
Lingulata, *Tongue-shaped.*
Lobata, *gashed.* pl. 7. f. 19.
Lobum, *Lobe.* pl. 7. f. 17.
Loculamentum, *Cell.* pl. 3. f. 4.
Longum, *long.*
Lucida, *transparent.*
Lunata, *Crescent-shaped* pl. 7. f. 11
Lyrata, *Lyre-shaped,* pl. 8. f. 62
Magnitudo, *size.*
Marcescens, *shrivelling.*
Marginatum, *bordered.*
Masculus, *barren,* pl. 1. f. 21. (*a*) 22 (*a*) 23.
Membranacea, *membranaceous*
Monadelphia, *Threads united,* pl. 1. f. 16.
Monandria, *one Chive,* pl. 1. f. 1
Monoecia, *Chives and Pointals separate,* pl. 1. f. 21.

Monogynia, *one Pointal*, pl. 1. f. 1. (c) f. 2. (a)
Monopetala, *one Petal*, pl. 1. f. 1, 2, 3, 4.
Monophyllum, *one leaf.*
Mucronatum, *sharp-pointed.*
Multifidum, *many Clefts.*
Multiflori, *many Flowers.*
Multipartitum, *deeply divided into many parts.*
Muricata *covered with sharp points.*
Musci, *Mosses*, pl 1. f. C. D.
Mutica, *without Awns.*
Natans, *floating.*
Navicularis, *boat shaped*, pl. 5. f. 13, (a, a)
Nectarium, *Honey-cup*, pl. 3. f. 3. (k) pl. 5, f. 1; (a) f 2, 3, 4.
Nervosa, *stringy*, pl 8, f. 46.
Nidulantia, *dispersed in Pulp.*
Nitida, *shining.*
Nuda, *naked.*
Nutans, *nodding.*
Nux, *Nut*, pl. 5, f. 21. (b, b)
Obcordatum, *inversely heart shaped*, pl. 8, f. 69
Obliqua, *oblique.*
Oblonga, *oblong*, pl. 7. f. 5
Obsolete, *indistinctly.*
Obtusus, *blunt.*
Octandria, *eight Chives*, pl. 1, f 8
Operculum, *lid*, pl. 1, f D, (b)
Operculatum, *covered with a lid.*
Oppositifolia, *opposite the leaves.*
Oppositus, *opposite in Pairs*, pl. 9, f. 1.
Orbiculata, *round and flat*, pl. 7. f. 1.
Ore Perianthii, *Rim of the Cup.*
Ossea, *hard as bone.*
Ovale, *oval*, pl. 7. f. 4.
Ovata, *Egg-shaped*, pl. 7, f. 3.
Pagina, *Surface.*
Palatum, *Palate*, pl. 4, f. 10, (c)
Palea, *Chaff.*
Paleacea, *chaffy.*
Palmata, *hand-shaped* pl. 7, f. 22
Panduriformis, *Fiddle-shaped*
Panicula, *Panicle*, pl. 6. f. 50.
Paniculatus, *panicled.*
Papilionacea, *Butterfly-shaped*, pl. 4. f. 13. f. 14
Papillosa, *pimpled.*
Pappus, *Feather*, pl. 4. f. 22. (l) pl. 6. f. 2. (a. b)
Parallelum, *parallel.*
Parasiticus, *parasitical.*
Partialis, *partial.*
Partita, *divided*, pl. 7. f. 28.
Patens, *expanding*, pl. 9. f. 5. (c, c)
Patulus, *open.*
Pedatum, *Birdsfoot*, pl. 7. f. 49.
Pedicellus, *little Fruit-stalk*, pl. 6. f. 7. (a, a, a, a, a, a)
Peduncularis, *belonging to a Fruit-stalk.*
Pedunculati; *growing on Fruit-stalks.*
Pedunculus, *Fruitstalk*, pl. 9. f. 8. (c) f. 5. (m.)
Peltatis, *Leaves with Leaf-stalks fixed in the Center*, pl. 9. f. 4. (a)
Peltatum, *Target-shaped.*
Pencilliformis, *Pencil-shaped*, pl. 2. f. 11. (c, c)
Pendula, *pendant.*
Pentagonus, *five-sided.*
Pentagynia, *five Pointals*, pl. 1. f. 22: (b)
Pentandria, *five Chives*, pl. 1. f. 5.
Pentapetala, *five Petals*, pl. 4. f. 6. pl. 5. f. 2.
Pentaphyllus, *five-leaved.*
Perennis, *perennial.*
Perfoliatum, *perforated*, pl. 9. f. 60. (g.)
Perianthium, *Cup*, pl. 3. f. 1. f. 10. f. 5. (a) pl. 4. f, 12. (b) f. 14, (a) f 18, (a)
Pericarpium, *Seed vessel*, pl. 5, from f. 5 to f 21.
Perichætium, *Receptacle of Mosses.*
Persistens, *permanent.*

Personata,

Perſonata, *gaping*, pl 4, f. 8. 9, 10.
Petalum. *Petal*, pl. 4. f. 18, (*b*, *b*, *b*, *b*).
Petaliformia, *reſembling Petals*.
Petiolaris, *fixed to the Leaf-ſtalk*.
Pet olaris, *fixed to the Leaf-ſtalk*.
Petiolatus, *with Leaf-ſtalks*.
Petiolus, *Leaf-ſtalk*, pl. 9, f. 4, (*c*)
Pileus, *Hat*, pl. 1. f. H. (*c*)
Pili, *Hair*.
Piloſa, *hairy*.
Pinnatifidum, *with winged Clefts*, pl. 7. f. 23.
Pinnatum, *winged*, pl. 8, f. 52. f. 53. f. 54.
Piſtillum, *Pointal*, pl. 3. f. 7. f. 2, (*d*, *e*, *f*.)
Plana, *flat*.
Plenus flos, *double Bloſſom*.
Plicata, *plaited*, pl. 7. f. 37.
Plumata, *plumed*.
Plumoſus, *downy*, pl. 4, f. 22, (*l*)
Plumula, *the aſcending part of the Heart*, pl. 6, f. 3, (*d*)
Pollen, *Duſt*, pl. 3, f 5, (*f*) f. 8. (*a*)
Polyadelphia, *Threads in many Sets* pl. 1. f. 18.
Polyandria, *many Chives*.
Polygamia, *various diſpoſitions*, pl. 1. f. 23.
Polyginia, *many Pointals*.
Polyphyllum, *many leaved*.
Polyſtachius, *many Spikes*.
Pomum, *Apple*, pl. 5, f. 20.
Pori, *Pores*, pl. 3, f. 3, (*k*)
Poſticus, *hinder part*.
Præmorſus, *bitten*, pl. 7. f. 18.
Priſmaticus, *Priſm-ſhaped*.
Procumbens, *trailing*.
Prolifer, *headed Stem*.
Proliferi flores, *one growing out of another*.
Prominulum, *prominent*.
Propago, *off-ſet*.
Proprium, *Individual Bloſſom*, pl. 4. f. 21. f. 26.
Pubes, *Cloathing*.
Pulpoſa, *pulpy*, pl. 5, f. 21
Pulveratum, *duſted*.
Punctata, *dotted*, pl. 4, f. 23, (*a*)
Racemus, *Bunch*, pl. 6, f. 8.
Rachis, *Spike-ſtalk*, pl. 2, f. 23, f. 24.
Radiata, *radiate*, pl. 4, f. 24.
Radicalia, *Root-leaves*, pl. 9. f. 7.
Radicans *ſtriking Root*, pl. 10, f. 7.
Radius *Circumference*.
Radii, *Spokes*, pl. 6. f. 9. (*e.e.e.e.*)
Rameum, *Branch Leaf*.
Ramoſiſſimus, *greatly branched*.
Ramoſus, *branching*.
Ramus, *branch*.
Rameum, *growing on the branches*.
Receptaculum, *Receptacle*, pl. 4, f. 11, (*c*) f. 23. (*a*)
Reclinatum, *curved*, pl. 9. f. 5, (*e*, *e*.)
Recurvatum, *bent backwards*.
Recta, *ſtraight*.
Reflexa, *reflected*, pl. 4, f. 5.
Regularis, *regular*.
Remotus, *remote*.
Reniformis, *Kidney-ſhaped*, pl. 7. f. 9.
Repandus, *ſerpentine*, pl. 7, 29.
Repens, *creeping*, pl. 10, f. 7, f. 8.
Reſupinatus, *lying on its back*.
Retrorſum ſinuatum, *barbed*, pl. 7. f. 27.
Retrorſum ſerratum, *inverſely ſerrated*.
Retrofractus, *bent back as if broken*.
Retuſus, *dented*.
Revoluta, *rolled back*.
Rhombea, *Diamond ſhaped*.
Rigidus, *inflexible*.
Rimoſus, *abounding with Chinks*.
Ringens, *gaping*, pl. 4, f. 8, f. 9, f. 10.
Roſtellatum, *the deſcending part of the heart*, pl. 6, f. 3 (*c*.)
Roſtrum, *Bill*. pl. 5, f. 15. (*a*)
Rotata, *Wheel-ſhaped*, pl. 4, f. 6.
Rugoſa, *wrinkled*.
Runcinata, *notched*.
Sagittata, *Arrow-ſhaped*, pl. 7. f. 13.

Sarmen-

Sarmentosus, *Runners.*
Scabra, *rough.*
Scandens, *climbing.*
Scapus, *Stalk*, pl. 6. f. 4.
Scariosa, *Skinny.*
Scrotiforme, *Purse-shaped.*
Scutellum, *a Saucer*, pl. 1. f. F.
Scyphifer, *cup bearing*, pl. 1. f. E.
Secunda, *pointing one way.*
Securiformis, *Hatchet-shaped.*
Semen *Seed.*
Semiteres, *half cylindrical.*
Sempervirens, *evergreen.*
Senis, *by sixes.*
Sericea, *silky.*
Serratus, *serrated*, pl. 7. f. 31.
Sessilibus, *sitting*, pl. 9. f. 4. (*d.*)
Setacea, *bristly.*
Setæ, *Bristles.*
Silicula, *Pouch*, pl. 5, f. 10. f. 11.
Siliqua, *Pod*, pl. 5. f. 17.
Simplex, *simple.*
Simplicissimus, *undivided.*
Sinuata, *indented*; pl. 7. f. 25.
Solida, *solid.*
Solitarius, *solitary.*
Spadix, *sheathed Fruit-stalk*, pl. 3. f. 9. (*d*)
Sparsus, *scattered.*
Spatha, *Sheath*, pl. 3. f. 9. (*a. a.*)
Spathulata, *Spatula-shaped*, pl. 8. f. 64.
Spica, *Spike*, pl. 6. f. 5.
Spicula, *a little Spike*, pl. 6. f. 5. (*a. b. c. d.*)
Spina, *Thorn*, pl. 10. f. 1.
Spinescens, *thorny.*
Spinosa, *thorny.*
Spiralis, *spiral.*
Squamata, *scaly*, pl. 4. f. 20. f. 25. pl. 11. f. 4.
Squamosus, *scaly.*
Squarrosus, *scurfy.*
Stamina, *Chives*, pl. 3. f. 2. (*c. b.*) f. 6. (*g. h.*) f. 3. (*h. i.*)
Stamineus flos. *barren Flower*, pl. 1. f. 21. (*a.*) f. 22. (*a.*) f. 23. (*a.*)
Stellata *starry*, pl. 9. f. 3. (*b. b.*)
Sterilis, *barren.*
Stigma, *Summit*, pl. 3. f. 2. (*f.*) f. 5. (*e.*) f. 7. (*l.*)
Stimuli, *Stings.*
Stipes, *Pillar*, pl. 1. f. H. (*b.*) pl. 6. f. 22. (*i.*)
Stipitatus, *standing on a Pillar.*
Stipula, *Prop*, pl. 10. f. 6. (*b: b.*)
Stoloniferus, *with Suckers.*
Striatus, *scored.*
Strictus, *stiff and straight.*
Strigosa, *strong Lance shaped Bristles*
Strobilus, *Cone*, pl. 5. f. 18.
Stylus, *Shaft*, pl. 3. f. 2. (*e.*) f. 5. (*d.*) f. 7. (*k.*)
Subdivisus, *subdivided.*
Submersum, *growing beneath the Surface of the Water.*
Subramosus, *a little branched.*
Subrotundum, *circular*, pl. 7. f. 2.
Subulatum, *Awl-shaped*, pl. 7. f. 8.
Suffruticosus, *somewhat woody.*
Sulcata, *furrowed.*
Superflua, *superfluous.*
Superum, *superior.*
Supradecomposita, *more than doubly compound*, pl. 8. f. 57 to 61.
Sutura, *Seam.*
Syngenesia, *Tips united*, pl. 1. f. 19, pl. 4. f. 21.
Teres, *cylindrical*, pl. 8. f. 68.
Tergeminum, *double Twinfork*, pl. 8. f. 57.
Terminalis, *terminating.*
Ternatum, *threefold*, pl. 7. f. 51.
Ternis, *growing by threes*, pl. 7. f. 47.
Tetradynamia, *four Chives longer.* pl. 1. f. 15.
Tetragonus, *four edged.*
Tetragynia, *four Pointals.*
Tetrandria. *four Chives*, pl. 1. f. 4.
Thyrsus, *Cluster.*
Tomentosum, *downy.*
Tomentum, *Down.*
Torosum, *protuberating.*
Tortilis, *twisted.*
Transversum, *transverse.*

Trapeziformis, *irregular square.*
Triandria, *three Chives*, pl. 1. f. 3.
Triangularis, *triangular*, pl. 7. f. 12.
Tricocca, *three Seeds in three Cells.*
Tricuspidata, *three pointed.*
Trigona, *three edged.*
Trigynia. *three Pointals*, pl. 1. f. 23. (*c.*)
Trinervata, *three fibred.*
Trinervis, *with three Fibres.*
Tripartitum, *with three Divisions.*
Tripinnatum, *triply winged*, pl. 8. f. 60, 61.
Triplinervis, *triple Fibres.*
Triquetra, *three cornered.*
Triternatum, *triply three-fold*, pl. 8. f. 59.
Truncatus, *lopped*, pl. 8. f. 63.
Truncus, *Trunk.*
Tuberculus, *Tubercle.*
Tuberosa, *tuberous.*
Tubulosa *tubular*, pl. 4. f. 26.
Tubus, *Tube*, pl. 4. f. 1. (*a.*) f. 7. (*a.*)
Tunicata, *coated.*
Turbinata, *Turban-shaped.*
Turgidum, *swollen.*
Umbella, *Rundle*, pl. 6. f. 9.
Umbellula, *Rundlet*, pl. 6. f. 9. (*b. b. b. b.*)
Umbilicatum, *dimpled.*
Uncinatum, *hooked.* pl. 4. f. 25. (*b.*)
Undata, *waved*, pl. 8. f. 66.
Unguis, *Claw*, pl. 4. f. 11. (*a. a.*)
Unicus, *single.*
Uniflora, *having but one Flower.*
Unilateralis, *growing only from one Side.*
Universale, *general.*
Urens, *stinging.*
Utriculus, *Bag.*
Vaginans, *sheathing*, pl. 9. f. 4. (*i.*)
Valvula, *Valve*, pl. 5. f. 6. (*a.a.a.*) f. 12. (*a. a.*) f. 13. (*a. a.*) f. 16. (*a. b.*)
Venosæ, *full of Veins*, pl. 7. f. 44.
Ventricosa, *distended*, pl. 4. f. 4.
Verrucosa, *warty.*
Versatilis, *vanelike.*
Verticillatus, *whorled.*
Verticilli, *Whorls*, pl. 6. f. 11. (*a. a. a.*)
Vexillum, *Standard*, pl. 4. f. 15. f. 13. (*b.*) f. 14. (*b.*)
Villi, *soft Hairs.*
Villosa, *woolly.*
Virgatus, *Rod-shaped.*
Viscida, *clammy.*
Viscositas, *Clamminess.*
Vivipara, *viviparous.*
Volubilis, *twining*, pl. 10. f. 5.
Volva, *Cap*, pl. 1. f. H. (*c.*)

AN

A N

Explanation of the Plates.

P L A T E III.

PARTS composing a FLOWER.

Fig. 1. A back view of a Rose to shew the *Empalement*, or *flower Cup*. *a. a. a. a. a.* the Segments of the Cup.

Fig. 2. A figure of the Crown Imperial, to shew

a. a. a. a. a. a. the Petals.

b. b. b. b. b. b. the Chives.

c. c. c. c. c. c. The Tips.

d. the Seed-bud.

e. the Shaft.

f. the Summit.

Fig. 3. *g.* a Petal of the Crown Imperial separated from the Flower.

h. i. A Chive. *h.* the Thread. *i.* the Tip.

k. A Honey-cup Pore.

Fig. 4. The Seed-vessel of the Crown Imperial cut a-cross, to shew the three Cells. During the existence of the Blossom this was called the Seed-bud.

PLATE III.

FIG. 5. A Flower with the Empalement, the Chives and the Pointal; but the *Petals* taken away.

a. The Empalement, or Cup.

b. b. b. b. b. b. The Tips of the Chives.

c. The Seed-bud.

d. The Shaft.

e. The Summit.

f. One of the Tips diſcharging its duſt.

FIG. 6. *g. h.* A Chive taken out of a flower.

g. The Thread. *h.* The Tip, which in this inſtance is double.

FIG. 7. *i. k. l.* A Pointal taken out of a flower. *i.* The Seed-bud. *k.* The Shaft. *l.* The Summit.

FIG. 8. *a.* A Particle of Duſt greatly magnified. *b.* The vapour eſcaping from it, which is ſuppoſed to to paſs through the Pointal to fertilize the Seed bud.

FIG. 9. A Daffodil and its ſheathing Empalement. *a. a.* the Sheath. *d.* The ſheathed fruit-ſtalk.

FIG. 10 A Cup which is the Empalement of a Polyanthus, with five ſharp teeth in the rim.

Plate III.

PLATE IV.

BLOSSOMS.

FIG. 1. A Bloſſom of one Petal, ſalver-ſhaped.
a. The Tube. *b. b.* The Border.

FIG. 2. A bell-ſhaped Bloſſom.

FIG. 3. A tubular bell-ſhaped Bloſſom.

FIG. 4. A Bloſſom bell-ſhaped but diſtended.

FIG. 5. A Bloſſom with ſix reflected Segments.

FIG. 6. A back view of a wheel-ſhaped Bloſſom, to ſhew the ſhortneſs of the Tube.

FIG. 7. A funnel-ſhaped Bloſſom. *a.* The Tube. *b.* The Border. *c.* The Cup.

FIG. 8. 9. Gaping Bloſſoms.
a. a. The upper Lip.
b. b. The lower Lip.
c. c. The Tube.
d. d. The Mouth.

FIG. 10. A gaping Bloſſom. *a.* The upper Lip. *b.* the lower Lip. *c.* The Palate.

FIG. 11. A croſs ſhaped Bloſſom with the Cup taken away, to ſhew *a. a.* the Claws of the Petals. *b. b. b. b.* The Limbs of the Petals. *c.* The Receptacle.

FIG. 12. A croſs ſhaped Bloſſom with the Empalement or Cup. *a. a. a. a.* The Petals. *b.* The Cup, hunched at the Baſe.

FIG. 13, 14. Two views of butterfly-ſhaped Bloſſoms. *a. a.* The Cups. *b. b.* The Standards. *c. c.* The Wings. *d.* The Keel.

FIG. 15. The Standard of a butterfly-ſhaped Bloſſom ſeparated from the other Petals. *c.* The Claw.

FIG. 16. One of the Wings of a butterfly-ſhaped Bloſſom ſeperated from the other Petals. *m.* The Claw.

FIG. 17. The Keel, or lowermoſt petal of a butterfly-ſhaped Bloſſom ſeparated from the other Petals.

PLATE IV.

FIG. 18. The Cup. Chives and Pointal of a butterfly-shaped Blossom after the Petals are taken away. *a*. The Cup *h*. The Chives. *i*. the Pointal.

COMPOUND FLOWERS.

FIG. 19. A Flower of DANDELION, as an example of a compound Flower in which all the Florets are strap-shaped.

FIG. 20. The common Empalement of a compound Flower, composed of upright Scales *d d*; and reflected Scales *c. c.*

FIG. 21. A strap-shaped Floret taken out of a compound Flower. *e*. the Blossom. *f*. the Seed-bud. *g*. the Tips forming a hollow Cylinder, through which passes the Pointal, with the two reflected Summits, *h*,

FIG. 22. *k*. the Seed of a compound Flower. *i*, the Pillar supporting the downy Feather *l*.

FIG. 23. A naked, dotted Receptacle of a compound Flower. *a*. the Receptacle. *b*. the Empalement reflected.

FIG. 24. The Flower of a DAISIE, as an example of a *Radiate* compound Flower, *a. a. a. a.* the strap-shaped Florets in the Circumference, *b*. the tubular Florets in the Center.

FIG. 25. The Flower of *Burdock*, as an example of a compound Flower in which all the Florets are tubular. *a*. the scaly tiled Empalement. *b*. one of the scales with its hooked Point. *c. c.* the tubular Florets.

FIG. 26. One of the tubular Florets separated from the rest. *d*. the Blossom. *c*. the Seed-bud. *f*. the Pointal.

FIG. 27. One of the seeds. *d*. the pyramidal seed, crowned by the short Feather *h*.

PLATE

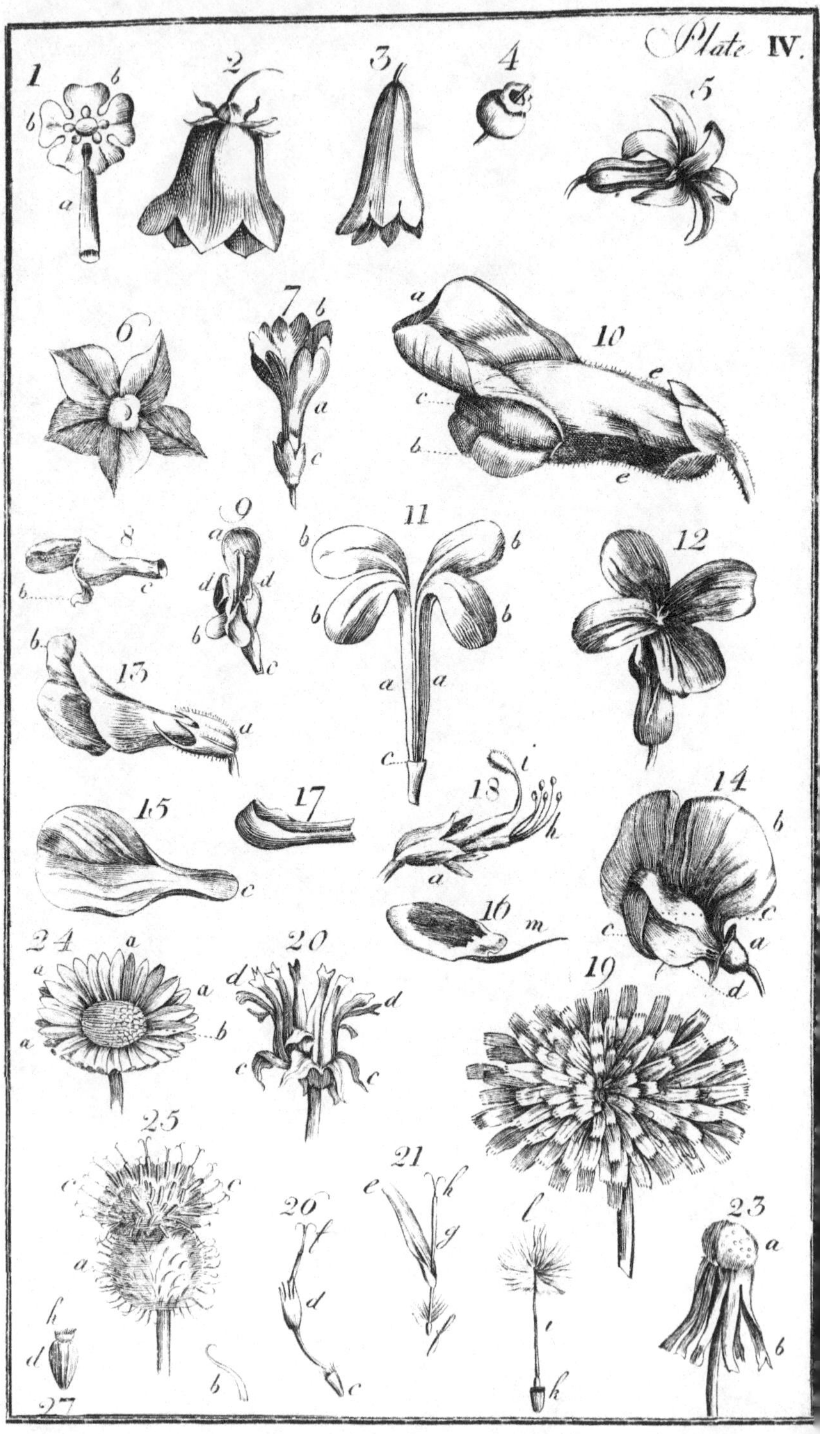
Plate IV.

PLATE V.

HONEYCUPS.

Fig. 1. The Blossom of a Daffodil, with the bell-shaped Honeycup, *a*.

Fig. 2. The Blossom of the Parnassus to shew the Honeycups *a. a. a. a. a.* which are little Globes supported upon Pillars, thirteen in each place.

Fig. 3. *a. a.* The Horned Honeycups of the Wolfs bane. *b. b.* the Foot-stalks that support them.

Fig. 4. *a.* The horn-shaped Honeycup of the Larkspur. *b. c. d. e. f* the Petals.

SEED-VESSELS.

Fig. 5. *c. c.* The globular Capsule of a Poppy. *a. a.* the holes through which the Seeds escape. *b.* the radiate Summit.

Fig. 6. A Capsule with three Valves, opening at the top. *a. a. a.* the Valves.

Fig. 7. A Capsule cut open length-ways, to shew the the Receptacle, with the Seeds fixed to it.

Fig. 8. A Capsule opening by holes at the sides, *a. a.* holes through which the Seeds escape.

Fig. 9. A Capsule that opens like a snuff box, or as if it was cut round. *a.* the Capsule entire. *b.* the Capsule open. *c.* the Receptacle as it appears after the Seeds are removed.

Fig. 10. An inversely heart-shaped Pouch, notched at the end.

Fig. 11. A circular Pouch notched at the end.

Fig. 12. A Pouch opened a little to shew *a. a.* the Valves. *b. b.* the Partition betwixt the Valves.

Fig. 13. A Capsule with two boat-shaped Valves, and one cell. *a. a* the Valves opening length-ways.

Fig. 4.

PLATE V.

Fig. 14. A Capsule cut open horizontally to shew *c.c.c.c.* the Valves. *b. b. b. b.* the Partitions. *d.* the Column in the Center to which the Partitions are connected. *a.a.a.a.* the Receptacles and Seeds.

Fig. 15. Seeds of Geranium, with a long Bill. *b.* the Seeds. *a.* the Bill.

Fig. 16. A Shell, or Seed-vessel of two Valves, in which the Seeds are fixed to the upper Seam only. *a. b.* the Valves.

Fig. 17. A Pod, or Seed-vessel of two Valves, in which the Seeds are fixed to the two Seams alternately.

a. b. the Valves. *d. d. d. d. c. c. c,* the Seeds.

Fig. 18. A Cone, cut through length-ways, to shew the Scales and the Seeds.

Fig. 19. A Berry cut across to shew *a. a.* the Seeds. *b. b.* the Pulp. *c. c.* the Coat.

Fig. 20. A fleshy Capsule, or Apple cut across to shew *b. b. b. b. b.* the five Cells.

Fig. 21. A pulpy Seed-vessel cut across. *a. a.* the pulpy part. *b. b.* the Nut or Stone.

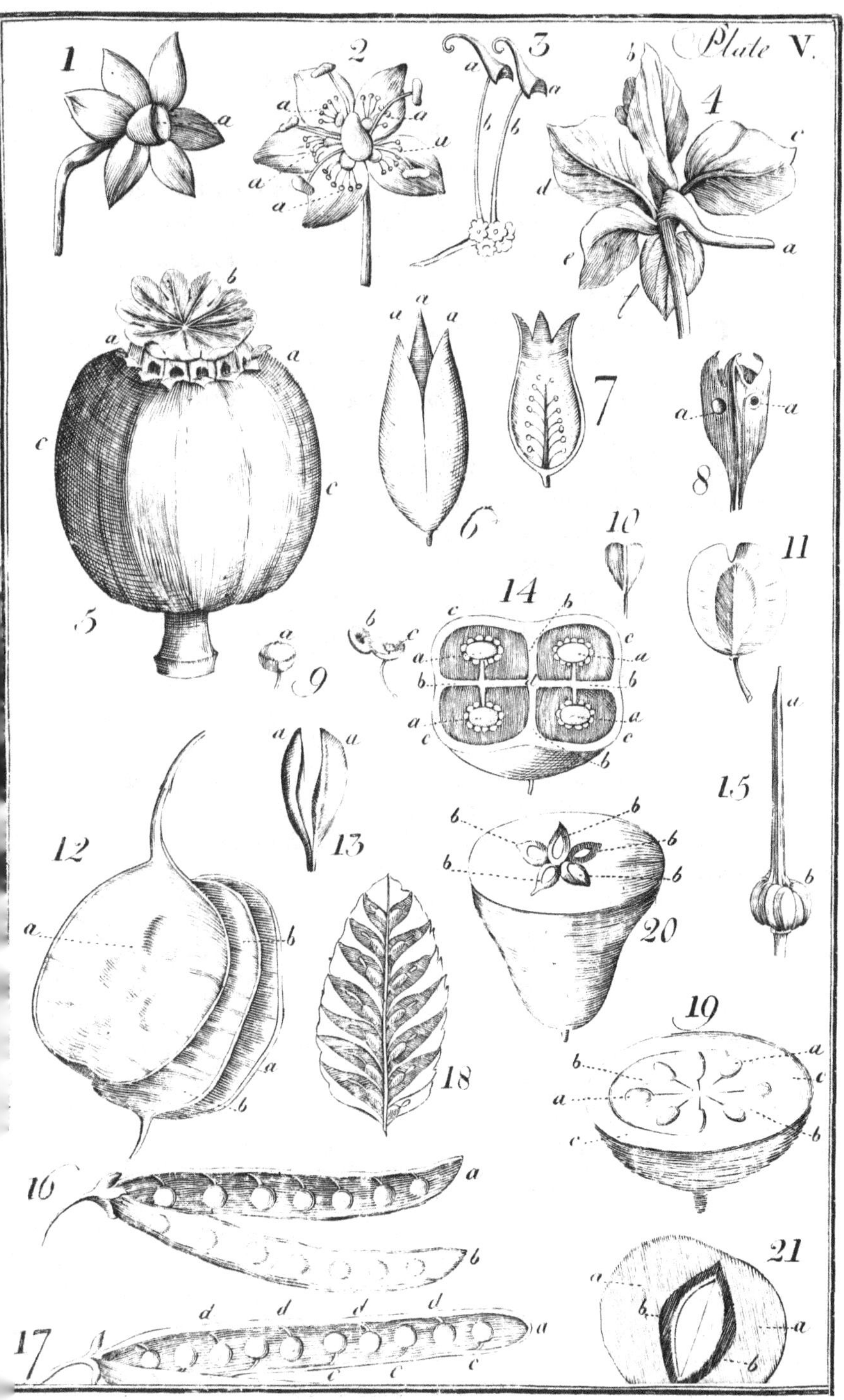
Plate V.
1
2
3
4
5
6
7
8
9
10
11
12
13
14
15
16
17
18
19
20
21

PLATE VI.

SEEDS.

FIG. 1. The Seed-vessel of the Spindle to shew the seed-coat.. *a. a.* The Valves of the Capsule. *b.* a Seed. *c. c.* the Seed-coat opened to shew the Seed.

FIG. 2. A Seed with its Feather.

a. A hairy feather. *b.* a downy Feather.

d. The pillar supporting the Feather. *c.* the Seed.

FIG. 3. The Seed of a Bean split in two, after being soaked a little while in water.

a. a. The Seed-lobes.

b. The Heart.

c. The descending part of the Heart.

d. The ascending part of the Heart..

e. The Eye.

FRUITSTALKS.

FIG. 4. A Stalk. It supports the Flowers, and springs directly from the Root.

FIG. 5. A Spike. *a. b. c. d.* the little Spikes.

FIG. 6. A Panicle.

FIG. 7. A broad topped Spike. *a. a. a. a. a. a.* the little Fruit-stalks.

FIG. 8. A Bunch.

FIG. 9. A Rundle. *b. b. b. b.* Rundlets, *c. c.* the General Fence. *d. d. d. d.* the Partial Fence. *e. e. e. e.* The Spokes of the Rundle.

Fig. 10. A Tuft.

FIG. 11. Whorls of Flowers. *a. a. a.* the Whorls.

FIG. 12. A Catkin.

LEAVES

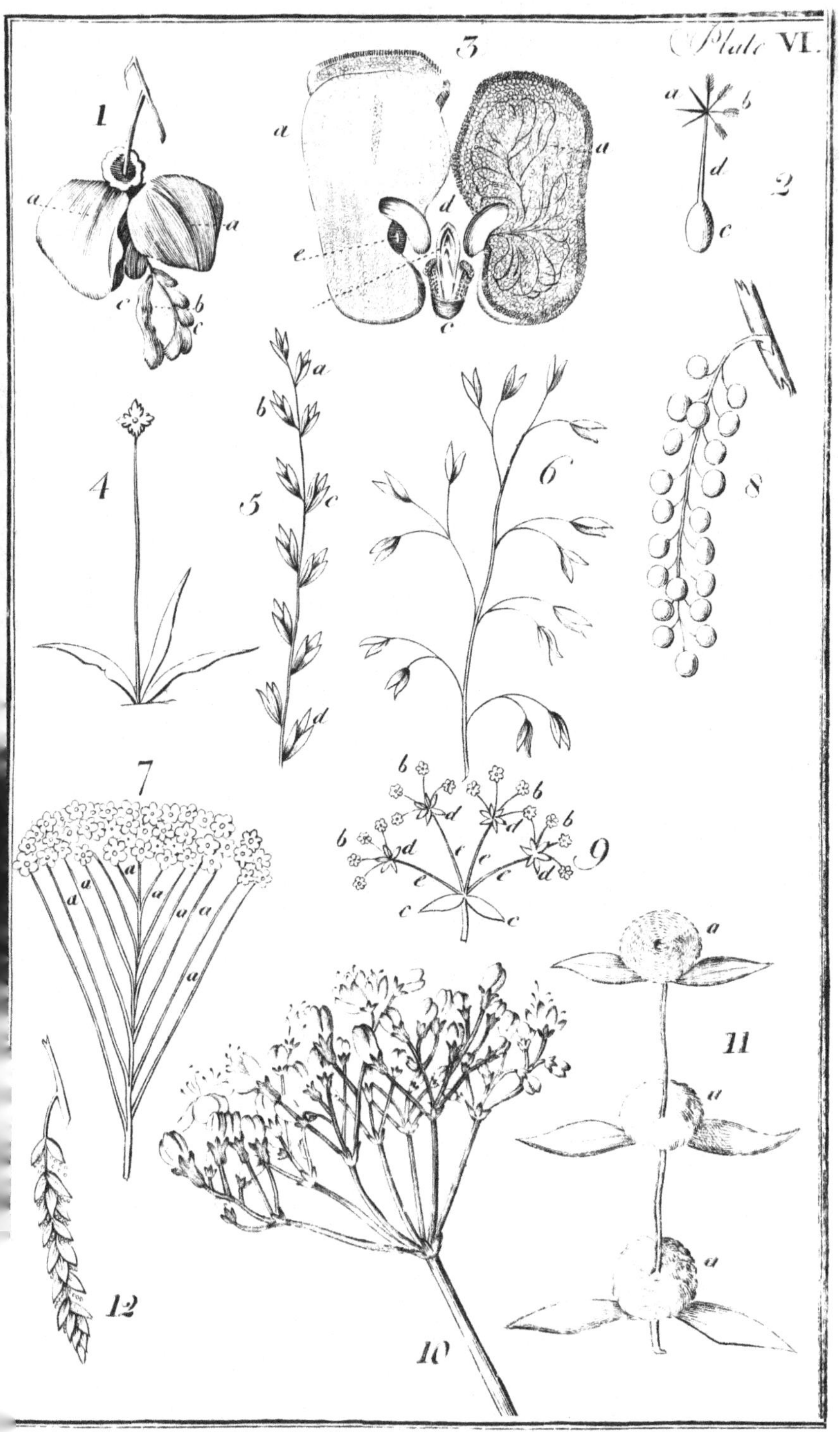
Plate VI.
1
2
3
4
5
6
7
8
9
10
11
12

AUTHORS *and* EDITIONS *referred to.*

Baub pin.	Caſpari Bauhini Pinax. Baſil 1671, quarto.
Gerard.	Johnſon's Gerard. Fol. London 1633.
Park.	Parkinſon's Theatrum Botanicum. London. 1640. Folio.
Ray's Syn.	Joannis Raii Synopſis methodica ſtirpium Britannicarum. Ed. 3d. London 1724. 8vo. Ed. 1ſt. 1690. 2d 1696.
Ray's Hiſt. Plant.	Joannis Raii Hiſtoria plantarum. 3 vol. fol. London. 1696.
Dillenius.	Joannis Jacobi Dillenii Hiſtoria muſcorum, 4t°. Oxon. 1741.
Hudſon's Flor. Angl.	Gulielmi Hudſoni Flora Anglica. 8v°. London, 1762.
Cat. Cant.	Thomæ Martyn Catalogus horti botanici Cantabrigienſis 8v° 1771.
Flor. Suec.	Caroli Linnæi Flora Suecica. 8v° Stockholmiæ, 1755.
Iter Oeland.	*Ibid.* Iter Oelandicum et Gotlandicum. 8v° Stockholmiæ 1745.
Iter Scan.	*Ibid.* Iter Scanicum. Stockholmiæ, 1751. 8v°.
Iter Weſtrog.	*Ibid.* Iter Weſtrogothicum, Stockholmiæ, 1745 8v°.
Flor. Lapp.	*Ibid.* Flora Lapponica. Amſtelodam. 8v° 1736.
Gen. Plant.	*Ibid.* Genera Plantarum. 8v° Holmiæ, 1764.
Sp. Pl.	*Ibid.* Species Plantarum. 2 vol. 8v° Vindobonæ, 1764.
Syſt. Nat.	*Ibid.* Syſtema Naturæ. 8v° Holmiæ, 1767.
Mantiſs. Plant.	*Ibid.* Mantiſſa Plantarum. 8v° Vindobonæ, 1770.
Philoſ. Tranſ.	Philoſophical Tranſactions.
Gent. Mag.	Gentleman's Magazine.

LEAVES.

PLATE VII.

LEAVES.

Fig.

1 Round.
2 Circular.
3 Egg ſhaped.
4 Oval.
5 Oblong.
6 Spear-ſhaped.
7 Strap-ſhaped.
8 Awl-ſhaped.
9 Kidney-ſhaped.
10 Heart-ſhaped.
11 Creſcent-ſhaped.
12 Triangular.
13 Arrow-ſhaped.
14 Betwixt heart and arrow ſhaped.
15 Halberd-ſhaped.
16 Notched at the end.
17 Conſiſting of 3 Lobes.
18 Bitten.
19 Gaſhed.
20 With five angles.
21 Gnawed.
22 Hand-ſhaped.
23 With winged Cleſts.
24 Jagged.
25 Indented.
26 Toothed and indented.

Fig.

27 Barbed.
28 Divided.
29 Serpentine at the edge.
30 Toothed.
31 Serrated.
32 Doubly-ſerrated.
33 Doubly ſcolloped.
34 Sharply ſcolloped.
35 Bluntly ſcolloped.
36 Sharply notched at the end.
37 Plaited.
38 Scolloped.
39 Blunt.
40 Sharp.
41 Tapering to a point.
42 Blunt but ending in a point
43 Fringed.
44 Venous, or full of veins.
45 Triangularly ſpear ſhaped
46 Stringy.
47 Growing by threes upon leaf-ſtalks.
48 Fingered.
49 Shaped like a bird's-foot
50 In pairs.
51 Three-fold.

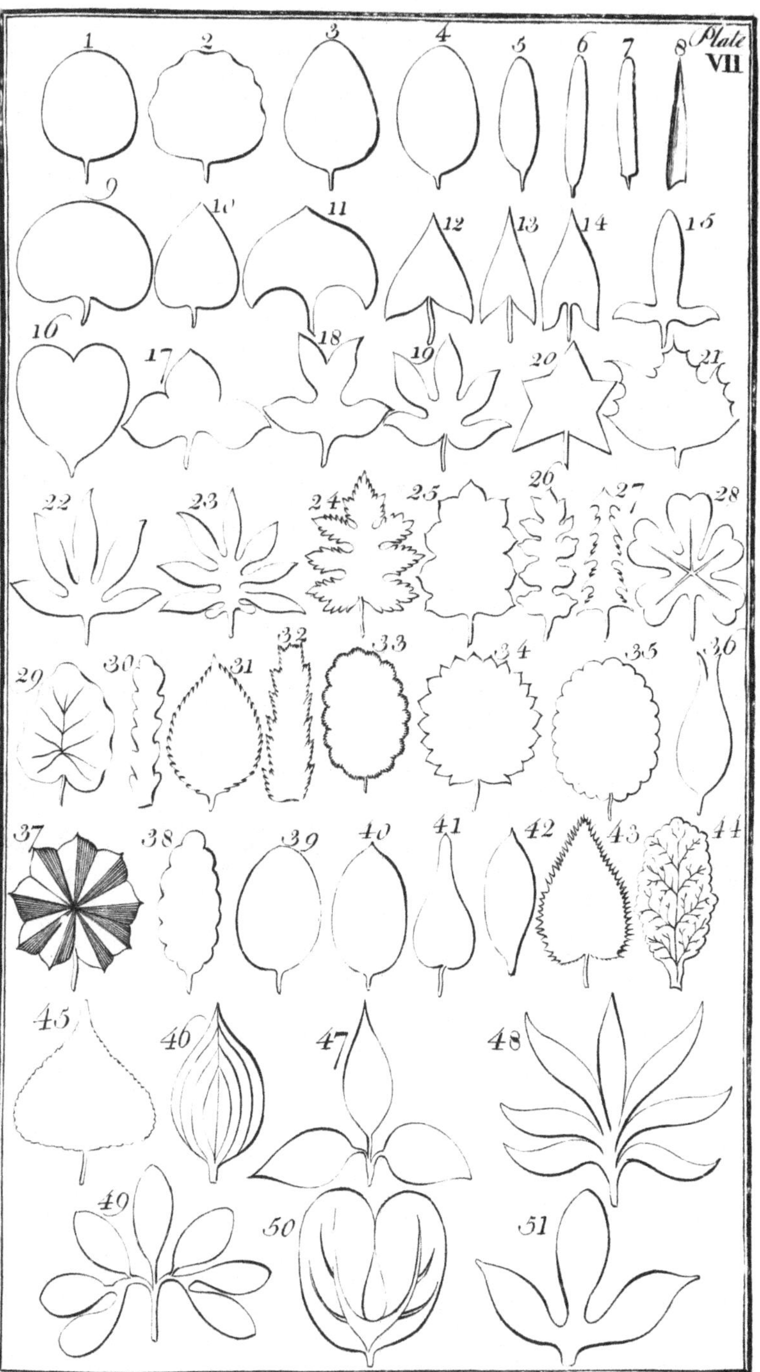

Plate VII
1
2
3
4
5
6
7
8
9
10
11
12
13
14
15
16
17
18
19
20
21
22
23
24
25
26
27
28
29
30
31
32
33
34
35
36
37
38
39
40
41
42
43
44
45
46
47
48
49
50
51

ERRORS *of the Press.*

Page 32, line 5, for *constitutes*, read *constitute.*
From page 36, to p. 69, add one to each number prefixed to the English Generic name.
Page 40, line 4, for *Phaloris*, read *Phalaris.*
- - - 50, in the side notes, for the upper *annua*, r. *annual.*
- - - 67, at the bottom for DOGSTAL, r. DOGSTAIL.
- - - 70, line 6, for 42 r. * 42.
- - - 74, line 32, for *radicea*, r. *radice.*
- - - 81, the last line, for *Spinx*, r. *Sphinx.*
- - - 94, at the top, for THREE CHIVES, r. FIVE CHIVES.
- - - 99, line 4, for HARESPONG, r. HARESTRONG.
- - - 180, line 17, for *pastinaca*, r. *pastinaca.*
- - - 188, after the bottom line, add *Honeycups.*
- - - 193, at the top, for TWO POINTALS, read FIVE POINTALS.
- - - 196, for 159 HYACINTH r. 149 HYACINTH
- - - 197, the lowermost line but one, for WATER PLANTAIN, r. THRUMWORT.
- - - 236, line 40, for *Pylygonum* r. *Polygonum.*
- - - 297, line 36, for *Barbara*, r. *Barba.*
- - - 340, line 2, for *Glecoma*, r. *Glechoma.*
- - - 388, No. 275, for CALBAGE, r. CABBAGE.
- - - 474, line 17, for *titled*, r. *tiled*
- - - 539, line 7, for *unsual*, r. *unusual.*
- - - 702, line 19, for *anthoceris* r. *anthoceros.*
- - - 712, in the side notes, for *orbicularus*, r. *orbicularis.*

PLATE VIII.

LEAVES.

FIG.

52 Winged, with an odd little leaf at the end.

53 Abruptly winged.

54 Winged with the little leaves alternate.

55 Interruptedly winged.

56 Doubly winged.

57 Doubly three-fold.

58 Winged and terminated by a tendril.

59 Triply threefold.

60 Triply winged, without an odd little leaf at the end.

FIG.

61 Triply winged, with an odd little leaf at the end.

62 Lyre-ſhaped.

63 Lopped at the end.

64 Spatula ſhaped.

65 Wedge-ſhaped.

66 Waved at the edge.

67 Curled.

68 Cylindrical.

69 Inverſely heart ſhaped.

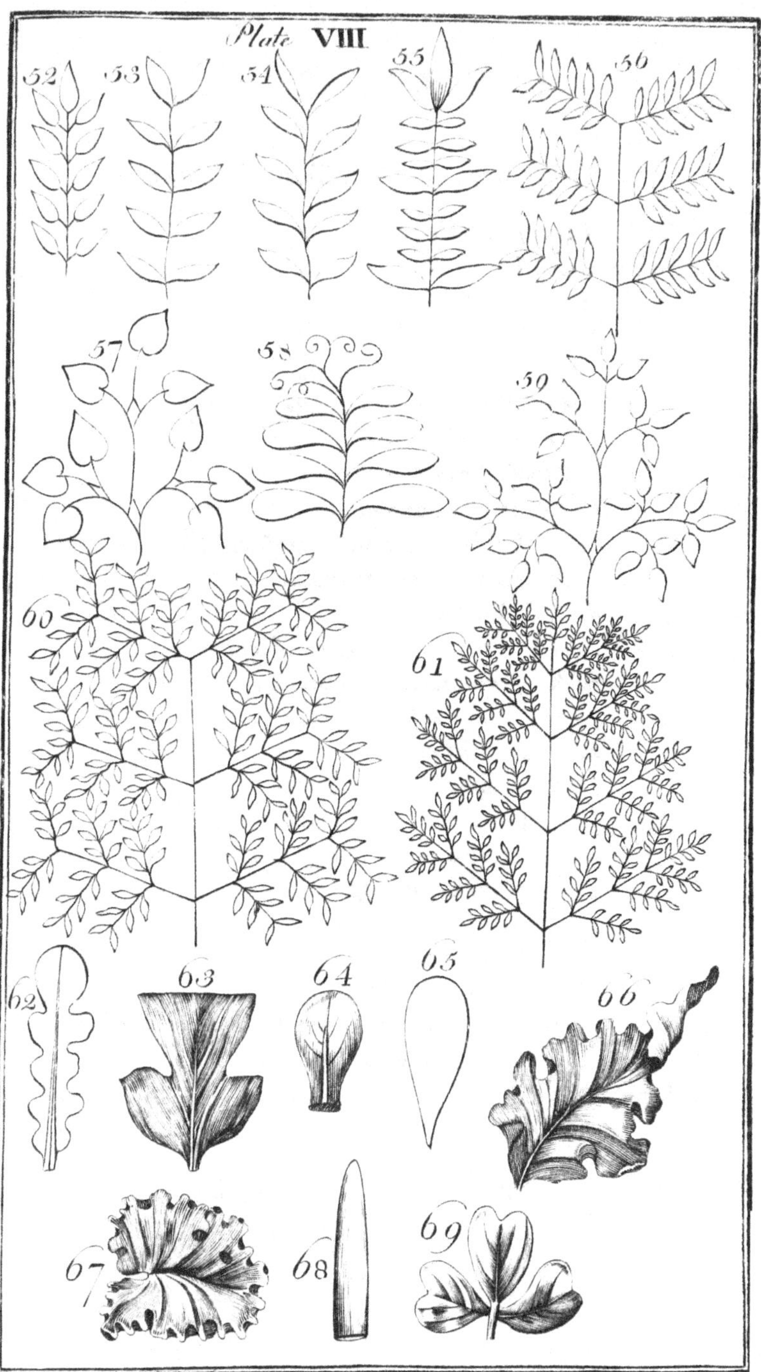
Plate VIII
52
53
54
55
56
57
58
59
60
61
62
63
64
65
66
67
68
69

PLATE IX.

Diſpoſition and Direction of Leaves.

FIG. 1. Leaves in croſs-pairs.

FIG. 2. Tiled Leaves.

FIG. 3. *a.* a jointed Leaf.

b. b. Starry Leaves.

c. c. Leaves growing by fours.

d. d. d. d. d. Leaves alternate. In fig. 5. all the Leaves are oppoſite.

e. Chaffy Leaves.

f. Leaves in a bundle.

FIG. 4. *a.* A Leaf with a central Leaf-ſtalk.

b. A Leaf with its Leaf-ſtalk. *c.*

d. A ſitting Leaf.

e. A Leaf running along the Stem.

f. A leaf embracing the Stem.

g. A perforated Leaf.

h. h. Twin Leaves.

i. A leaf ſheathing the Stem.

FIG. 5. *a. a.* Leaves bent inwards.

b. b. Leaves upright.

c. c. Leaves expanding.

d. d. Leaves horizontal.

e. e. Leaves curved.

f. f. Leaves rolled back.

m A Fruit-ſtalk riſing from the baſe of the Leaf.

FIG. 6. Leaves contiguous to the Stem.

FIG. 7. Root-leaves. *a.* the root. *b. b. b.* the leaves riſing immediately out of it, without the intervention of any Stem.

FIG. 8. *a. a.* Floral Leaves, different from the other Leaves of the plant, *b. b.* A Fruit-ſtalk. *c.*

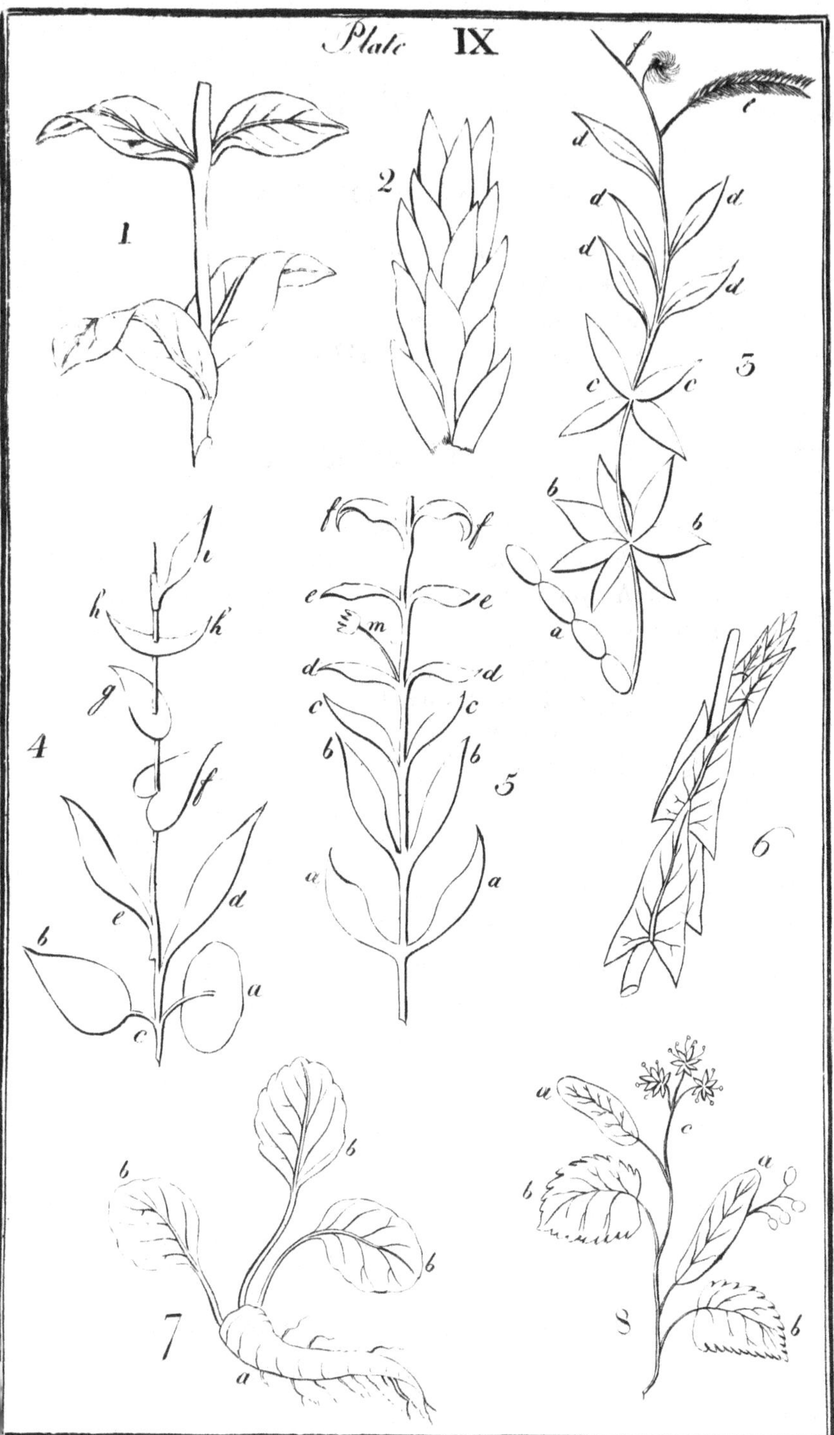
Plate IX
1
2
3
4
5
6
7
8

PLATE X.

WEAPONS.

Fig. 1. *a. a. a. a.* Simple thorns.
b. b. b. A triple thorn.
Fig. 2. *a. a.* Simple Prickles.
b. b, Forked or triple Prickles.

STEMS, &c.

Fig. 3. A jointed Straw. (*a. a. a.*) The Joints.
Fig. 4. A forked Stem.
Fig. 5. A twining Stem.
Fig. 6. *a. a.* A Tendril.
b. b. Props.
c. c. Concave Glands.
Fig. 7. A creeping Root.
Fig. 8. A creeping Stem.

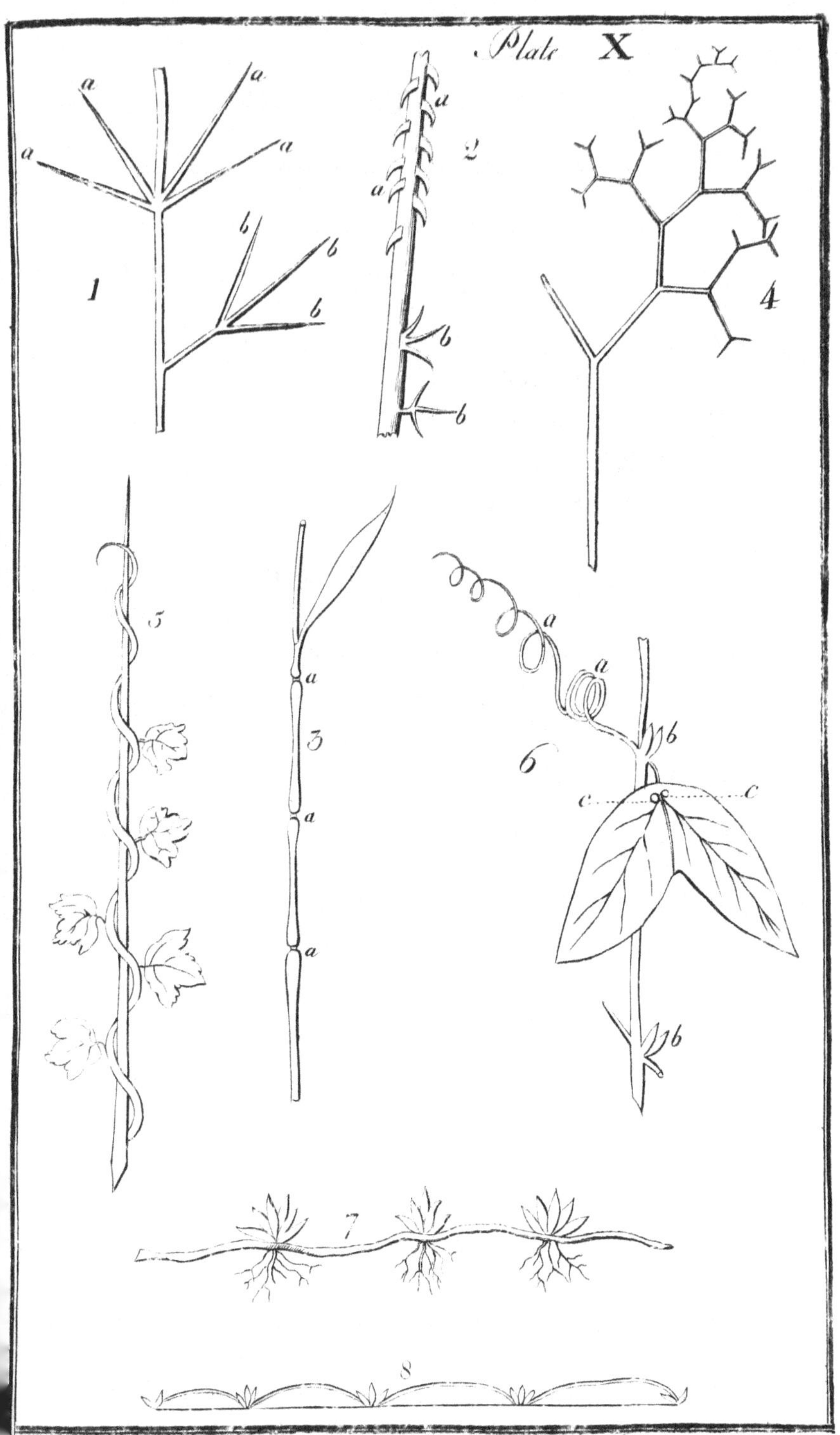
Plate X
1
a
a
a
a
b
b
b
2
a
a
b
b
4
5
3
a
a
a
6
a
a
b
c
c
b
7
8

PLATE XL.

FIG. 1. *a.a.a.a.* Glands ſupported upon Foot-ſtalks.

ROOTS.

FIG. 2. A coated bulbous Root, cut a-croſs to ſhew the Coats which compoſe it.

FIG. 3. A ſolid bulbous Root.

FIG. 4. A ſcaly bulbous Root.

FIG. 5. A branching Root.

FIG. 6. A ſpindle-ſhaped Root.

FIG. 7. A tuberous Root.

INSTRUMENTS.

FIG. 8. Two Diſſecting Needles, with ivory handles, belonging to the Botanical Microſcope.

FIG. 9. A pair of Spring Plyers for diſſection, belonging to the Botanical Microſcope.

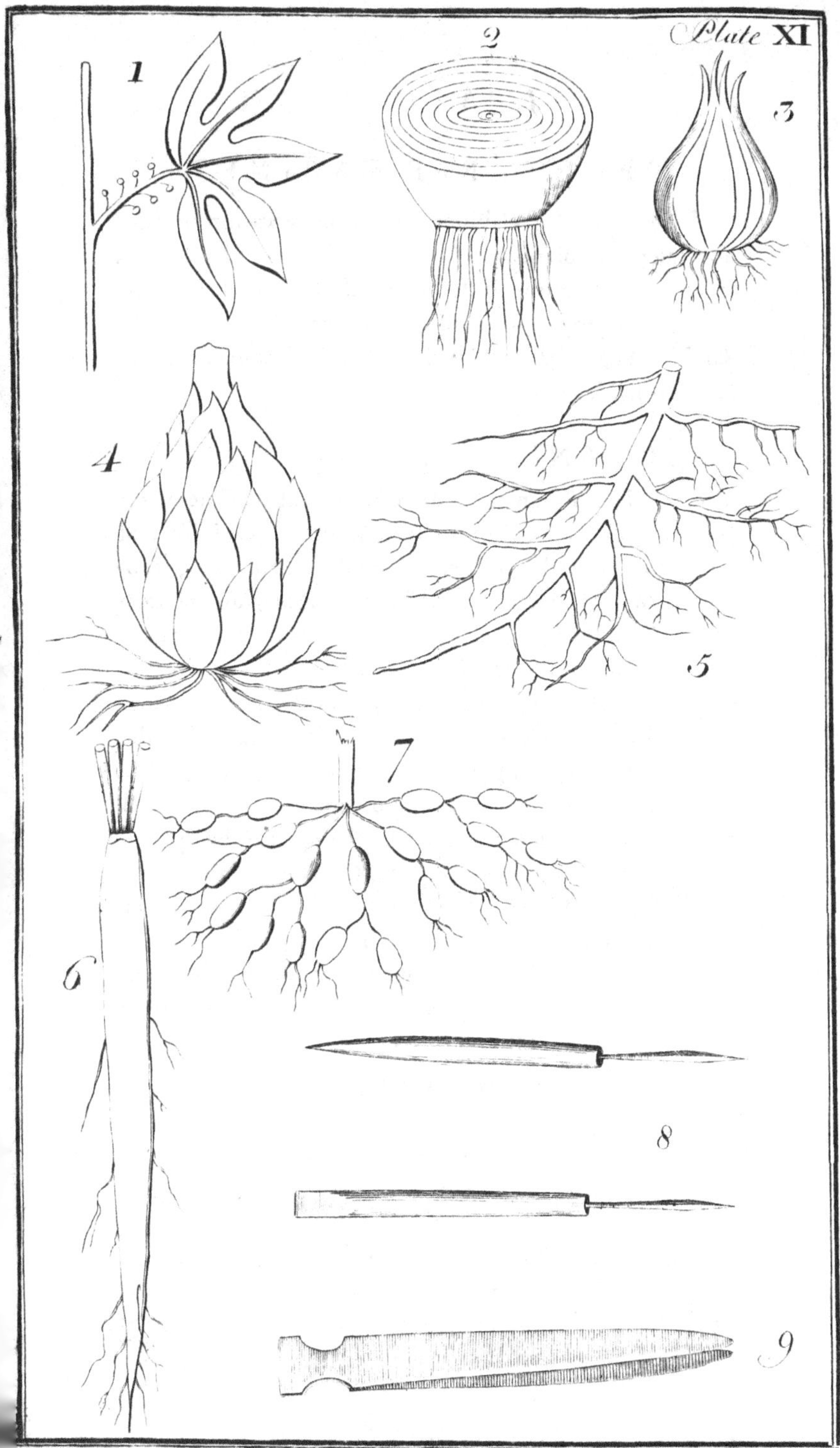
Plate XI
1
2
3
4
5
6
7
8
9

PLATE XII.

VEGETABLE CABINET.

A Section of a Cabinet for the preſervation of dried ſpecimens of plants. The numbers denote the drawers appropriated to the different Claſſes. The ſize of the drawers, is proportioned to the number of plants in each Claſs. They are calculated to contain ſpecimens of all the Britiſh Vegetables.

BOTANICAL MICROSCOPE.

Fig. 2. The Botanical Microſcope. *a. a.* The ſtages upon which the objects to be viewed and diſſected are placed. *b. b. c. c.* Circular braſs cells, containing lenſes of different magniſying powers. Theſe lenſes ſlide higher or lower, to adapt the focus to diſtinct viſion. Either of the lenſes may be taken out occaſionally and held in the hand. In the ſtage *a. a.* are the holes to contain the inſtruments figured in the preceding plate. The beſt way to uſe the microſcope is to ſet it upon a table, of ſuch a height that the eye can be applied with eaſe, almoſt cloſe to the lens. The elbows reſting upon the table, the two hands will be ſteady, and at liberty to uſe the diſſecting inſtruments. The Microſcope ſtands upon either end, according as you want to uſe the greater or the leſſer magnifying power.

FINIS.

DIRECTIONS *to the* BOOK-BINDER.

Pl. I.	To front title Vol. I.		Pl. VII.	To front	— 828
II.	To front page	31	VIII.	——	— 830
Pl. III.	Vol. II. ——	820	IX.	——	—— 832
IV.	——	822	X.	——	— 834
V.	——	— 824	XI.	——	—— 836
VI.	——	— 826	XII	——	—— 838

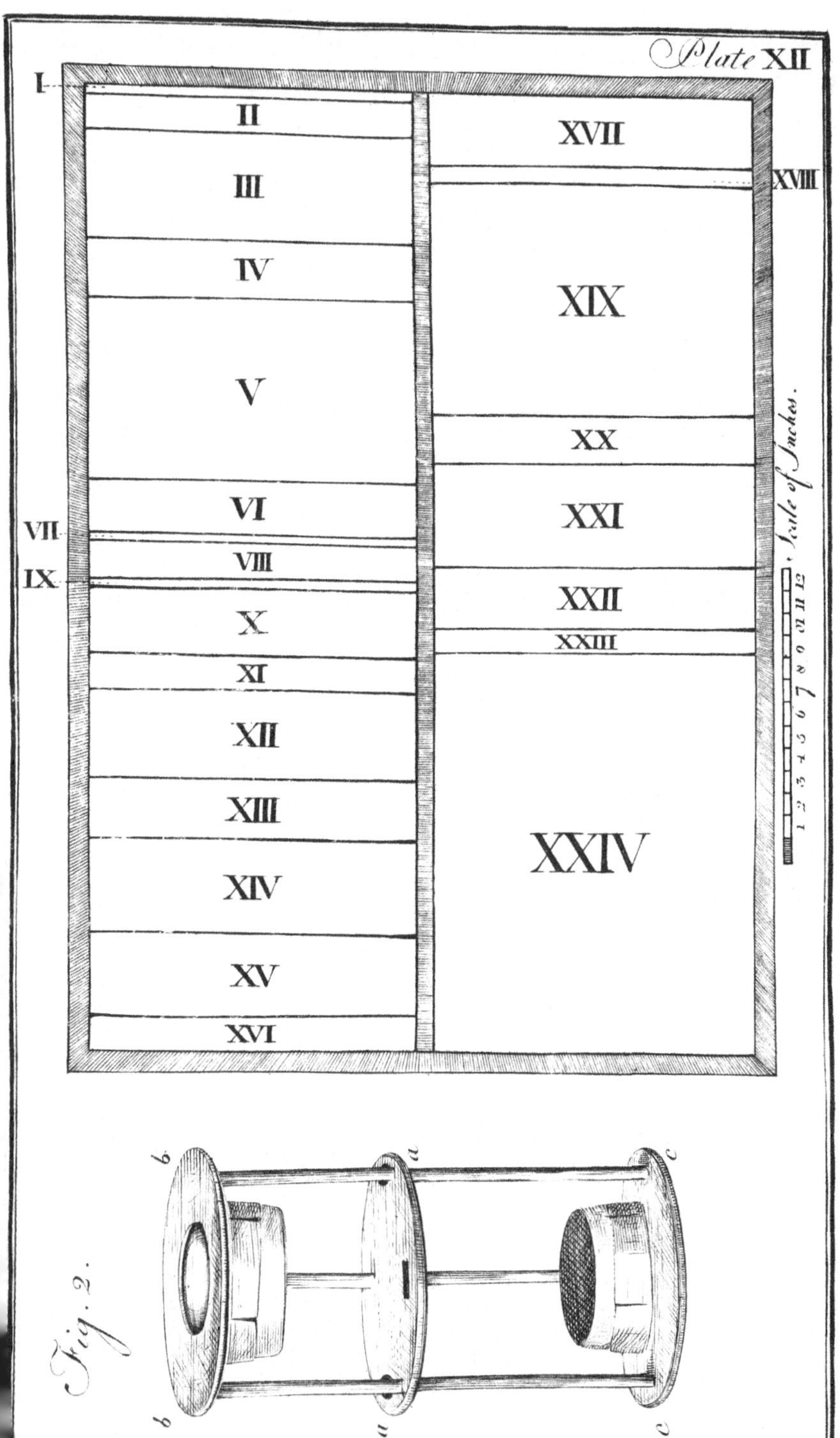

Plate XII
I
II
III
IV
V
VI
VII
VIII
IX
X
XI
XII
XIII
XIV
XV
XVI
XVII
XVIII
XIX
XX
XXI
XXII
XXIII
XXIV
Scale of Inches.
1 2 3 4 5 6 7 8 9 10 11 12
Fig. 2.
b
a
c
b
a
c

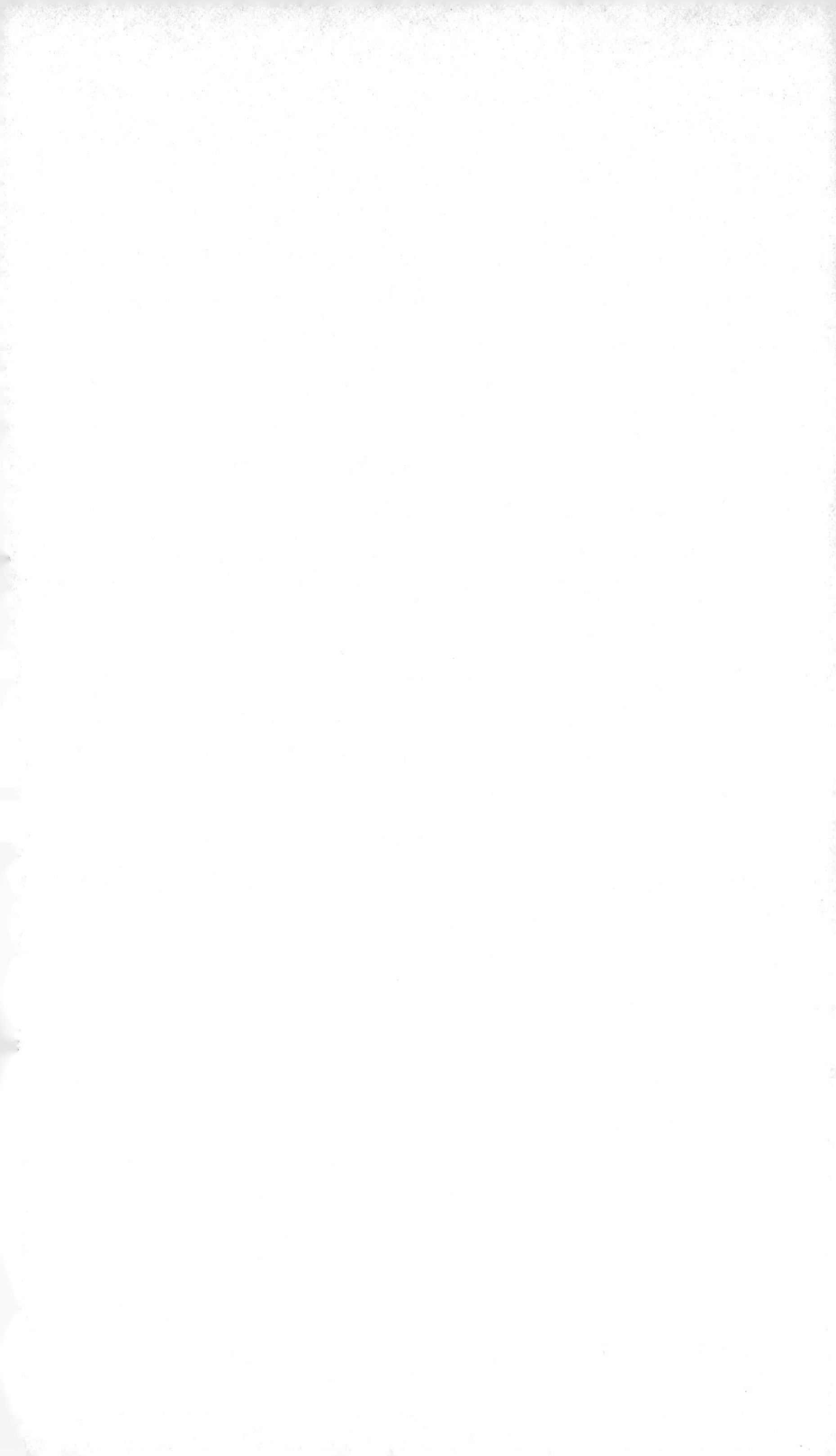

Printed in the United States
By Bookmasters